Wastewater Management: Theory and Practice

Wastewater Management:
Theory and Practice

Editor: Gabriel Craig

CALLISTO REFERENCE

www.callistoreference.com

Callisto Reference,
118-35 Queens Blvd., Suite 400,
Forest Hills, NY 11375, USA

Visit us on the World Wide Web at:
www.callistoreference.com

ISBN: 978-1-64116-605-8 (Hardback)

Cataloging-in-Publication Data

Wastewater management : theory and practice / edited by Gabriel Craig.
     p. cm.
Includes bibliographical references and index.
ISBN 978-1-64116-605-8
1. Sewage. 2. Sewage--Management. 3. Sewage--Purification. 4. Sewage disposal.
5. Water reuse. I. Craig, Gabriel.
TD745 .W37 2022
628.36--dc23

Table of Contents

Preface

Wastewater management is a procedure that is used to remove contaminants from sewage or waste water and convert it into an effluent that can later be reused. The main objective of waste water treatment is disposing or reusing water. The treatment of wastewater takes place in wastewater treatment plants where the pollutants are broken down or removed. The wastewater treatment process involves various steps like phase separation, sedimentation, filtration, oxidation and polishing. The biological processes which are used for treating wastewater include aerated lagoons, activated sludge and slow stand filters. The treatment of wastewater also involves the separation of solids from liquids, by the process of sedimentation. This book provides comprehensive insights into the field of wastewater management as well as its theory and practice. It also presents researches and studies performed by experts across the globe. This book will serve as a valuable source of reference for graduate and post graduate students.

The researches compiled throughout the book are authentic and of high quality, combining several disciplines and from very diverse regions from around the world. Drawing on the contributions of many researchers from diverse countries, the book's objective is to provide the readers with the latest achievements in the area of research. This book will surely be a source of knowledge to all interested and researching the field.

In the end, I would like to express my deep sense of gratitude to all the authors for meeting the set deadlines in completing and submitting their research chapters. I would also like to thank the publisher for the support offered to us throughout the course of the book. Finally, I extend my sincere thanks to my family for being a constant source of inspiration and encouragement.

Editor

Experience of Suckling Perfection of Secondary Clarifier of Aeration Station in Almaty, Kazakhstan

Kairat T. Ospanov[1*], Aleksandr V. Demchenko[2], Zhanar Kudaiberdi[1]

[1] Kazakh National Research Technical University Named After K.I. Satpayev, Satpayev's Street 22, Almaty, Kazakhstan

[2] State Municipal Enterprise "Tospa Su", Shubazeva's Str. 128, Alamty Region, Yhapek Batyr Village, Kazakhstan

* Corresponding author's e-mail: ospanovkairat@mail.ru

ABSTRACT

This article presents the results of experimental studies pertaining to the suckling improvement of secondary clarifier in an aeration station in Almaty. According to the obtained research results, the work of secondary clarifiers was evaluated for the removal of suspended solids, concentration of returned sludge and sediment moisture. The technical specifications of aerations, secondary settling tanks and cesspool emptier in the aeration station of Almaty were defined. Additionally, it was shown that the use of the articulated sucker with movable scrapers can increase the efficiency of cleaning the bottom of the circular tanks. In this practical test, the proposed suckling radial settler in aeration station of Almaty during the year showed that the sucker reduces the costs for the installation and maintenance of cesspool emptier and has no damaging effects on the bottom of the sump.

Keywords: water disposal, waste water, secondary sedimentation tank, activated sludge, cesspool emptier, sucker

INTRODUCTION

The treatment of sewage water in the cities of Kazakhstan is mainly performed by using biological aeration tanks with secondary clarifier. A forecast for the future and future developments in the field of wastewater treatment show that this method will remain the primary one in the future, but it is important to pay attention to the improvement of the facilities related to the biological method. The efficiency of biological wastewater treatment is determined by the work of secondary settling tanks, which should ensure the allocation of the activated sludge from the purified water. As the secondary clarifier runs efficiently, greater concentration of returned sludge and lesser degree of recycling should be used while operating the aeration tank [Andreev et al. 2006, Burger et al. 2011, Zhmur 2003]. The main parameters of the biological cleaning process must be interconnected, namely: the volume of the aeration tanks, the amount of oxidation and contaminants in the wastewater, the contact time of wastewater with activated sludge. In addition, the sedimentation properties of the activated sludge, which are defined by all of the above-mentioned parameters must comply with the technological capabilities applied by secondary clarifiers – satisfactorily separation of the effluent from the sludge [Dziubo and Alferova 2014, Hunze 2005, Zhmur 2003]. Activated sludge is constantly forming new cells, carrying the biochemical oxidation of organic pollutants. The intensity of the sludge on the growth of cells is regulated by several factors and depends on: the nature of the oxidized substrate; the temperature of treated wastewater, which determines the degree of the assimilation processes; self-oxidizing capacity of the activated sludge, depending on the load and the period of aeration on structures; sedimentation characteristics of the sludge and the removal of suspended solids from the secondary clarifiers; the presence of toxicants (lowering increase) or mutagens-growth stimulants [Al-Bastaki 2004, Hunze 2005, Pijuan et al. 2012].

The majority of activated sludge in the secondary settling tank, should be pumped back into the aeration tank. This constitutes the circulating active sludge, which falls into the aeration tank through the regenerator. Usually, the amount of sludge in the secondary settling tank is greater than needed for the circulation; therefore, the excess is sent for recycling.

The volume of the returned sludge removed from the secondary settling tanks and supplied to the regenerator constitutes from 30 to 70% of the volume of the treated wastewater. For each treatment plant, the figure is individual and determined by the calculation of the degree of recycling. The excessive sludge with a moisture content of 99.2% was 4 dm^3 / day per resident, and has a greater moisture content than the raw sludge from a primary sedimentation tank, which increases the total amount of sludge [de Clercq 2003, Zhmur 2003].

The work of secondary clarifiers is estimated by the removal of suspended solids, concentration of returned sludge and sediment moisture. These parameters characterize their main functions such as separating the purified water from the activated sludge and silt seal [Ni and Yu 2012, Zhmur 2011].

Control of the secondary clarifiers operation is a very important task of the operating service, because the efficiency of the secondary settling directly affects the course of biochemical oxidation in the aerotanks and, to a large extent, determines the content of suspended solids in purified water, i.e. loss of active biomass sludge and, accordingly, its growth.

The effectiveness of the secondary clarifiers is influenced by other causes such as hydrodynamic flows; secondary clarifiers are more sensitive to load and unevenness in volume of inflow of sewage than the primary ones, since they are loaded from the circulating flow of returned sludge, and the sludge precipitates easier; the type of sump and sludge collection system used; characteristics of the activated sludge and a variety of biological processes [Sanin et al. 2011, Solovieva 2008, Zhmur 2003].

In contrast to the raw sludge, the active sludge is more sensitive to fallow lands, which requires the use of more sophisticated sludge collection and pumping system from the bottom of the sump and from pits [Mishukov and Solovieva 2001, Zhmur 2003]. The activated sludge in the sumps is more susceptible to the process of rotting in a packed bed, where anoxic conditions are present. The height of the packed bed in the sump depending on the mode of shipping sludge may range from 0.5 to 1.0 m in the radial sumps [Ivanov et al. 2012].

During the exploitation of secondary clarifiers, it is very important to establish and maintain an optimal state of the bed height of sludge. In winter, the height of the sludge layer can be 25% from the depth of the settler, and in summer – no more than 10%.

With the accumulation of sludge in the sumps, and exceeded the optimum layer height of standing sludge, the humidity of returned sludge decreases, but its concentration is increased, which may contribute to an excessive removal of suspended solids. Excessively frequent release of excess sludge from the settling tank and overly active circulation of returned sludge lead to an increase in the excess sludge moisture, which increases the volume of the necessary facilities for the processing and disposal of sludge [Hunze 2005, Vaxelaire et al. 2004].

In other words, if more than optimal amount of sludge is withdrawn from the secondary clarifier , then the excess volume returns to the aeration tank of water, if less, then most of the settled sludge collects in the sump and the quality of the treated water reduces.

Therefore, technological operation mode of the secondary clarifier was given to ensure adequate parameters of the envisioned projects.

The efficiency of the secondary clarifier depends on the compliance of real hydraulic load and its design values and the uniformity of its distribution, as well as the timely and continuous uniform flow of sediment removal. The optimal level of standing sediment can be controlled by the values of returned sludge doses. Experience has shown that with the returned sludge at a dose of 4–6 g/dm^3, theremoval of suspended solids from the secondary clarifiers is about 15 mg/dm^3, whereas returned sludge at a dose greater than 6 g/dm^3 increased the removal from 15 to 20 mg/dm^3.

A significant increase in the removal of suspended solids from the secondary clarifiers (40 mg /dm^3) occurs with the return sludge concentration of 8 g/dm^3, which, apparently, is the threshold for typical structures, treating municipal wastewater [Pijuan et al. 2012, Vaxelaire et al. 2004].

At the same time, the major factors contributing to the excessive removal of suspended solids from the secondary clarifiers are hydraulic overload, which is caused by excess amount of wastewater entering the treatment, the structural imperfections of secondary settling tanks or unsatisfactory operation of secondary settling tanks, the formation of sludge deposits on the bottom of the secondary settling tank, which can be due to the irregularities in the bottom of the settling tank, poor performance of the cesspool emptier, untimely removal and the sludge system without delay in its shipment for recycling [Canales et al. 1994, Deleris et al. 2002, Zhmur 2003].

The secondary clarifiers differ radically from the primary ones in respect to the properties of substances. While the sludge in primary settlers may lie for some time without decay, in the secondary one, even small deposits of sediment decay and deteriorate the aeration mode throughout the system. Rotting returned sludge treatment upsets the system of treatment and as a result, its effect significantly reduces.

Therefore, the sludge removal system of secondary settling tanks should include work under daily peak load, rather than around the clock.

The humidity of shipped and returned sludge can vary within wide limits from 99.2 to 99.7%, which corresponds to the content of dry matter in the sludge from 3 to 8 g/ m³.The results pertaining to determination of moisture and solids of returned sludge should be consistent with each other, which is an indirect validation of the measurements [Guest et al. 2009, Ivanov et al. 2012, Yasui and Shibata 1994].

The most perfect system of sediment collecting is found in the radial secondary clarifiers, which divide into a system with scraper and cesspool emptier. The residence time of the sludge in the bottom of the sump depends on the radial velocity of the scraping device, number of wings or cesspool emptier, and the distance between the scrapers to radial mud pit or the length of the wings cesspool emptier.

The presence of mud deposits on the bottom peripheral part of the settler effects on increasing the removal of suspended solids, as hydraulic wastewater streams coming from the central pipe, are directed to the periphery of the sump and washed sludge settled in the sump to the surface of the walls [Auerbach et al. 2007, Denisov et al. 2016].

In the radial sumps, due to unevenness or the bottom or because of the presence of zones with sediment, deposits may unevenly accumulate at the bottom of the settler in certain areas, which

is the reason for its fermentation and buoyancy, and cause deterioration of the properties of return sludge. In such cases, the sump is emptied, cleaned reservoir, and the causes are eliminated as far as possible [Denisov et al. 2016, Radjenovic et al. 2009, Zhmur 2003].

Sometimes, the elimination of sludge deposits requires serious reconstruction settlers. For example, if the cesspool emptier bead does not reach several meters, only the center well of the settler precipitate is removed and observed at the periphery of the permanent sludge deposits. This deficiency can be corrected by replacing the scrapers on the cesspool emptier [David et al. 2009, Ratkovich et al. 2013].

MATERIALS AND METHODS

The Almaty sewage system works on incomplete separate system, one of them the storm – with the tap water in the small river, the other citywide – for industrial and domestic wastewater.

The drains of the city received in citywide sewer system are cleaned in the wastewater treatment plant. High-rise building's location provides gravity mode movement of the main mass of the waste water, using the natural terrain.

The station carries out mechanical and complete artificial biological wastewater treatment, with additional purification to storage and bioponds as well as subsequent disinfection with chlorine at the discharge of effluent into the Ili river. The design capacity of aeration station is 640 000 m³ per day.

The actual volume of waste water in Almaty arriving at the wastewater treatment plant in 2016 has averaged 395 000 m³ per day.

The wastewater enters the treatment plant by three collectors, the diameter of the two of them in front of the aeration station is d = 1500 mm and d = 1000 mm.

A special receiving chamber which is used to equalize the rates and even distribution of waste on working grates is found here.

The chamber on the ferro-concrete drainage channels are directed to the grid. Metal gates with electric drive to switch off from the work of the individual arrays are installed in the connecting channels [Technological regulations 2005].

The scum trapped on lattices was collected in a special container and was disinfected with a solution of bleach; then, they were transported to the Aeration station sludge beds together with the municipal solid waste.

Detention of heavy solids occurs in the horizontal sand catchers. The precipitated solids and sand were transported to the pits with the hydraulic system, where sand is pumped to the hydro elevator platform.

After sand traps drain the common tray, which enabled their quantitative measurement, it enters to the distribution bowl of primary clarifiers. The removal of suspended solids wastewater, which is capable of settling or floating under the gravity, occurs in radial primary sedimentation tanks. The crude residue precipitated in each sump scrapers installed on the farm scraper is shifted to the pit from which it is pumped to the sludge beds. After settling in a settler, the clarified effluent is collected in a common channel and sent to biological purification facilities. The wastewater enters the airlift pump chambers by receiving reinforced concrete channel, where airlifts pump it into the aeration tanks for biological treatment [Technological regulations 2005].

Technical characteristics of aeration station are given in Table 1.

After mixing in the aeration tank, the waste water of the activated sludge from the supply line

is sent into the secondary clarifiers. The technical characteristics of the secondary clarifiers are presented in Table 2.

In the secondary settling tanks, the activated sludge settles and using the cesspool emptier brand, PSI-40 is removed from the bottom of sumps. [Technological regulations 2005] Technical Specifications of cesspool emptier RWI-40 shown in Table 3.

The cesspool emptier design consists of a rotating mechanism with a suckling and a periph-

Table 1. Technical characteristics of aero tanks in the aeration station in Almaty

The name of indicators	The value of indicators
Number of aero tanks	2 blocks
Block of aero tank #1	experimental, deep, consisting of the sections A and B
Volume of section	$V = 63140 \ m^3$
Dimensions (inplan) block	110 x 164 m
Sizes of one section	110 x 80 m
Workingdepth	H = 7.0 m
The number of corridors in section	8
Project (planned) performance of block	320 thousand.m^3per day
Block of aero tank #2	Sample consisting four sections, consisting of aerotank №1, №2, №3, №4
Volume of one aero tank	$V = 25000 \ m^3$
Volume of one regenerator	$V = 6250 \ m^3$
Dimensions (inplan) block	144 x 144 m
Sizes of one aero tank	144 x 36 m
Workingdepth	H = 5.0 m
The number of corridors in section	4 ;
Project (planned) performance of block	320 thousand.m^3per day

Table 2. Technical characteristics of the secondary sedimentation tanks of aeration station in Almaty

The name of indicators	The value of indicators
Type of sump	typical secondary radial settling tank
Thenumberofsumps	12
The diameter of the sump	D = 40 m
Hydraulicdepth	H = 4.35 m
The depth of the flow in the settler	H_1 = 3.65 m
The height of the sludge zone	H_2 = 0.7 m
The diameter of the supply line	D = 1500 mm
The diameter of the discharge pipe	D = 1200 mm
Diameter of sludge conductor	D = 800 mm
Cesspoolemptier	ИВР-40
Number of suckling in a sump	4
Locationof suckling	one suckling for four lines of different lengths, with different radius of rotation
Rotational speed of the movable truss to the sump	one turn of 35 to 45 minutes
Sludgereturnmethod	transfer pumps in the return sludge channel, then transmission of gravity in the aerotank regenerators
Volume of the settling zone	$V = 4580 \ m^3$

Table 3. Specifications of cesspool emptier RWI-40

The name of indicators	The value of indicators
Productivity, m^3 / h	1728
Frequency of rotation of cesspool emptier, vol/ hour	1÷3
Electricmotorpower, kW	1.5
Overall dimensions, mm, no more: – length – width – height	40850 6000 7100
The mass of the rotating parts, kg, no more	14170
Weight of the fixed leg portions kg, no more	6330
Total weight, kg, not more	20500
Ratedmainsvoltage, V	380

eral drive. The working bodies are suckling cesspool emptier, which are connected to the collector pipe. The secondary settling tank is designed according to a standard project 902–2-377.83 [1983] in which suckling cesspool emptier looks as shown in Figure 1.

The activated sludge enters the chamber sludge shield and the movable weir gate through conduit, and then is pumped into the axial distribution channel return sludge and from there it is transported to aeration.

Excess activated sludge is pumped to the sludge beds with the help of the main pumping station. The sewage treatment plant wastewater treatment plant operates continuously, day and night, treating all the waste water of the city and its suburbs to the required degree of purification.

After complete biological treatment, the purified water, is sent to the drive Sorbulak across the earthen channel (49 kilometers) or, through a special water divider sent to bioponds for deep cleaning. From bioponds, the water after disinfection on chlorinator is discharged into the Ili river.

The secondary clarifiers of the Almaty aeration station have significant drawbacks, leading to a decrease in the efficiency of wastewater treatment. Additionally, the cesspool emptier 902–2-377.83 model project for secondary radial sumps is produced with little or no design changes. However, the requirements for the quality of wastewater after secondary clarifiers have significantly tightened since then, and sample design cesspool emptier is no longer able to fulfill their tasks. The analysis of the aeration station in Almaty showed that the so-called "dead zones" formed in the settler due to non-parallelism of the planes and the side of the settler bottom part, uneven bottom of the settler, various construction and installation of deviations between the bottom of the settling tank and the suckling.

In these zones, rotting and subsequent flotation ascent of the sludge not collected from the bottom occurs and, as a consequence, results in an increase in the content of suspended substances in the purified water.

At the bottom of the settler zone, the average concentration of sludge is 15–22 g/l and it significantly reduced by a distance greater than 15 cm, which is comparable to the distance from the bottom of the settling tank to the intake slot suckling in compliance with all technical requirements for the installation cesspool emptier. Thus, a significant amount of compacted mud is not going to suckling, and is stirred up when driving along the circumference of suckling. This results in an is inevitable dilution of the sludge by trapping boundary layers, up to a breakthrough in the water sucking, which leads to a sludge collection with high humidity.

RESULTS AND DISCUSSION

In order to eliminate the above-mentioned drawbacks, we developed and implemented a sucker for cleaning the bottom of the circular tanks on the aeration station of Almaty. There is a patent for the invention of the Republic of Kazakhstan [Ospanov et al. 2014].

Figure 2 shows the suckling scheme for cleaning the bottom of the circular tanks.

The proposed sucker for the mechanical purification of bottom circular tanks is made of reinforced metal and rubber. The fixed part 1 of the suckling is made from sheet metal with the thickness of 6 mm, the frame 4 on which are fixed

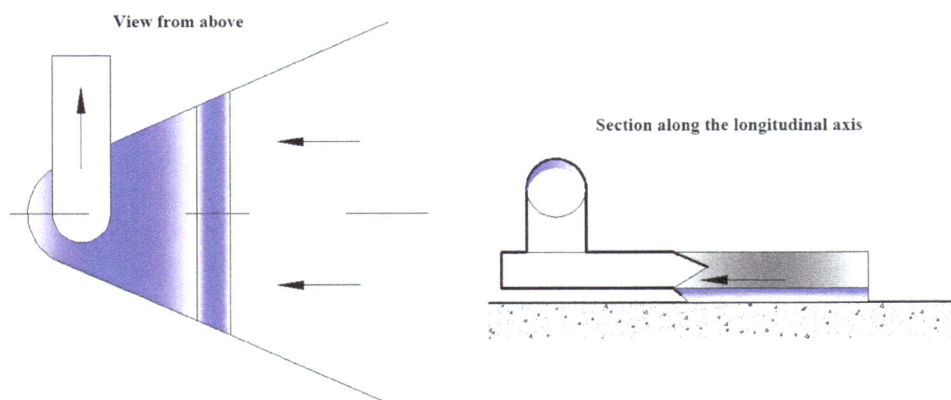

Figure 1. Scheme of sucking on a standard project 902–2-377.83 [1983]

View from above

Side view

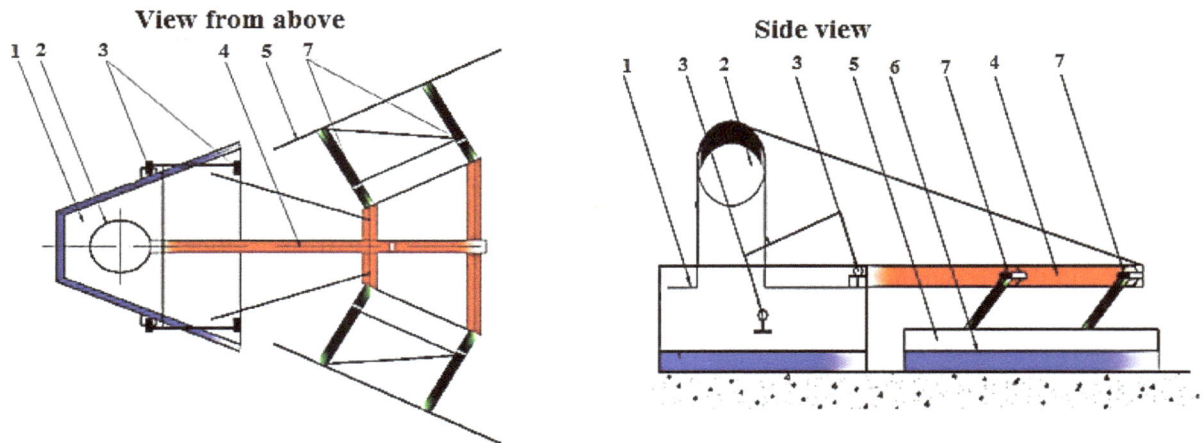

Figure 2. Scheme of suckling for cleaning the bottom of the circular tanks [Ospanov et al. 2014]:
1 – fixed part of the suckling; 2 – cesspool emptier pipeline for the removal of sludge; 3 – the hinges of the moving part of suckling; 4 – mounting (frame) scoops scraper; 5 – scoops scraper; 6 – flexible component scoops scraper; 7 – hinges scoops scraper.

mounted scraper scoops of pipes has a diameter of 100 mm, whereas the scoops scrapers 5 are made from sheet metal with the thickness of 3 mm. In order to provide stiffness in the longitudinal direction, reinforced metal area of the scrapers is 50×50 mm. A 20 mm thick reinforced rubber is mounted under scraper with M16bolts. Each of the two wipers is pivotally 3 secured to the frame tubes of the arms of the corner 63×63 mm. In order to stiffen the levers, a further reinforcement with Ø16mm was used.

The device operates as follows: When wastewater is present in the sump, radial cleaning processes by deposition of suspended particles on the bottom of the settler occur, so the bottom of the sump forms a radial layer of sludge. The sludge formed at the bottom is removed by suction through a fixed part of the suckling in the cesspool emptier pipeline 2.

The use of the proposed mobile scrapers allows efficient and reliable operation of the sucker, as the size and location of the scoops scraper provide them passage across the surface to be cleaned the bottom of the circular tanks. This flexible component scoops scraper 6 completely in contact with the bottom of the circular tanks, covering all surface irregularities. Compound scraper scoops hinged 3, 7 allowing them to rotate relative to each other in a vertical plane and the best fit to the uneven floor, which also improves the quality of cleaning.

This sucker for cleaning the bottom of the circular tanks successfully works in Almaty aeration station and removes sludge from the entire area of

circular tanks, improving the state of the sludge pumped to the aeration tanks. The photos of suckling for cleaning the bottom of the circular tanks during the reconstruction of the city of Almaty aeration station are shown in Figure 3.

The economic benefit was also provided because the sucker can be repaired on site, resulting in minimum operating costs. Figure 4 shows the process of replacing the flexible component scoops scraper that is reinforced rubber.

In general, this sucker was introduced in the secondary clarifier aeration station in Almaty in April 2014 and works successfully to this day.

In order to evaluate the effectiveness of suckling works for the secondary clarifier, we defined the concentration of suspended solids and BOD_5 were determined before and after reconstruction. The analyses of the comparison results before and after reconstruction are given in Table 4.

The table shows that after the reconstruction of the secondary clarifier suckling Almaty aeration station concentrations of suspended solids and BOD_5 as compared with the reconstruction to declined.

CONCLUSIONS

The analysis of the secondary clarifier in Almaty aeration station showed that sludge accumulation zone is formed at the bottom of the settling tank. Thus, there is a decay of the sludge as a consequence, reduced overall efficiency of the biological purification, and also increases in the

Figure 3. Photo of suckling of aeration station in Almaty [Ospanov 2017]

Figure 4. Picture showing replacement of flexible component scoops scraper [Ospanov 2017]

Table 4. Analysis comparing the degree of purification of waste water before and after reconstruction of the secondary clarifier suckling Almaty aeration station [Ospanov 2017]

Indicators, mg/l	Beforereconstruction			After reconstruction		
	Incoming water	Mechanical purified water	Biological purified water	Incoming water	Mechanical purified water	Biological purified water
Suspended solids	387.8	101	12.7	375.6	100.5	8.1
BOD_5	311.8	85.3	9.6	391.2	88.7	5.7

content of suspended solids in the purified water occur. In order to eliminate the above-mentioned drawbacks, improving the design of the secondary clarifier suckling Almaty aeration station can be applied. This problem is solved with the use of a hinge suckling with mobile scrapers. The use of the proposed mobile scrapers allows covering all the irregularities of the surface of the bottom of the secondary settling tank. The design of the proposed sucker for cleaning the bottom of the settler was tested in circular tanks of Almaty aeration station during the year. At the same time, the following observations were pointed out:

- possible swelling of sludge in the settling tank;
- providing the silt fence with a maximum concentration;

- the degree of water purification increases;
- construction has no damaging effects on the bottom of the sump;
- depreciation of the flexible components does not exceed the permissible limits for a year of continuous operation.

Acknowledgements

The authors are grateful to the employees of the State municipal enterprise "Tospa Su" of the city of Almaty for their assistance in carrying out this work.

REFERENCES

1. Al-Bastaki N.M. 2004. Performance of advanced methods for treatment of wastewater: UV/TiO2, RO and UF. Chemical Engineering and Processing. 43 (7), 935–940.

2. Andreev S., Grishin B., Maksimov S., Titov A., Nikolaev V. 2006. Intensification of the process of mass exchange in aeration structures of biological sewage treatment as a factor affecting the improvement of secondary sedimentation tanks. News of higher educational institutions. Construction. No. 11–12. 56–60.

3. Auerbach E.A., Seyfried E. E., McMahon K.D. 2007. Tetracycline resistance genes in activated sludge wastewater treatment plants. Water Research. No. 41. 1143–1151.

4. Burger R., Diehl S., Nopens I. 2011. A consistent modelling methodology for secondary settling tanks in wastewater treatment. Water Research. 45(6). 2247–2260.

5. Canales A., Pareilleux R.J.L., Goma G., Huyard A. 1994. Decreased sludge production strategy for domestic wastewater treatment. Water Science and Technology. No. 30 (8) 97–106.

6. David R., Saucez P., Vasel J.L., Vande Wouwer A. 2009. Modeling and numerical simulation of secondary settlers: A Method of Lines strategy. Water Research. 43(2). 319–330.

7. de Clercq B. 2003. Computational fluid dynamics of settling tanks e development of experiments and rheological, settling, and scraper submodels. University of Gent. 324 p.

8. Deleris S., Geaugey V., Camacho P., Debellefontaine H., Paul E. 2002. Minimization of sludge production in biological processes: an alternative solution for the problem of sludge disposal. Water Science and Technology, 46(10). 63–70.

9. Denisov A.A., Rozaeva A.V., Shamanova L.A., Koroleva E.A., Pavlenko A.L., Canarskaya Z.A. 2016. Method of calculating the optimal size and operating mode of the secondary settler in biological wastewater treatment technology. Vestnik Kazanskogo tehnologicheskogo universiteta [Bulletin of Kazan Technological University], 19(3), 101–104.

10. Dziubo V.V., Alferova L.I. 2014. Modernization of air-lift unit for active sludge recirculation. Vodoochistka. No. 10. 29–33.

11. Guest J.S., Skerlos S.J., Barnard J.L., Beck M.B., Daigger G.T., Hilger H., Jackson S.J., Karvazy K., Kelly L., Macpherson L., Mihelcic J.R., Promanik A., Raskin L., Van Loosdrecht M.C.M., Yeh D., Love N.G. 2009. A new planning and design paradigm to achieve sustainable resource recovery from wastewater. Environmental Science & Technology. 43(16). 6126–6130.

12. Hunze M. 2005. Simulation in der kommunalen Abwasserreinigung. Oldenburg Industrieverlag. Munich. Germany, pp. 380.

13. Ivanov V.G., Amelichkin S.G., Medvedev A.N. 2012. Constructive solutions for the modernization of existing secondary sedimentation tanks using thin-layer modules. Voda i Ecologiya: Problemy i Resheniya. No. 2–3 (50–51). 62–69.

14. Mishukov B.G., Solovieva E.A. 2001. The results of secondary radial sedimentation of sewage treatment plants and their mathematical interpretation. Voda i Ecologiya: Problemy i Resheniya. 2 (7). 42–45.

15. Ni B.J., Yu H.Q. 2012. Microbial products of activated sludge systems in biological wastewater treatment systems: a critical review. Critical Reviews in Environmental Science and Technology. No. 42. Pp. 187–223.

16. Ospanov K.T. 2017. Integrated technology of sewage sludge treatment. Monograph. Almaty: KazNRTU,– 171 p.

17. Ospanov K.T., Demchenko A.V., Tamabaev O.P. 2014. A device for mechanical cleaning of the bottom of a radial settler. Patent 2014/0748.1 dated May 30, 2014.

18. Pijuan M., Wang Q., Ye L., Yuan Z. 2012. Improving secondary sludge biodegradability using free nitrous acid treatment. Bioresource Technology. No. 116. 92–98.

19. Radjenovic J., Petrovic M., Barcelo D. 2009. Fate and distribution of pharmaceuticals in wastewater and sewage sludge of the conventional activated sludge (CAS) and advanced membrane bioreactor (MBR) treatment. Water Research. No. 43, 831–841.

20. Ratkovich N., Horn W., Helmus F.P., Rosenberger S., Naessens W., Nopens I., Bentzen T.R. 2013. Activated sludge rheology: A critical review on data collection and modelling. Water Research. No. 47. 463–482.

21. Sanin F.D., Clarkson W.W., Vesilind P.A. 2011. Sludge Engineering: the Treatment and Disposal of Wastewater Sludges, first ed. DEStech Publications Inc, Pennsylvania.

22. Solovieva E.A. 2008. Improving the design of secondary sedimentation tanks at sewage treatment plants. Promyshlennoe I grazhdanskoestroitel'stvo. No. 10. 37–38.

23. Technological regulations for the operation of treatment facilities at the Aeration station in Almaty. 2005, pp. 128 .

24. Typical project 902–2-377.83 Sewage sewage secondary sedimentation tanks made of prefabricated reinforced concrete with a diameter of 40 m. Album 6, part 2. Moscow 1983, pp. 74.

25. Vaxelaire J., Ce´zac P. 2004. Moisture distribution in activated sludge: a review. Water Research. No. 38. 2215–2230.

26. Yasui H., Shibata M. 1994. An innovative approach to reduce excess sludge production in the activated sludge process. Water Science and Technology. 30 (9). 11–20.

27. Zhmur N.S. 2011. Analysis of the causes of development, and methods for suppressing filamentous swelling of activated sludge and silt foaming. Part 1. Vodosnabzhenie i Kanalizatciya. No. 1. 94–107.

28. Zhmur N.S. 2003. Technological and biochemical processes of sewage treatment at facilities with aerotanks. Moscow: Akvaros, pp. 512.

2

Enhancement of Dairy Wastewater Treatment in a Combined Anaerobic Baffled and Biofilm Reactor with Magneto-Active Packing Media

Marta Kisielewska[1*], Marcin Dębowski[1], Marcin Zieliński[1], Mirosław Krzemieniewski[1]

[1] University of Warmia and Mazury in Olsztyn, Department of Environmental Engineering, ul. Warszawska 117, 10-950 Olsztyn, Poland

[*] Corresponding author's e-mail: jedrzejewska@uwm.edu.pl

ABSTRACT

In this study, a new reactor concept was designed for combining the advantages of anaerobic baffled reactors and biofilm reactors for treating dairy wastewater. The magneto-active microporous packing media manufactured by extrusion technology and modified by the addition of relevant amounts of metal catalysts and magnetic activation were used. The effects of active packing media placing in the different functional areas (hydrolysis or methanogenic) on the reactor performance (organic matter and nutrients removal, biogas production) were studied. The highest biogas production of 337 L/d and biogas yield of 415 mL/g $COD_{removed}$ were achieved when the packing media with magnetic properties were placed in the methanogenic tanks. A stimulatory effect of placing the active packing media in methanogenic tanks on the organic matter removal (86% as COD) and suspended solids elimination from wastewater were noted; however, the magnetic properties did not contribute towards higher organic matter and nutrients removal. Incorporation of metals into the plastic packing media enhanced the phosphorus removal (85–87%).

Keywords: biofilm reactor; compartmentalized reactor design, dairy sewage; magneto-active support medium; biogas; phosphorus removal

INTRODUCTION

Anaerobic digestion (AD) has been successfully applied in full-scale for the treatment of dairy wastewater (Tiwary et al., 2015). Several bioreactors types have been used; however, the biofilm-based reactors are the most common reactors employed (Rajagopal et al., 2013). Anaerobic dairy wastewater treatment in biofilm reactors has been conducted by using inert carriers such as seashells, plastic materials, ceramic, natural stones, gravel, pumice, sintered glass (Karadag et al. 2015). However, the use of packing media that enhance the sorption, precipitation and the binding of biogenic compounds constitute an alternative to the currently used support materials. Hawkes et al. (1995) used sand and activated carbon as support media for ice-cream wastewater treatment, and activated carbon ex-

hibited better performance than sand. Other authors showed that the corrosion of metal elements used as the packing media in an anaerobic reactor enhanced Chemical Oxygen Demand (COD) and phosphorus removal from dairy wastewater (Jędrzejewska-Cicińska, Krzemieniewski, 2010). As a constant magnetic field enhanced the efficiency of dairy wastewater treatment (Zieliński et al. 2014), the magneto-active packing media prepared by coating plastic media with iron and copper powder for dairy wastewater treatment were investigated as well (Dębowski et al. 2014). The authors obtained significantly improved COD removal efficiency, biogas production and methane content in biogas.

The biochemical processes in anaerobic bioreactors are accomplished by the cooperation of four different groups of microorganisms: fermentative, syntrophic, acetogenic, and methanogenic bacteria (Venkiteshwaran et al. 2015). The mutualistic behavior of various anaerobic microorganisms results in the decomposition of complex organic substances into simple, chemically stabilized compounds, mainly methane and CO2. However, each microbial group has different optimal environmental conditions and kinetics (Karadag et al. 2015). In order to provide the optimum conditions for satisfying the requirements of each group of anaerobic microorganisms, a reactor chamber could be divided into compartments (Gulhane et al. 2017). Anaerobic Baffled Reactors (ABRs) consist of alternating hanging and standing baffles, which compartmentalizes the reactor resulting in partial separation of acidogenesis and methanogenesis (Plumb et al. 2001). It has been shown that the compartmentalized design of anaerobic reactor provided a higher resilience to the hydraulic and organic shock loads, higher treatment rate, longer biomass retention times, and lower sludge yields than other anaerobic treatment systems (Reynaud, Buckley 2016).

Combining the advantages of ABF and biofilm reactors, an anaerobic reactor with a vertical labyrinth wastewater flow filled with the magneto-active packing media has been designed. This work aims to study the treatability of dairy wastewater in a new model of reactor and the effects of the active packing media placing in the different functional reactors areas (hydrolysis or methanogenic tank) on the reactor performance (organic matter and nutrients removal, biogas production).

MATERIALS AND METHODS

Wastewater

Dairy wastewater was prepared by dissolving 20 g milk powder manufactured by Dairy Plant in Ostrowia Mazowiecka (Poland) per litre of distilled water to achieve the COD concentration of 20,000 mg COD/L. The composition of dairy wastewater with a high content of COD was determined according to literature data, where the COD values were varying from 0.1 to 100 g/L (Slavov 2017). The dairy wastewater characteristics were as follows: total nitrogen (TN) 620±29.2 mg/L, ammonia nitrogen (AN) 33±5.7 mg/L, total phosphorus (TP) 109±9.2 mg/L, total suspended solids (TSS) 26±3.6 mg/L.

Biofilm support active media preparation

The active packing media (APM) used in the experiment were manufactured by microporous extrusion of granulated transparent and plasticised polyvinyl chloride (PVC). PVC, commercially available as Alfavinyl GFM/4–31–TR, was manufactured by Alfa Sp. z o.o. (Poland). Major parameters of the PVC used in the experiment are as follows: density – 1230 kg/m^3, Young's modulus – 2600 MPa, tensile strength – 21 MPa, elongation at break – 300%, shore A-hardness – 80 °Sh. While manufacturing the packing media, PVC was modified by the introduction of 0.8% wt granulated blowing agent Hydrocerol 530 (Clariant Masterbatch). Additionally, the PVC was blended with chemically pure copper and iron powder (Cometox). Each time, 5.0% wt. of metal additives was introduced to PVC used for the production of packing elements. The copper to iron weight ratio was $1_{Cu}:9_{Fe}$ and was set following the previous studies carried out by the authors (data not published). The characteristics of the manufactured APM were as follows: density – 810±0.6 kg/m^3, shore A-hardness – 24.8±0.4°Sh, tensile strength – 140.5±2.2 N, porosity – 39.13±0.4%.

The magnetic properties of the magneto-active packing media (M-APM) were achieved by the introduction of magnets to 40% of plastic elements placed inside the reactor. The characteristics of the neodymium magnets used in the experiments are as follows: diameter – 10±0.1 mm, height – 5±0.1 mm, magnetic flux ~3952·10^{-3} mWb, magnetic moment ~450.528·10^{-6} mWb·m, magnetic field in geometrical center of surface of the magnetic pole with a distance of 0.7 mm ~0.368 T, magnetic field next to the edge of surface of the magnetic pole (max.) with a distance of 0.7 mm ~0.384 T, pull force – 0.85 kg.

Experimental research installation setup and operation

A 100 L anaerobic reactor with a vertical labyrinth wastewater flow was used in the experiments. The reactor comprised three cylindrical stainless steel containers; that vertical channels and weirs pushed the wastewater through a series of chambers. The influent was pumped to the bottom part of the internal chamber serving as the hydrolysis tank with a working volume of 20 L. The internal recirculation with the yield of 20 L/h inside the hydrolysis tank ensured a complete mixing and circulation of the anaerobic sludge and wastewater. The outer chambers served as methanogenic tanks with an active volume of 70 L and downward wastewater flow in the central tank and 10 L and upward flow in the external. Additionally, the external tank served as the clarification chamber. The temperature in reactor was maintained at 35°C, the hydraulic retention time (HRT) was 2 days, while the organic loading rate (OLR) was maintained at the level of 10 g COD/L·d. The reactor was initially inoculated at a ratio of 40% (by volume). The inoculum originated from the closed fermentation tanks of a local municipal wastewater treatment plant. The concentration of volatile solids (VS) seeded into the reactor was 62.7% total solids (TS).

The experiment was divided into two stages, in which the packing media were placed in the hydrolysis tank (H) or the methanogenic tanks (M). The entire volume of the tanks was filled with the active packing media. In both cases, the experiments were conducted in the two parallel series with the APM (H1 and M1) and with M-APM (H2 and M2), (Fig. 1).

Analytical methods

The chemical oxygen demand (COD), total nitrogen (TN) and total phosphorus (TP) in dairy wastewater and effluent were analysed once every 24 h using a DR 5000 spectrophotometer (Hach Lange). The content of total solids (TS) and volatile solids (VS) in anaerobic sludge, dairy wastewater and effluent were determined according to the gravimetric method.

The biogas flow rate was measured continuously using a XFM17S digital gas flow meter (Aalborg Instruments & Controls, Inc., USA). The quality of biogas was analysed using a GMF 430

meter (GasData, England) and a gas chromatograph (GC, 7890A Agilent, USA) equipped with a thermal conductivity detector (TCD). The GC was fitted with the two Hayesep Q columns (80/100 mesh), two molecular sieve columns (60/80 mesh) and Porapak Q column (80/100) operating at a temperature of 70°C. The temperatures of the injection and detector ports were 150°C and 250°C, respectively. Helium and argon were used as the carrier gases at a flow of 15 mL/min.

The statistical analysis of results was carried out using the STATISTICA 10.0 PL software package. The hypothesis on the distribution of each analysed variable was verified using the W Shapiro–Wilk test. One-way analysis of variance (ANOVA) was used to determine the significance of differences between variables. The homogeneity of variance in the groups was analysed using Levene's test. The RIR Tukey test was used to determine the significance of differences between the analysed variables. The statistical significance was adopted at p = 0.05.

RESULTS

Organic matter and nutrients removal

The study showed that the packing media placement had a significant effect (p<0.05) on the COD removal efficiency from the dairy wastewater (Fig. 2, Fig. 3). When the active packing media were placed in the methanogenic tanks, the COD removal was over 84%, while the effects obtained with the active media placed in the hydrolysis tank did not exceed 79% (Fig. 2). A similar trend was noted for COD load removal (Fig. 3). The difference between H1/M1 and H2/

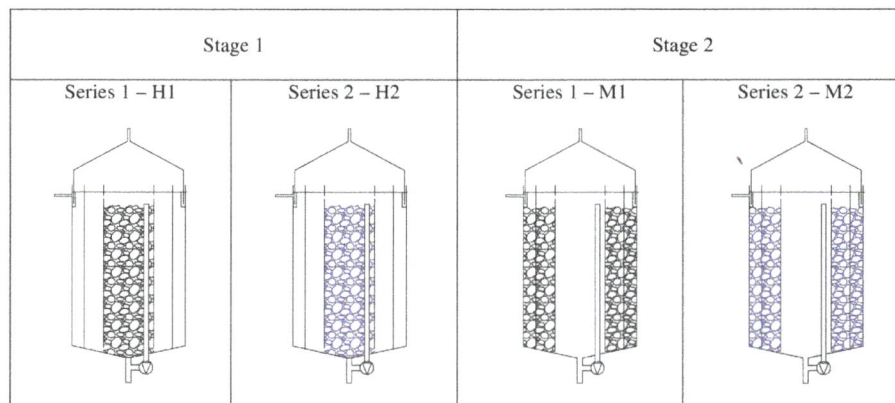

| Stage 1 | | Stage 2 | |
| Series 1 – H1 | Series 2 – H2 | Series 1 – M1 | Series 2 – M2 |

Fig. 1. The study organization: H1 – APM placed in the hydrolysis tank, H2 – M-APM placed in the hydrolysis tank, M1 – APM placed in the methanogenesis tanks, M2 – M-APM placed in the methanogenesis tanks

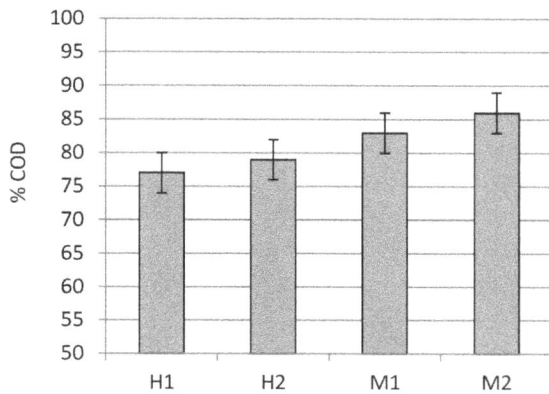

Fig. 2. The effects of active packing media placement and magnetic properties on the COD removal

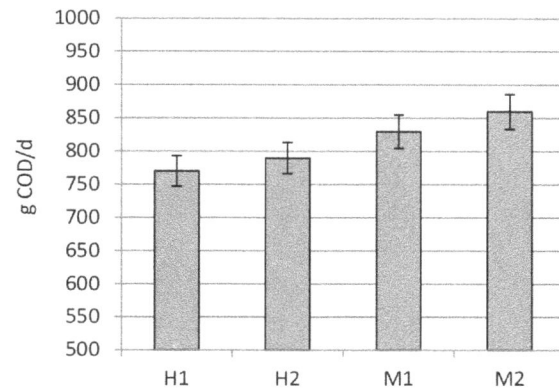

Fig. 3. The effects of active packing media placement and magnetic properties on the COD loading removal

M2 was 60 g/d and 72 g/d, respectively (p<0.05). The type of the applied medium had no effects on the COD removal (p>0.05). The use of M-APM in the hydrolysis tank (H2) allowed to remove 791 g COD/d from dairy wastewater, which was only 16 g COD/d higher than using the APM (H1), (Fig. 3). Similarly, using M-APM in the methanogenic tank (M2) allowed to remove 863 g COD/d, which was 28 g/d more than in M1 (p>0.05).

Neither the type of the active media, nor the placement of the media in anaerobic digester had a significant effect (p>0.05) on the TN removal. During the study, the TN removal ranged from 22% in H1 to 26% in M2 (Fig. 4). Both active media (APM and M-APM) contributed to a high TP removal ranging from 85% to 87% in all series (Fig. 5). However, there were no differences (p>0.05) in the series with different placement of the media in the reactor chamber (M and H). A slightly higher (2 – 3%, p>0.05) TP removal was noted in H2 and M2 series using M-APM. The placement of the active media in methano-

genic tanks affected the concentration of suspended solids in the effluent, which was significantly lower in M1 and M2 series (p<0.05), (Fig. 6). The magnetic properties of M-APM in H-series did not affect (p>0.05) the reduction of the suspended solids in the effluent; however, in the M-series, the differences were significant (p<0.05), (Fig. 6).

Biogas and methane production

The magnetic properties of the active packing media contributed towards higher biogas production and biogas yield, both in the H-series and M-series (Fig. 7, Fig. 8). In the M-series, the increase in daily biogas production was about 10%, while in H-series it was 8% (p<0.05). Moreover, the placement of the active packing media in methanogenic tanks contributed towards a better biogas production and biogas yield (p<0.05), (Fig. 7, Fig. 8). The highest biogas production of 337 L/d and yield of 415 mL/g $COD_{removed}$ were achieved in M2. Neither magnetic properties of

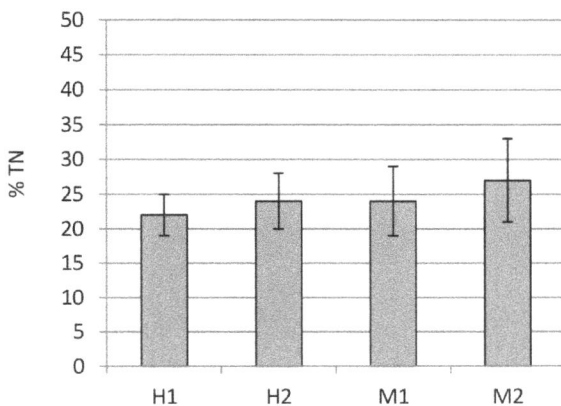

Fig. 4. The effects of active packing media placement and magnetic properties on the TN removal

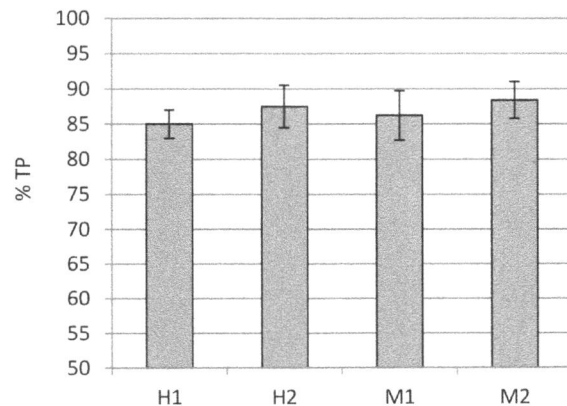

Fig. 5. The effects of active packing media placement and magnetic properties on the TP removal

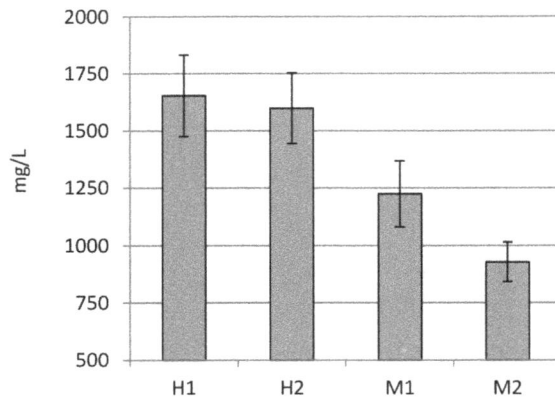

Fig. 6. The effects of active packing media placement and magnetic properties on the suspended solids removal

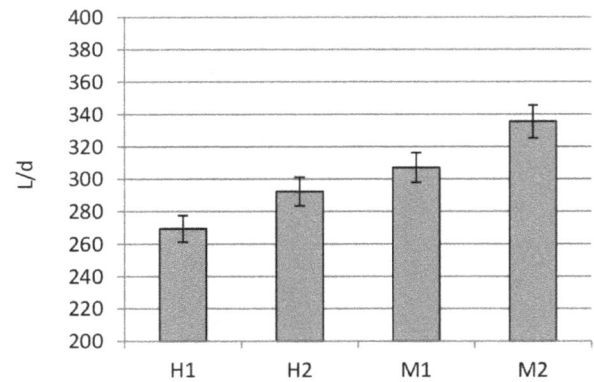

Fig. 7. The effects of active packing media placement and magnetic properties on the daily biogas production

the active media nor their placement did affected the methane content in the biogas (p>0.05), (Fig. 9). The methane concentration was similar in H-series and M-series, and ranged from 66% to 67% during the study (Fig. 9).

DISCUSSION

A new reactor concept was designed for combining the advantages of ABR and biofilm reactors. Our study results confirmed that the use of the active media in a combined anaerobic baffled and biofilm reactor enhanced the organic matter and phosphorus removal from dairy wastewater, reduced the the suspended solids concentration in the effluent and increased the methane production.

In biofilm reactors, the support materials are responsible for the microbial biomass and suspended solids retention within reactors. The mi-

crobial cells concentrations in biofilm reactors can achieve 74 g/L by adhering and absorbing of cells to the support material (Qureshi et al. 2005). High biomass concentration contributed to the high treatment performance. In our study, the COD removal efficiency ranged from 77% to 86% at OLR of 10 kg COD/m^3·d, and the highest COD removal was achieved when the packing media were placed in methanogenic tanks. During the treatment of dairy wastewater, Rajinikanth et al. (2009) obtained 80% COD removal efficiency at OLR of 17 kg COD/m^3·d in an up-flow anaerobic filter. Kundu et al. (2013) operated an anaerobic fluidized bed reactor for the treatment of simulated milk wastewater. They achieved the COD removal of 78% at OLR of 8 kg COD/m^3·d. In turn, Rodgers et al. (2004) studied whey treatment in anaerobic moving biofilm reactor with plastic support media and obtained 89% COD removal at OLR of 11.6 kg COD/m^3·d.

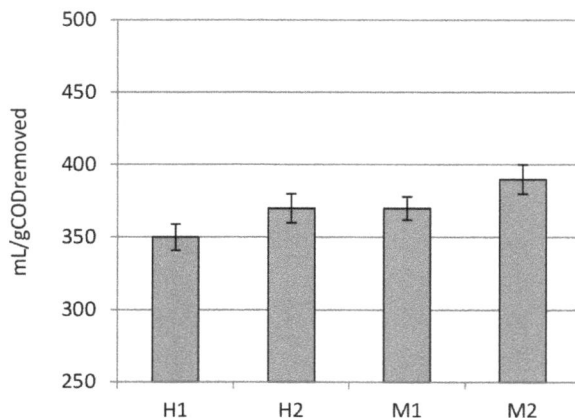

Fig. 8. The effects of active packing media placement and magnetic properties on the biogas yield

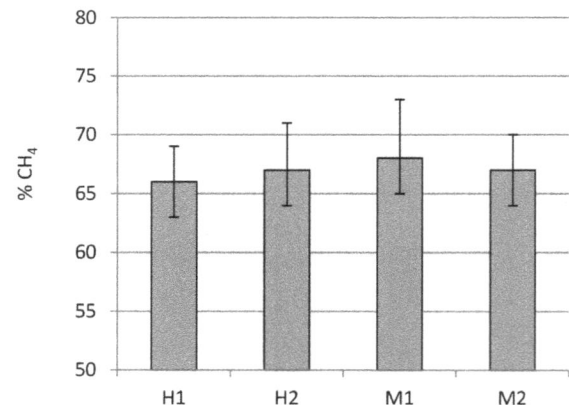

Fig. 9. The effects of active packing media placement and magnetic properties on the methane concentration in biogas

The use of packing media that enhance the sorption, precipitation and the binding of biogenic compounds constitutes an alternative to the currently used support materials. In the presented study, an innovative active media with metal additives placed in anaerobic reactors allowed for the effective phosphorus removal (85–87%) and suspended solids elimination. The corrosion of metals stimulates the biochemical degradation of organic compounds by modification of anaerobic environment inside the reactor through a reduction in the oxidation-reduction potential (ORP) and an increase in the buffer capacity (Zhang et al., 2011; Wu et al., 2015). It also enhances the abiotic processes of contaminant removal, because the formed colloids (i.e. $Fe(OH)_2$ and $Fe(OH)_3$) help to eliminate the organic suspended solids in flocculation, adsorption and precipitation reactions (Noubacept, 2008). Moreover, insoluble vivianite compounds are formed in the reaction of metal ions with phosphates, which reduces the phosphorus concentration in the treated wastewater (Shi et al., 2011; Wu et al., 2015). Similar process may favour the nitrogen compounds removal via forming complexes with iron ions. However, during the study, the use of packing media did not enhance the TN removal.

Placing the active media in methanogenic tanks significantly enhanced the COD removal and suspended solids elimination from dairy wastewater. The magnetic properties of the active media did not affect the COD, phosphorus and nitrogen removal, but it enhanced the suspended solids elimination from the effluent. The enhancement of suspended solids coagulation as well as better sedimentation of activated sludge by magnetic field (Zaidi et al. 2014).

According to the literature, the corrosion of metals intensified methanogenesis (Shi et al., 2011; Liu et al., 2015). During the study, the placement of the active packing media in methanogenic tanks significantly enhanced the biogas production and yield. During AD, metal catalysts such as zero-valent iron are electron-donors for methanogenic and denitrifying bacteria (Karri et al. 2005). Under anaerobic conditions, iron corrosion is triggered by gaseous H_2, and the formed CO_2 is reduced to CH_4 by hydrogenotrophic methanogens such as *Methanococcus thermolithotrophicus*, *Methanobacterium thermoautotrophicum*, or *Methanospirillum hungatei* (Wu et al. 2015). Shi et al. (2011) used iron nanoparticles to supplement the UASB reactor and obtained an increase in the biogas productivity and

the methane content in biogas (by 12.9–17.9% and 10.7–12.9%, respectively). In a study on the UASB reactor filled with iron elements Zhang et al. (2011) achieved a 66.8% content of methane in biogas versus 47.9% in biogas generated in an UASB reactor without elements.

The magnetic properties of M-APM used in the presented study enhanced the daily biogas production and yield. According to Ji et al. (2010), a suitable level of magnetic induction, not greater than 17.8 mT stimulates the metabolic activity of microorganisms, thus improving the efficiency of wastewater treatment. Zieliński et al. (2014) showed that the value of magnetic induction of 0.38 T positively affected the organic matter removal and biogas production. Krzemieniewski et al. (2004) applied a stable magnetic field with induction of 0.4–0.6 T and achieved an increased efficiency of the COD, ammonia nitrogen and orthophosphates removal from municipal, household and dairy wastewater (25–55%, 50–66%, 87–90%, respectively) when compared to the removal rates in the system without magnetic field assistance.

CONCLUSIONS

This study demonstrated that incorporation of metals into the plastic packing media used in anaerobic reactors could enhance the organic matter, phosphorus and suspended solids removal, and also contributed to a better biogas production. Placing the active packing media in the methanogenic functional areas of an anaerobic reactor significantly enhanced the COD removal form dairy wastewater, suspended solids elimination, as well as the daily biogas production and biogas yield. The magnetic properties of the active media did not affect any COD and phosphorus removal, but it greatly contributed towards higher biogas production and suspended solids elimination.

Acknowledgements

The research was conducted under Project POIG.01.01.02–14–034/09, entitled: Catalytic fillings for bioreactors for industrial wastewater treatment technology, sponsored by the Programme Innovative Technical Support Systems for Sustainable Development Task IV.5.2. and was also supported by Project No. 18.610.008–300 from the University of Warmia and Mazury in Olsztyn.

REFERENCES

1. Dębowski M., Zieliński M., Krzemieniewski M., Brudniak A. 2014. Effect of magneto-active filling on the effectiveness of methane fermentation of dairy wastewaters. International Journal of Green Energy, doi.org/10.1080/15435075.2014.909362.

2. Gulhane M., Pandit P., Khardenavis A., Singh D., Purohit H. 2017. Study of microbial community plasticity for anaerobic digestion of vegetable waste in Anaerobic Baffled Reactor. Renewable Energy, 101, 59–66.

3. Hawkes F.R., Donnelly T., Anderson G.K. 1995. Comparative performance of anaerobic digesters operating on ice-cream wastewater. Water Research, 29, 525–533.

4. Jędrzejewska-Cicińska M., Krzemieniewski M. 2010. Effect of corrosion of steel elements on the treatment of dairy wastewater in a UASB reactor. Environmental Technology, 31, 585–589.

5. Ji Y., Wang Y., Sun J., Yan T., Li J., Zhao T., Yin X., Sun C. 2010. Enhancement of biological treatment of wastewater by magnetic field. Bioresource Technology, 101, 8535–8540.

6. Karadag D., Köroğlu O.E., Ozkaya B., Cakmakci M. 2015. A review on anaerobic biofilm reactors for the treatment of dairy industry wastewater. Process Biochemistry, 50, 262–271.

7. Karri S., Sierra-Alvarez R., Field J.A. 2005. Zero valent iron as an electron-donor for methanogenesis and sulfate reduction in anaerobic sludge. Biotechnology and Bioengineering, 92, 810–819.

8. Krzemieniewski M., Dębowski M., Janczukowicz W., Pesta J. 2004. Effect of the Constant Magnetic Field on the Composition of Dairy Wastewater and Domestic Sewage. Polish Journal of Environmental Studies, 13, 45–53.

9. Kundu K., Bergmann I., Hahnke S., Klocke M., Sharma S., Sreekrishnan T.R. 2013. Carbon source – A strong determinant of microbial community structure and performance of an anaerobic reactor. Journal of Biotechnology, 168, 616–624.

10. Liu Y., Wang Q., Zhang Y., Ni B.J. 2015. Zero valent iron significantly enhances methane production from waste activated sludge by improving biochemical methane potential rather than hydrolysis rate. Scientific Reports, 5, 8263.

11. Noubactep C. 2008. A critical review on the process of contaminant removal in Fe^0-H_2O systems. Environmental Technology, 29, 909–920.

12. Plumb J.J., Bell J., Stuckey D.C. 2001. Microbial Populations Associated with Treatment of an Industrial Dye Effluent in an Anaerobic Baffled Reactor. Applied and Environmental Microbiology, 67, 3226–3235.

13. Qureshi N., Annous B.A., Ezeji T.C., Karcher P., Maddox I.S. 2005. Biofilm reactors for industrial bioconversion processes: employing potential of enhanced reaction rates. Microbial Cell Factories, 4, 24.

14. Rajagopal R., Saady N.M.C., Torrijos M., Thanikal J.V., Hung Y.T. 2013. Sustainable Agro-Food Industrial Wastewater Treatment Using High Rate Anaerobic Process. Water, 5, 292–311.

15. Rajinikanth R., Ganesh R., Escudie R., Mehrotra I., Kumar P., Thanikal J.V., Torrijos M. 2009. High rate anaerobic filter with floating supports for the treatment of effluents from small-scale agro-food industries. Desalination and Water Treatment, 4, 183–190.

16. Reynaud N., Buckley C.A. 2016. The anaerobic baffled reactor (ABR) treating communal wastewater under mesophilic conditions: a review. Water Science Technology, 73, 463–478.

17. Rodgers M., Zhan X.M., Dolan B. 2004. Mixing characteristics and whey wastewater treatment of a novel moving anaerobic biofilm reactor. Journal of Environmental Science and Health Part A: Toxic/hazardous Substances & Environmental Engineering, 39, 2183–2193.

18. Shi R., Xu H., Zhang Y. 2011. Enhanced treatment of wastewater from the vitamin C biosynthesis industry using a UASB reactor supplemented with zero-valent iron. Environmental Technology, 32, 1859–1865.

19. Slavov A.K. 2017. General Characteristics and Treatment Possibilities of Dairy Wastewater – A Review. Food Technology and Biotechnology, 55, 14–28.

20. Tiwary A., Williams I.D., Pant D.C., Kishore V.V.N. 2015. Emerging perspectives on environmental burden minimisation initiatives from anaerobic digestion technologies for community scale biomass valorization. Renewable and Sustainable Energy Reviews, 42, 883–901.

21. Venkiteshwaran K., Bocher B., Maki J., Zitomer D. 2015. Relating Anaerobic Digestion Microbial Community and Process Function. Microbiology Insights, 8, 37–44.

22. Wu D., Zheng S., Ding A., Sun G., Yang M. 2015. Performance of a zero valent iron-based anaerobic system in swine wastewater treatment. Journal of Hazardous Materials, 286, 1–6.

23. Zaidi N.S., Sohaili J., Muda K., Sillanpää M. 2014. Magnetic field application and its potential in water and wastewater treatment systems. Separation & Purification, 43, 206–240.

24. Zhang Y., Jing Y., Quan X., Liu Y., Onu P. 2011. A Built-In Zero Valent Iron Anaerobic Reactor to Enhance Treatment of Azo Dye Wastewater. Water Science and Technology, 63, 741–746.

25. Zieliński M., Dębowski M., Krzemieniewski M., Dudek M., Grala A. 2014. Effect of the constant magnetic field (CMF) with various values of magnetic induction on the effectiveness of dairy wastewaters treatment under anaerobic conditions. Polish Journal of Environmental Studies, 23, 255–261.

3

Study on the Influence of Selected Technological Parameters of a Rotating Biological Contactor on the Degree of Liquid Aeration

Joanna Szulżyk-Cieplak[1*], Aneta Tarnogórska[1], Zygmunt Lenik[1]

[1] Faculty of Fundamentals of Technology, Lublin University of Technology, ul. Nadbystrzycka 38, 20-618 Lublin, Poland

* Corresponding author's e-mail: j.szulzyk-cieplak@pollub.pl

ABSTRACT

The subject of the research involves a rotating biological contractor with a bi-directional longitudinal flow as an element of a synchronized system of disposal and biological treatment of domestic wastewater in small-bore sewerage. The rotor design is based on a system of corrugated protective pipes, arranged in coils wound around its rotation axis. The pipes are wound in a way that enables a bi-directional flow of liquids. During the rotor rotation in wastewater, the contactor is simultaneously emptied and filled with wastewater. The role of corrugated protective pipes is twofold; on the one hand, they constitute a surface for the biofilm development and on the other, they enable the flow of liquids, thus ensuring its aeration. The contactor design aims to achieve intensive aeration of transported wastewater, which will allow for greater development of microorganism populations participating in hybrid wastewater treatment, i.e. the methods involving activated sludge and trickling filter. An analysis on the influence of rotor operation kinematics on the efficiency of liquid aeration was conducted. The aeration capacity for variable rotational speed (0.7 rpm, 1.5 rpm, 2.0 rpm, 3.0 rpm) and direction of the contactor rotating element were calculated. In the considered case, oxygen transfer coefficient K_{La} was within the range of 0.011÷0.023 1/min. The obtained results indicate a clear connection between the system kinematics and the degree of liquid aeration in the contactor.

Keywords: biological rotating contactor, domestic wastewater, aeration, oxygen transfer coefficient

INTRODUCTION

The water and wastewater management constitutes one of the main factors influencing the living standard of residents as well as the condition and quality of the natural environment. Management of water resources should conform to the principle of sustainable development, which aims to ensure the possibility of satisfying the basic needs of both the contemporary and future generations, without upsetting the natural balance. This principle holds in the case of creating water use conditions, water protection and water resources management and it aims at achieving good ecological state of waters in a country. Proper water management must be based on a rational and modern wastewater management. As indicated in the literature data (Szulżyk-Cieplak et al. 2015, Duda et al. 2016, Błażejewski, 2012), one of the main problems of water and waste-

water management in Poland is connected with the low degree of sewage disposal services characterizing the rural areas. Approximately 70% of country population uses sewerage systems, including only 37.3% or countryside residents. The basic problem of building a sewerage system in rural areas is the necessity of constructing and operating numerous small wastewater treatment plants or constructing extremely long collection and transportation systems which supply wastewater to a collective treatment plant. As relatively few residents benefit from these solutions, they are ineffective in respect to economy, ecology, and energy consumption. A solution to the issue of water and wastewater management in rural areas may involve the application of non-conventional systems. One of such systems includes simultaneous disposal and treatment of wastewater in a rotating biological contactor (Szulżyk-Cieplak et al. 2016).

STUDY AREA

Rotating biological contactors (RBCs) are widely applied in the treatment of both the domestic and industrial wastewater. They constitute an alternative to the conventional method involving the activated sludge technology of wastewater treatment (Patwardhan 2003). Rotating contactors are characterized by the stability of treatment processes, low electric energy consumption, short retention time, low maintenance costs and ease of operation (Ghawi et al. 2009, Pathan et al. 2015). There are numerous solutions related to the construction and technology of rotating biological contactors. In her work, Ryschka (Ryschka et al. 2014) proposed four main criterions for categorization of rotating biological contactors: type of media (discs, cylinders), media submergence level (classic or submerged rotating biological contactor), type of drive (mechanical or air), and treatment method (conventional and hybrid rotating biological contactors). On the other hand, Żubrowska-Sudoł (Żubrowska-Sudoł et al. 2007) distinguished three criteria for the division of rotating biological contactors: type of wastewater flow: Moving Bed Biofilm Reactor (MBBR) and Moving Bed Sequencing Batch Biofilm Reactor (MBSBBR), share of biomass in the form of biofilm and activated sludge (conventional and hybrid MBBR), and the type of treatment process involved (removal of organic pollutants – COD, BOD_5, nitrification process, denitrification process, integrated removal of C, N, and P compounds, anaerobic ammonia oxidation). As can be noted, in both cases there is a common classification criterion of rotating biological contactors into conventional and hybrid ones. One of the most common types of classic rotating biological contactors, in which the treatment occurs as a result of wastewater coming into contact with biofilm microorganisms, is a rotating disc biological contactor (Patwardhan 2003, Cortez et al. 2013, Hassard et al. 2015). It is made of discs centred around a horizontal shaft, submerged in wastewater to about 40% of their surface area. The surface of these discs is suitable for the development of biofilm. During their rotation, the biofilm attached to disc surface is carried into the air and aerated. Oxygen is supplied to wastewater in two ways: in the process of mixing resulting from the revolving shaft and due to the diffusion from the biofilm. As opposed to the conventional RBCs,

the hybrid ones combine two factors of biological pollutant break-down in one reactor: biomass in the form of activated sludge floc and biofilm (Staehler 2004, Lenik et al. 2016).

The efficiency of biodegradation in rotating biological contactors is dependent on a number of parameters, including: the content of oxygen dissolved in wastewater, flow intensity of wastewater, content of organic compounds, rotational speed of the rotor as well as the system configuration, including the method of liquid distribution and aeration conditions (Waskar et al. 2012, Mba et al. 2007, Celenza 1999). These parameters affect, among others, maintain aerobic conditions in the reactor. It is known that the effectiveness of oxygen transfer is a critical factor determining successful application of any bioreactor for the treatment of wastewater (Chavan et al.2008). The research results (Pradeep et al. 2011, Borghi et al. 1985) indicate the concentration of dissolved oxygen in wastewater, and thus the degradation efficiency of pollutants contained within, increase along with the rotational speed of the contactor.

The biological rotating contactor with bi-directional longitudinal flow under study constitutes a novel solution (Lenik et al. 2016). The rotor, which is the main element of the device, ensures beneficial aeration conditions necessary for the biochemical pollutant degradation process and serves as a mobile trickling filter. The paper discussed the studies which enabled the evaluation of the impact of the rotor kinematic parameters on the efficiency of liquid aeration process.

MATERIAL AND METHODS

Experimental set-up

The scheme of the laboratory rig is presented in fig. 1. The test rig includes a biological contactor with a rotating element, i.e. a rotor (1), which is powered by a drive system comprising a electric geared motor (7) with a stepless adjustment of revolutions 0 ÷ 10 rpm. The test rig was equipped with an electromagnetic flow meter (6) which enables to control the flow rate of the influent and a RDO/DO dissolved oxygen concentration meter (3).

The rotating biological contactor with bi-directional longitudinal flow (Photo 2.1) is placed in a tank made of polyester resin (1). The central element of the RBC is a rotor (2) built of mul-

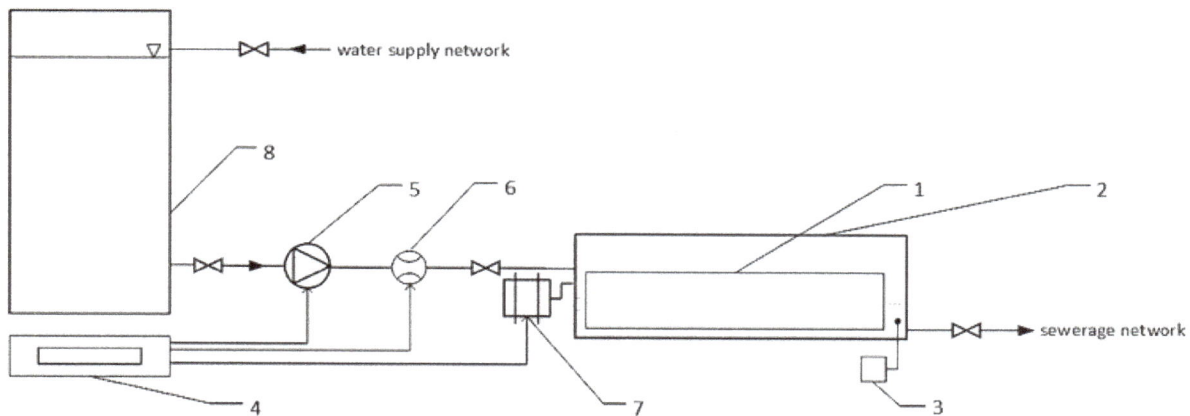

Figure 1. Schematic diagram of the laboratory rotating biological contactor: 1 – rotor, 2 – RBC housing, 3 – RDO/DO meter, 4 – control panel, 5 – peristaltic pump, 6 – electromagnetic flow meter, 7 – drive system, 8 – tank containing the test medium

tiple layers of corrugated pipes (3) wound cylindrically around its axis housing a shaft with a float (4). The amount of liquid supplied to the tank with rotor should enable its free rotation on its own axis, which enables a constant flow of the medium through the system of corrugated protective pipes. The level of medium in the tank should not exceed the height of the height of rotor drive shaft axis and – simultaneously – a half of the contactor diameter.

On the basis simulations employing the MES method, an appropriate angle of pipe winding was selected, with alternating direction and sense. This ensured the assumed flow rate and aeration of the medium (Lenik et al. 2013). The winding of corrugated pipes (Figure 2) enables a bi-directional flow of the liquid medium. The flow in layers is alternating, i.e. the fluid in the first layer flows in one direction, whereas in the next layer, the flow direction is opposite. Aeration of

the liquid medium is intensified by the alternating sense and direction of corrugated channels; their number and interlacing have a significant impact on the hydrodynamics of the process involving mixing air with wastewater.

Research methodology

The degree of liquid aeration was investigated for various rotational speed of the rotor. Literature data indicate that the energy usage for motor drive increases exponentially with increasing rotational speed of the contactor, thus for the minimal operating expense the lowest rotor speeds should be selected (0,7 ÷ 2,0 rpm) (Hassard et al. 2015). Therefore, the following rotational speeds were selected: 3 rpm, 2 rpm, 1.5 rpm and 0,7 rpm. The rotor operation in counter-clockwise and clockwise rotation was compared as well. Each series was conducted in triplicate, whereas the

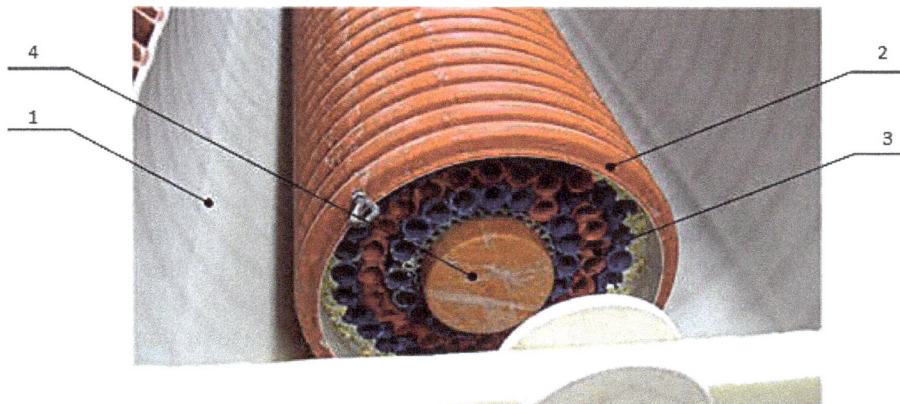

Photo 1. Laboratory model of a rotating biological contactor with bi-directional longitudinal flow:
1 – RBC housing, 2 – rotor, 3 – corrugated protective pipes, 4 – float

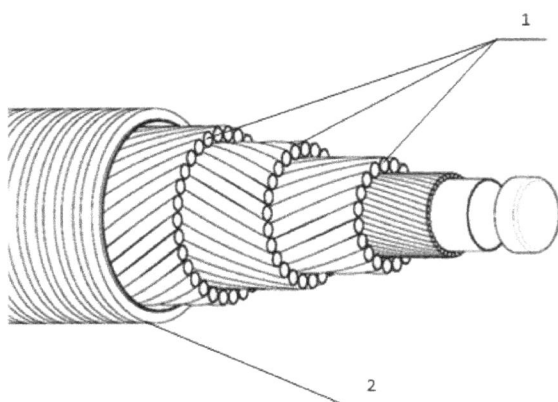

Figure 2. Arrangement of corrugated protective pipes in the rotor with bi-directional longitudinal flow: 1 – cylindrically wound layers of corrugated protective pipes, 2 – rotor housing

mean values from particular measurement series were used for the preparation of results.

The test medium comprised tap water with the temperature $t_0 \approx 16\ °C$. The studies were carried out with the water deoxidized by means of sodium sulfite addition (Na_2SO_3). In order to deoxidize the required amount of tap water, the initial concentration of dissolved oxygen in sample "0" was measured, which was necessary to calculate the theoretical amount of the oxygen-reducing agent. The tap water was deoxidized according to the reaction (1):

$$2\ Na_2SO_3 + O_2 \rightarrow 2\ Na_2SO_4 \qquad (1)$$

The required amount of the reducing agent, calculated on the basis of reaction (1), was supplemented with an extra 20%. A single series of experiment required deoxidization of 90 dm^3 of tap water in order to achieve appropriate rotor submersion. Water was considered deoxidized, when the content of dissolved oxygen was lower than 1 mg O_2/dm^3.

Prior to conducting the experiment, appropriate rotational speed and rotation direction were set by means of the control panel. When suitable kinematic parameters for each measurement series were selected, an oxygen sensor was placed in the tank containing the aerated medium. The content of dissolved oxygen was read in 30 second intervals. The liquid aeration was performed until three consecutive values of oxygen concentration with the difference lower than 1 mg O_2/dm^3 were read.

The obtained results enabled to calculate the aeration capacity (efficiency of contactor aera-

tion) for particular kinematic parameters of the system operation. The efficiency was calculated with the formula (2).

$$K_L a = \frac{1}{t_p} * ln\frac{C_s}{C_s - C_t} \qquad (2)$$

where: $K_L a$ – volumetric oxygen transfer coefficient [1/min],
t_p – Total aeration time,
C_s – saturation in a given temperature,
C_t – maximum dissolved oxygen concentration

RESULTS AND DISCUSSION

The results of conducted studies were presented graphically on graphs (Fig. 3÷5). In each of the analyzed cases, the changes in the oxygen concentration in the test medium were of certain character. An exemplary flow of the liquid aeration process was presented in Figure 3 (rotor rotational speed v = 3 rpm, counter-clockwise and clockwise rotation).

While analyzing the graph, one can observe a slightly faster increase pertaining to the concentration of dissolved oxygen in the test medium when clockwise rotation was applied. A similar trend was observed for the remaining rotation speeds, i.e. v = 2 rpm, v = 1.5 rpm and v = 0.7 rpm.

The influence of rotor rotational speed on the aeration degree of the test medium was analyzed. Fig. 4 presents the results of the studies conducted for variable rotational speed (v = 1.5 rpm, 2 rpm, 3 rpm) and clockwise rotation.

The data presented on the graph (Fig. 4) indicate the dependency of the liquid aeration degree on the rotor rotational speed. Increasing the rotor rotational speed simultaneously raises the efficiency of RBC operation in terms of the liquid aeration degree. The results obtained coincide with the results of other researchers on the subject of rotating disc contactors. Moreover, the conducted research shows that the increase in rotational speed reduces the time required for a liquid to stabilize, i.e. mixing of a reducing agent with the medium in order to carry out the aeration process properly. A similar dependency was observed in the case of counter-clockwise rotation. Taking into account operating expense of the process, the results obtained indicate that the optimal solution appears to be the rotor speed of v = 2 rpm and the aeration time of t = 60 min.

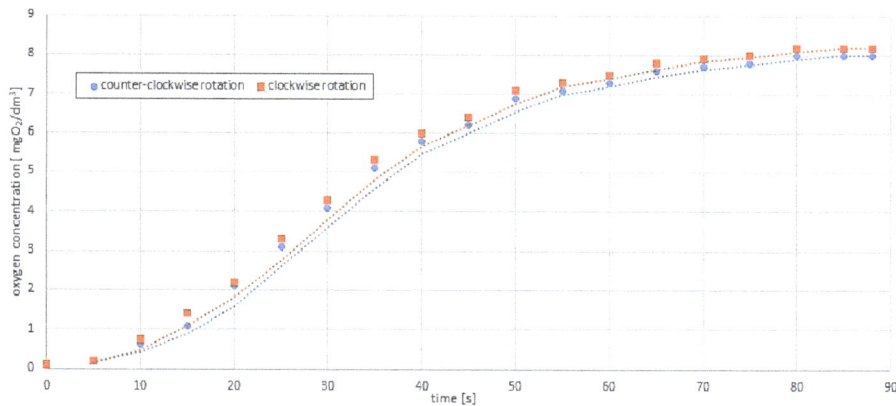

Figure 3. Changes in the concentration of dissolved oxygen in the test medium for the rotational speed of 3 rpm.

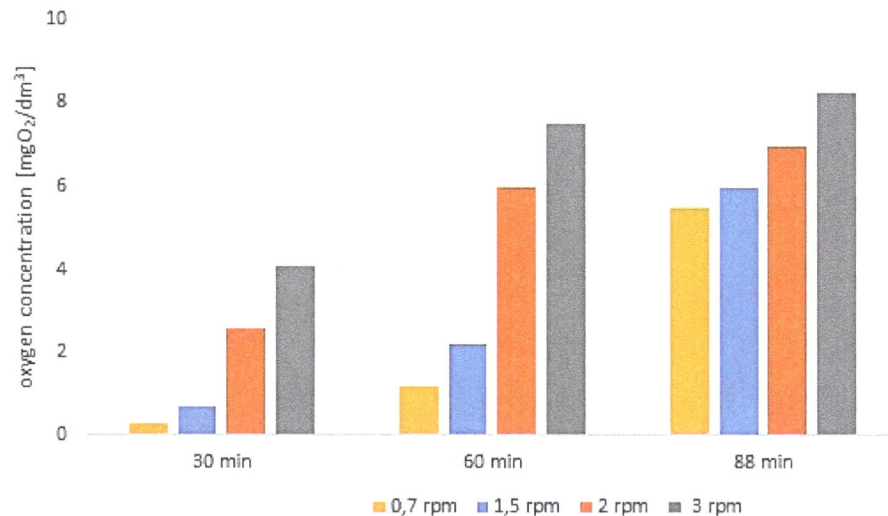

Figure 4. Oxygen concentration in the test medium following the aeration time t = 30 min, 60 min and 88 min for variable rotor rotational speed (v = 0.7 rpm, 1.5 rpm, 2 rpm, 3 rpm min) with clockwise rotation applied

The studies conducted enabled to determine the volumetric oxygen transfer coefficient (K_{La}) for variable kinematic parameters of the system. The calculations were performed on the basis of the dependency (2), whereas the calculations results were presented in Table 1 and a graph (Fig. 5).

The obtained results indicate an exponential dependency of the liquid aeration efficiency on the kinematic parameters characterizing the system, i.e. direction and rotational speed of rotor. The highest volumetric oxygen transfer coefficient was obtained for the rotational speed of v = 3 rpm, clockwise. Overall, comparing the results obtained with the results obtained by other researchers (Boumansour et al. 1998, Paolini 1986), it can be concluded that the applied constructional solution of the reactor in relation to the degree of liquid aeration gives satisfactory results. For rotational speed of v = 0, 7÷3.0 rpm, higher value of oxygen transfer coefficient was obtained as compared to the value obtained by Boumansour and Paolini. (Fig. 5).

CONCLUSIONS

1. Ensuring adequate aerobic conditions constitutes one of the main factors influencing the efficiency of reactor operation based on biochemical processes carried out under aerobic conditions. The obtained results indicate a possibility adjusting the aeration degree of liquids in a reactor – which is one of the decisive parameters connected with the efficiency and quality or rotating biological contactor operation – through rotor operation kinematics.

Table 1. Values of volumetric oxygen transfer coefficient for variable kinematic parameters in the system.

Rotor rotational speed [rpm]	Liquid aeration time [min]	Temperature of the test medium [°C]	Saturation in a given temperature [mg/dm³]	Concentration of dissolved oxygen. [mg/dm³]	Volumetric oxygen transfer coefficient [1/min]
Counter-clockwise rotation					
3	90.5	17.2	9.75	8.16	0.020
2	94.5	18.2	9.54	7.02	0.014
1.5	97.0	17.6	9.54	6.26	0.011
0.7	99.0	17.5	9.54	5.52	0.008
Clockwise rotation					
3	88.5	17.8	9.54	8.26	0.023
2	92.5	18.0	9.54	7.10	0.015
1.5	95.0	17.2	9.75	6.30	0.012
0.7	98.0	17.5	9.54	5.60	0.009

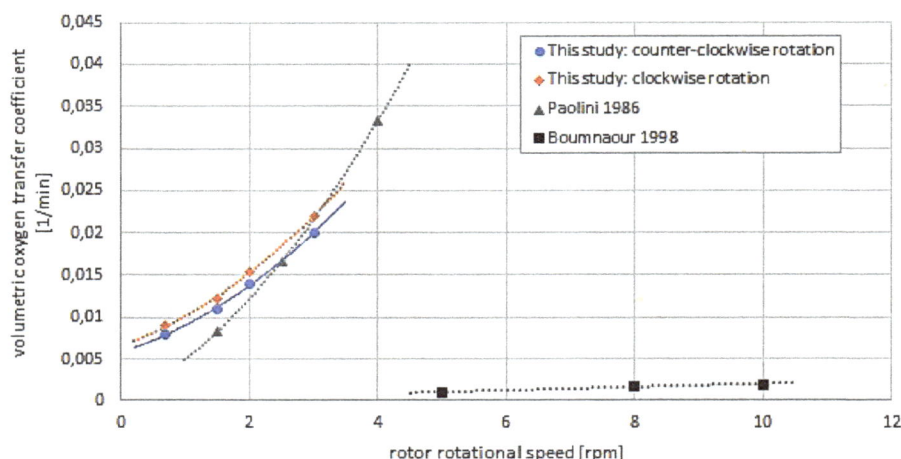

Figure 5. Dependence of volumetric oxygen transfer coefficient K_{La} on the rotor rotational speed

2. The volumetric oxygen transfer coefficient K_{La} was determined on the basis of the measurements pertaining to the dissolved oxygen, which were carried out until saturation was achieved. In the case considered, K_{La} was within the range of 0.011÷0.023 1/min, which is a satisfactory result as compared to the results of other researchers. Taking into account operating expense of the process, the results obtained indicate that the optimal solution appearss to be the rotor speed of v = 2 rpm and the aeration time of t = 60 min.

3. A comprehensive examination of various alternatives for the system kinematics allowed for selection of the suitable RBC configuration to be employed in a demonstrative sewerage network. This device ensures adequate aeration conditions, necessary for the biochemical pollutant breakdown process and simultaneously serves as a Moving Bed Biofilm Reactor. The solution employed does not require construction of the mechanical parts found in the conventional treatment plants and wastewater can be supplied directly to the tank with a rotor, immediately followed by the aeration. Therefore, the application of the rotor as an element of a simultaneous wastewater discharge and treatment system in small settlement units, may reduce the investment costs of the water and wastewater management.

REFERENCES

1. Błażejewski R., 2012. Status and opportunities for water and sewage infrastructure in Poland (In Polish), Gaz, Wodna i Technika Sanitarna, 2, 49–51.

2. Borghi M., Pallazi E., Parisi F., Ferraiolo G., 1985. Influence of process variables on the modelling and design of a rotating biological surface, Water Research, 19(5), 573–580.

3. Boumansour B.E., Vasel J.L., 1998. A new tracer gas method to measure oxygen transfer and enhancement factor on RBC, Water Research, 32(4), 1049–1058.

4. Celenza G., 1999. Industrial Waste Treatment Process Engineering: Biological Processes, Vol. II. CRC Press Taylor & Francis Group.

5. Chavan A., Mukherji S., 2008. Dimensional analysis for modeling oxygen transfer in rotating biological contactor, Bioresource Technology, 99(9), 3721–3728.

6. Cortez S.,Teixeira P., Oliveira R., Mota M., 2013. Bioreactors: Rotating Biological Contactors, Encyclopedia of Industrial Biotechnology, Press John Wiley & Sons, 1013–1030.

7. Duda A., Szulżyk-Cieplak J, Lenik, K., 2016. Innovative system of simultaneous transportation and treatment of sanitary wastewater in scattered dwelling areas, Journal of Ecological Engineering, 17(5), p. 84–89.

8. Ghawi A.H., Kris J., 2009. Use of rotating biological contactor for appropriate technology wastewater treatment, Slovak Journal of Civil Engineering, 3, 1–8.

9. Hassard F., Biddle J., Cartmell E., Jefferson B., Tyrrel S., Stephenson T., 2015. Rotating biological contactors for wastewater treatment – A review, Process Safety and Environmental Protection, 94, 285–306.

10. Lenik K., Korga S., Kozera R., Szalapko J., 2013. FEM and Flow Simulation Module for selecting parameters in rotors flow systems, Archives of Materials Science and Engineering, 2(59), 69–75.

11. Lenik K., Skubisz W., 2016. Urządzenie i sposób do biologicznego oczyszczania ścieków, Zgłoszenie nr (21) 411985, Biuletyn Urzędu Patentowego, 44(22), 19–19.

12. Mba D., Bannister R., 2007. Ensuring effluent standards by improving the design of Rotating Biological Contactors, Desalination, 208, 204–2015.

13. Paolini A.E., 1986. Effect of Biomass on Oxygen Transfer in RBC Systems, Water Pollution Control Federation, 58(4), 306–311.

14. Pathan A.A., Rasool B.M., Memon F.A., 2015. Effect of flow rate and disc area increment on the efficiency of rotating biological contactor for treating greywater, Mehran University Research Journal of Engineering & Technology, 2(34), 99–106.

15. Patwardhan, A.W., 2003. Rotating biological contactors: a review, Industrial & Engineering Chemistry Research, 42, 2035–2051.

16. Pradeep N.V., Hampannavar U.S., 2011. Biodegradation of phenol using rotating biological contactor, International Journal of Environmental Sciences, 2(1), 105–113.

17. Ryschka, J., Rak, A., 2014. Analysis the efficiency of biological wastewater treatment in air-driven rotating biological contactor (In Polish), Gaz, Woda i Technika Sanitarna, 8, 303–306.

18. Staehler, T. 2004. Aerobous continual method of biologically treating sewage (in Polish), Zgłoszenie patentowe PL 186625. 27.02.2004.

19. Szulżyk-Cieplak,J., Lenik, K., Ozonek, J., 2015. Assessment of rural sewerage system on the example of Lublin province (In Polish), In: Chmielewski J., Szpringer M. (Ed.), Zdrowie, praca, środowisko – współczesne dylematy, Warszawa, Instytut Ochrony Środowiska – PIB, 241–252.

20. Szulżyk-Cieplak J., Duda A., Lenik K., 2016. The selection of design parameters of the rotating biological contactor using tracer studies, Rocznik Ochrona Środowiska, 2(18), 897–908.

21. Waskar V.G., Kulkarni G.S., Kore V.S., 2012, Review on process, application and performance of rotating biological contactor (RBC). International Journal of Science and Research Publications, 2(7), 1–6.

22. Żubrowska-Sudoł, M., Jasińska, A., 2007, Division of reactors with mobile beds (In Polish), Gaz, Woda i Technika Sanitarna, 3, 29–31.

4

The Effect of Light-Emitting Diodes Illumination Period and Light Intensity on High Rate Algal Reactor System in Laundry Wastewater Treatment

Bieby Voijant Tangahu[1], Adhi Triatmojo[1], Ipung Fitri Purwanti[1], Setyo Budi Kurniawan[1*]

[1] Department of Environmental Engineering, Faculty of Civil, Environmental and Geo Engineering, Institut Teknologi Sepuluh Nopember, Jalan Raya ITS, Kampus ITS Sukolilo, Surabaya 60111, Indonesia

[*] Corresponding author's e-mail: setyobudi.kurniawan@gmail.com

ABSTRACT

Wastewater that contains high concentration of nutrients can create instability in water ecosystem if left untreated. Laundry wastewater contains nutrients in high concentration. The nutrients that commonly found in laundry wastewater are nitrogen and phosphorus. This study had a purpose to determine the effect of illumination period and light intensity for the removal of Chemical Oxygen Demand (COD), Nitrogen-ammonia (NH_3-N), and phosphate (P) content using *Chlorella vulgaris* in High Rate Algal Reactor (HRAR) treatment. Variables that used were exposure period of 12 and 24 hours and light intensity of 2000–3000 Lux, 4000–5000 Lux, and 6000–7000 lux. The parameters tested to determine the efficiency of nutrient removal were COD, Nitrogen-ammonia, phosphate and Chlorophyll α to determine the condition of algae development. The results showed that the highest nutrient removal were obtained by the reactor with 24 hours illumination period with light intensity of 6000–7000 Lux that was capable of removing 54.63% of COD, and 22.15% of P. The 12-hour illumination period was better in terms of NH_3-N removal, up to 50.07%. On the basis of the of statistic test result, the illumination period did not significantly influence the removal efficiency of COD, NH_3-N and P indicated by P-value >0.05, while the light intensity significantly affect the removal of COD and NH_3-N showed by P value <0.05.

Keywords: Algae, C. vulgaris, Laundry Wastewater, Illumination Period, Light-Emitting Diodes, Light Intensity, Microalgae.

INTRODUCTION

In large quantities, laundry wastewater can lead to eutrophication and algal blooming if left untreated [Tectona, 2011]. Laundry waste is one of the causes of algl blooming in water bodies because it has high organic content and nutrient concentrations if not controlled properly [Kurniawan et al., 2018]. One of the souces of wastewater containg high nutrient is the laundry industry. About 15 litres of clean water is needed for laundry industry to wash 1 kg of clothes. On average, the laundry industry washes 25 kg of clothes per day, producing high nutrient containing wastewater of about 400 litres/day [Ciabatti et al., 2009].

High Rate Algal Reactor (HRAR) is one of wastewater treatment systems using microalgae. This treatment aims to reduce the high nutrient contents from wastewater [Rawat et al., 2010]. In HRAR wastewater treatment, simbiotic relation between the heterotroph bacteria and microalgae cell that live in the water were used to remove pollutants [Hamouri., 2008]. The biological reaction in HRAR can reduce the organic and nutrient content in wastewater through decomposition by bacteria and nutrient conversion to microalgae biomass using the photosyntetic process [Polpraset, 1996; Purwanti et al., 2018]. HRAR is operated at average speed of 10–30 cm/sec to avoid sedimentation from microalgae cell [Fallowfield and Garret, 1985].

The assimilation of N by algae and floating aquatic plants accounted for about 65% of the total N removal, ammonia volatilization of 15% and about 20% of total removal in N content can be assumed based on the nitrification-denitrification process [Mostret and Grobbelar, 1987]. According to [Wang et al., 2010] the research with specific use of algae may allow for greater nutrient removal, in addition to specific species biomass byproducts for various purposes. *Chlorella Sp.* is a type of algae that has high tolerance levels of pollutants andis easy to obtain; hence, *Chlorella Sp.* is often used in wastewater treatment [Man et al., 2016].

According to our knowledge, the research on laundry wastewater treatment using HRAR system has not been widely conducted. The purpose of this study was to obtain the removal of nutrient content in laundry wastewater by using High Rate Algal Reactor (HRAR). Variation of the illumination period and light intensity were used to measure the removal efficiency of nutrient content, and the development of alga yielded. HRAR were performed at 6 days of retention time with variation of illumination period were 12 and 24 hours and variation of light intensity were 2000–3000 Lux, 4000–5000 Lux, and 6000–7000 Lux, This variation was chosen to understand the impact of light and dark reaction towards microalgae. The parameters used in this research were Chemical Oxygen Demand (COD), nitrogen-ammonia (NH3-N), and phosphate (P). Chlorophyll α will also be analyzed to comprehend the microalgae development. The removal efficiencies of each reactor were calculated by comparing the result of parameters analysis, before and after treatment process using HRAR.

MATERIALS AND METHODS

At the beginning of research, the COD, nitrogen ammonia (NH_3-N), and phosphate (P) content in laundry wastewater was analysed. Sampling was carried out in Keputih District, Sukolilo, Surabaya, Indonesia using integrated sample method on 4 different laundries. The results showed that the COD, NH_3-N and P values ranged between 600–829 mg/L, 4.96–8.66 and 8.53–12.62 mg/L, respectively. NH_3-N was analysed using Nessler Method. Phosphate in the orthophosphate form was analysed using 4500-P-Stannous Chloride Method and the COD analysis was performed by using close reflux method.

This experiment used 6 reactor containers consisting of 3 test reactors and 3 control reactors. The volumes of the employed reactors were ± 12 Liters by 30 cm diameter and 27 cm height. The lighting sources included 18 watts Light Emitting Diodes (LED) (Phillips, Indonesia). The light intensity was measured using Lux Meter (HS1010, Indonesia) on the surface of the reactor. The algae used in this study was *Chlorella vulgaris* that were previously filtered with 50 μ cloth filter to harvest microalgae with 94 ± 2% efficiency [Bejor et al., 2013].

Chlorella vulgaris was illuminated by LED with a variety of light intensities, i.e. 2000–3000 Lux, 4000–5000 Lux, and 6000–7000 Lux. The reactor retention time was 6 days with constant agitation by 12 or 24 hours of illumination period. The removal efficiencies of each reactor were analysed through the removal of COD, NH_3-N, and P parameters. Chlorophyll α was also analysed to understand the microalgae development under different conditions.

RESULTS AND DISCUSSION

Nitrogen-Ammonia removal

The light intensity experiment showed that 6000–7000 Lux was adequate for the algal growth. This observation was in line with [Ifeanyi et al., 2011] who reported that the light with higher intensity can boost the growth of algae. The growth of algae in a reactor can affect the HRAR nutrient removal efficiency [Mirquez et al., 2016], the higher light intensity, the higher of microalgae concentration, resulting in the better removal efficiency. The rate of ammonia removal in all batch reactors increased at higher light intensity, as showed in Figures 1 and 2. The effect of the illumination time was presented in Figures 3 and 4. A linear correlation between chlorophyll α and ammonia concentration in the reactor was also observed. It is suggested that ammonia was consumed by microalgae.

It can be seen that the ammonia content always decreases, but on day 3 and 4 ammonia increases again; afterwards, the trend on the next day continues to fall. This case occurred because some of the microalgae present in the reactor begin to die and decompose, thus releasing the nitrogen back into the reactor in the 3rd and 4th days. The optimum removal efficiency was obtained in the reactor with 6000–7000 Lux lighting, which is 50%, while the 4000–5000 Lux and 2000–3000

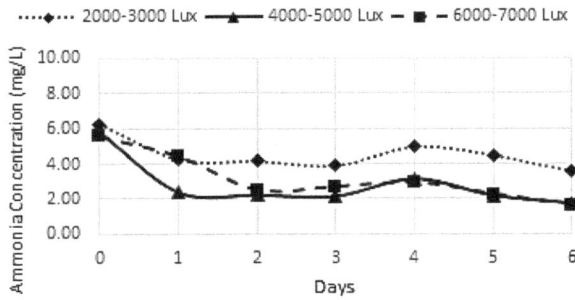

Figure 1. HRAR ammonia trend against light intensity

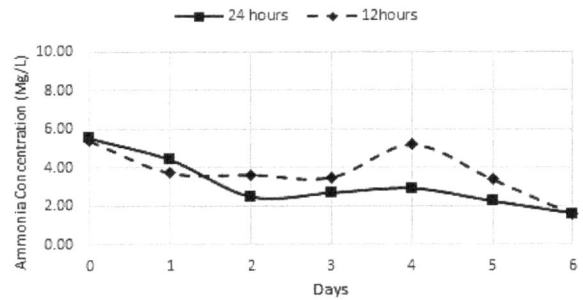

Figure 3. HRAR ammonia trend against illumination period

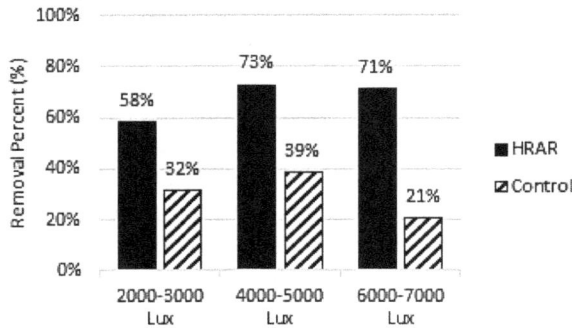

Figure 2. HRAR ammonia removal against light intensity

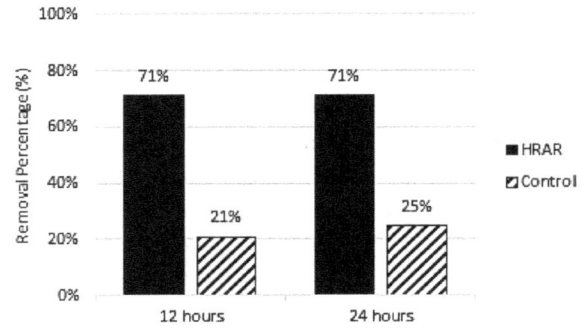

Figure 4. HRAR ammonia removal against illumination period

Lux reactors have 32% and 26% removal efficiencies, respectively. The control reactor has a lower efficiency than the reactor containing microalgae. The range of removal at the control reactor ranged from 21–39%, while the reactor containing microalgae showed that the NH_3-N removal ranged from 58–73%.

On the basis of Figures 3 and 4, the optimum NH_3-N removal efficiency was the same in all illumination period, which was 71%. This case indicated that the illumination time did not have a significant effect on the nitrogen removal. Different removal efficiencies was obtained on day 4, which amounted to 50.05% for 12 hours and 46.65% for 24 hours of illumination time. This result also suggests that the uptake of microalgae greatly affects the NH_3-N removal process, because in general, ammonia was the easiest nitrogen source to be taken up by microalgae [Garcia et al., 2000].

Phosphate removal

While correlating the removal percentage of phosphate with light intensity, the correlation coefficients were low, as we can see at Figures 5 and 6. This suggested that apart from the uptake by microalgae, other factors of have also played

a certain role. The mechanism that was important in the reduction of phosphate concentration was immobilization in the sediment with the precipitation of phosphorus-calcium and uptake bonds by microalgae [Chen et al., 2003].

Figure 5 showed that the overall phosphate concentration is decreasing. The phosphate concentration in the reactor tends to fluctuate, which is caused by microalgae excretion as a part of their metabolism [Vymazal, 1995]. On the basis of the Figure 6 it can be noticed that the reactor with 4000–5000 Lux had a removal of 30%, while 6000–7000 Lux had 27%. The reactor with 2000–3000 Lux had the lowest percentage of removal, which was 7%. The removal efficiency at the control reactor has the greatest removal efficiency, ranging from 19 to 36%. The phosphate removal was smaller when compared to NH_3-N because the required C: N: P ratio for microalgae growth is 106: 16: 1 [Chen et al., 2003]; thus, the phosphate uptake by microalgae will be smaller. Figures 7 and 8 show that the reactor with 12 hours of illumination period had 22.14% P removal, while 24 hours of illumination period had 13.24% removal. The results of the statistical analysis showed that the illumination period significantly affected the removal of phosphate in reactors.

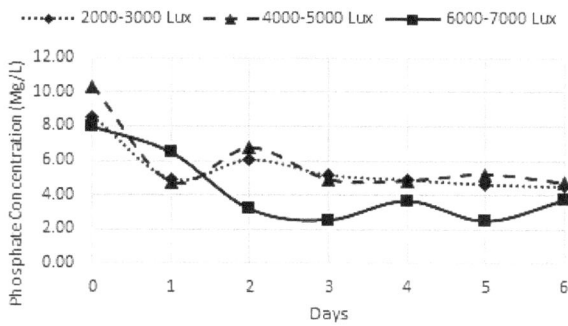

Figure 5. HRAR phosphate trend against light intensity

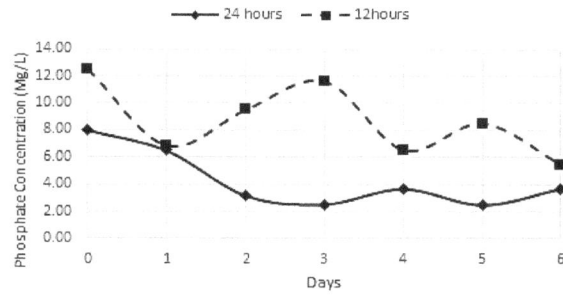

Figure 7. HRAR phospate trend against illumination period

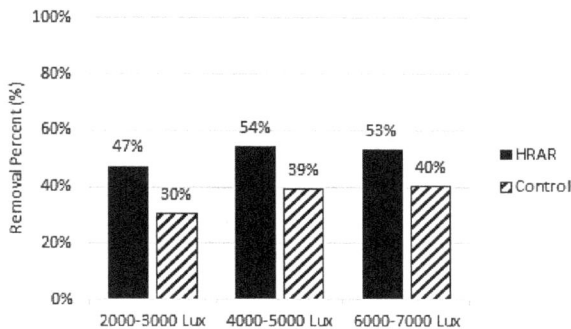

Figure 6. HRAR phosphate removal against light intensity

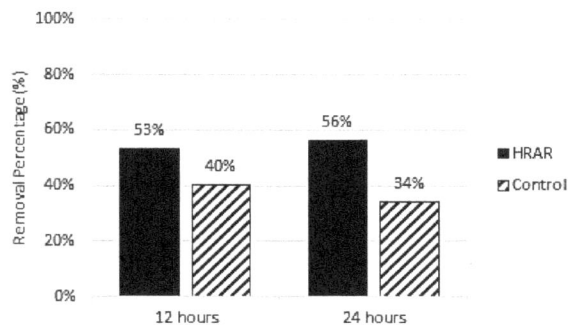

Figure 8. HRAR phospate removal against illumination period

Chemical Oxygen Demand removal

The correlation between the COD removal and light intensity were linear as shown in Figures 9 and 10. In the reactor with 2000–3000 Lux, the trend of COD concentration continues to decline and tend to be stable until the end of test period. In the reactor with 4000–5000 Lux and 6000–7000 Lux, the COD concentration increases occurred on days 2 and to 3. The fluctuations in the concentration occurred due to the decomposition of dead microalgae, counted as organic substances dissolved in wastewater. According to Figures 9 and 10, the greater the intensity of light, the higher the percentage of COD removal. This resulted from the control of the amount of light in light-dark cycle of algal metabolism. This cycle will break the organic substances down into simpler compounds and provide energy for metabolism.

The Chemical Oxygen Demand and illumination period had a linear correlation, as shown in Figure 11 and Figure 12. In the 12 hours of illumination period, the trend of COD concentration continues to decline and tends to be stable until the end of test period, In the 24 hours of illumination period, a concentration increase also

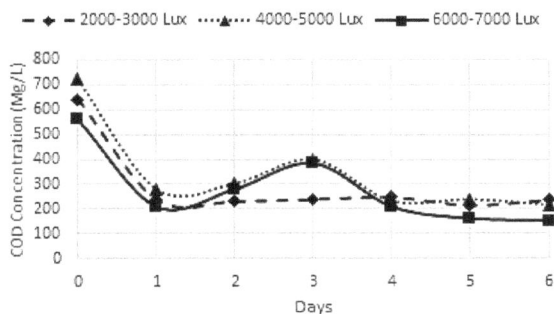

Figure 9. HRAR COD trend against light intensity

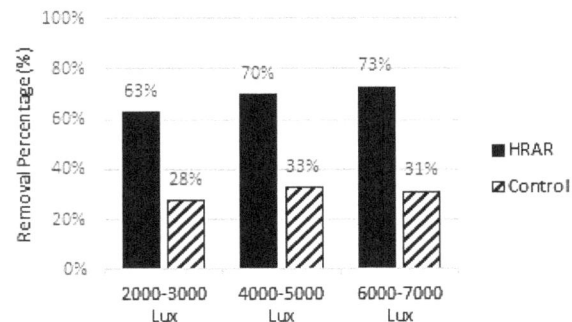

Figure 10. HRAR COD removal against light intensity

occurred on days 2 to 3, with the same reason suggested as the cause. Figure 11 showed that the 12 and 24 hours of illumination period had 55% and 42% removal percentages, respectively. The illumination period did not significantly affect the percentage of COD removal. This occurred due to the total of light intensity that actually affects the COD removal, not the illumination time. The light intensity was controlling the light-dark cycle in the algal metabolism to break the organic substances down, which resulted in COD reduction.

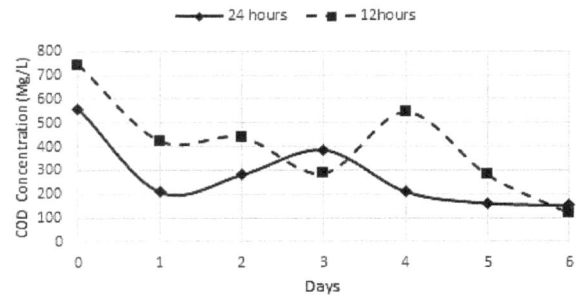

Figure 11. HRAR COD trend against illumination period

CONCLUSIONS

The illumination period had no significant effect on any of the parameters of the study. The largest removal percentage for COD is 54.62%, and Phosphate is 22.14% with light period exposure of 24 hours while NH_3-N was 50.07% with 12 hours of illumination period. Contrary to the illumination period, the light intensity had a significant effect on the removal of NH_3-N and COD. However, the phosphate removal was not significantly affected by light intensity. The largest removal percentage for COD was 55%, and for NH_3-N it amounted to 46% with 6000–7000 Lux of light intensity, while phosphate had 30% removal in the reactor with 4000–5000 Lux light intensity.

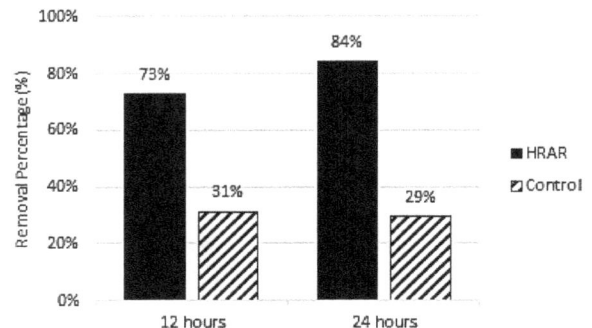

Figure 12. HRAR COD removal against illumination period

Acknowledgements

Authors would like to thank LPPM ITS for their financial support. In addition, we are grateful to the Department of Environmental Engineering, FTSLK, Institut Teknologi Sepuluh Nopember for their technical cooperation and the provision of chemical compounds.

REFERENCES

1. Bejor E.S., C. Mota, N.M. Ogarekpe, K.U. and Emerson, J. Ukpata. 2013. Low-Cost Harvesting of Microalgae Biomass from Water. International Journal of Development and Sustainability, 2 (1), 1–11.

2. Chen. P., Zhou, Q., Paing, J., Le, H. and Picot, B. 2003. Nutrient Removal by Integrated use of High Rate Algal Ponds and Macrophytes system in China. Water Science and Technology, 48(2), 251–257.

3. Ciabatti, I. F., L. Faralli, E. Fatrella and F. Togotti. 2009. Demonstration of Treatment System for Purification and Reuse of Laundry Wastewater. Bioresource Technology, 245 (1–3), 451–459.

4. Fallowfield, H. J. and Garret M. K. 1985. The Treatment of Wastes by Algal Culture. Journal of Applied Microbiology, 59 (14), 187S-205S.

5. Garcia, J., Mujeriego, R. and Marine, M.H. 2000. High Rate Algal Pond Operating Strategis for Urban Wastewater Nitrogen Removal. Journal of Applied Phycology, 12 (3), 331–339.

6. Hamouri, El B. 2008. How to Take Advantage of Combining High Rate Anaerobic Pre Treatment and High Rate Photosyntetic Post Treatment?. Department of Water Environment and Infrastructures Institute Agronomique et Veterinaire Hassan II, Rabat, Morroco

7. Ifeanyi V.O., Anyanwu B.N., Ogbulie J.N., Nwabueze R.N., Ekezie W., and Lawal O.S. 2011. Determination of the effect of light and salt concentrations on Aphanocapsa algal population, African Journal of Microbiology Research 5 (17), 2488–2492.

8. Kurniawan, Setyo Budi, Ipung Fitri Purwanti and Harmin Sulistiyaning Titah. 2018. The Effect of pH and Aluminium to Bacteria Isolated from Aluminium Recycling Industry. Journal of Ecological Engineering, 19 (3), 154–161.

9. Man, K. L., M. I. Yusoff, Y. and Uemura, J. Wei.

2016. Cultivation of Chlorella vulgaris using nutriens source from domestic wastewater for biodiesel production: Growth condition and kinetic studies. Renewable Energy, 103, 197–207.

10. Mirquez, Liliana Delgadillo, Behnam Taidi, Dominique Pareau, and Filipa Lopes. 2016. Nitrogen and Phosphate Removal from Wastewater with a Mixed Microalgae and Bacteria Culture. Biotechnology Reports, 11, 18–26.

11. Mostret, E.S., and Grobbelar, J.U. 1987. The Influence of Nitrogen and Phosphorus on Algal Growth and Quality in Outdoor Mass Algal Culture. Biomass, 13 (4), 219–233.

12. Polpraset, C. 1996. Organic Waste Recycling. Asian Institute of Technology, Thailand.

13. Purwanti, Ipung Fitri, Setyo Budi Kurniawan, Harmin Sulistiyaning Titah and Bieby Voijant Tangahu. 2018. Identification of Acid and Aluminium Resistant Bacteria Isolated from Aluminium Recycling Area. International Journal of Civil Engineering and Technology, 9 (2), 945–954

14. Rawat, I., Kumar, R. Ranjith, Mutanda, T., Bux, F. 2010. Dual Role of Microalgae: Phycoremediation of Domestic Wastewater and Biomass Production for Sustainable Biofuels Production. Applied Energy, 88 (10), 3411–3424.

15. Tectona, Johan. 2011. The utilization of Pterocarpus Indicus as Activated Carbon for Laundry Wastewater Treatment. Bachelor Thesis, Department of Environmental Engineering, Institut Teknologi Sepuluh Nopember, Surabaya, Indonesia.

16. Vymazal, J., 1995. Algae and Element Cycling in Wetlands. Lewis Publisher, California

17. Wang, L.A, Min M., Li Y.C., Chen P., Chen Y.F., Liu Y.H. 2010. Cultivation of green algae Chlorella sp. in different wastewaters from municipal wastewater treatment plant. Appl Biochem Biotechnol, 162, 1174–86.

5

Gravimetric Evolution during Sewage Sludge Biostabilization

Paola Posligua[1,2,3], Michelle Peñaherrera[1], Elvito Villegas[2], Carlos Banchón[4]

[1] Universidad de Las Américas (UDLA), Faculty of Engineering and Agrarian Sciences, Environmental Engineering, Av. de los Granados and José Queri, 59302, Quito, Ecuador

[2] Universidad Nacional Mayor de San Marcos (UNMSM), 07001, Lima, Perú

[3] Instituto Antártico Ecuatoriano (INAE), 59316, 9 de Octubre y Chile, Guayaquil, Ecuador

[4] Universidad Agraria del Ecuador (UAE), Environmental Engineering School, Faculty of Agrarian Sciences, Av. 25 de Julio and P. Jaramillo, 59304, Guayaquil, Ecuador

* Corresponding author's e-mail: paolaposligua@gmail.com

ABSTRACT

Sewage sludge is a by-product in the wastewater treatment and is an inherent hazardous issue because of the pathogenic contamination of natural resources. Therefore, in this study, domestic sludge was treated with premontane forest soil, macronutrients, and also pasteurization to reduce the content of volatile solids and pathogens. The best biostabilization treatment using premontane forest soil and pasteurization obtained a volatile solids reduction of 87% according to the environmental regulations, in which a biosolid is stable in a range of 38% of volatile solids reduction. In less than 30 days in a mesophilic range, the coliform count was reduced up to 71% when using forest soil and pasteurization. Thus, a biosolid-class B was obtained using gravimetric means as a platform to promote fast quality control.

Keywords: pathogens, biosolids, sewage, mesophilic, anaerobic.

INTRODUCTION

In the next 40 years, the world population will double and consequently, the capacity to treat the volume of wastewater according to the population increase will barely meet the demand; this would cause the waste to be discharged without known control to ecosystems. Municipal wastewater contains pathogenic bacteria, protozoa, viruses, and parasites, as well as oils, fats, detergents, soaps, nutrients, salts, and particles of hair, food, and paper. The treatment of municipal wastewater accounts for the worldwide production of approx. 48 million dry tons of sewage sludge, and its disposal has become an environmental problem because of the pathogenic risks (Fytili and Zabaniotou, 2008; Krüger et al., 2014; Li et al., 2007; Mu et al., 2016; Snowden-Swan et al., 2016; Wang et al., 2008a; Wei et al., 2003; Zahan et al., 2016). A typical drawback of the wastewater treatment system is the high-energy cost in the context of energy efficiency, carbon footprint, and recycling. Sludge treatment and its disposal account for up to 60% of the total operating costs, and the elimination of harmful pathogens and a new kind of contaminant, namely, emerging pollutants constitutes a critical step (Barrios et al. 2015; De Vrieze et al., 2016; Jenicek et al., 2012; Lewis et al., 1999; Ruffino et al., 2015; Wang et al., 2008b; Weemaes and Verstraete, 1998; Ye et al., 2014). Biological sewage sludge is mainly composed of organic matter (approx. 59–88%) from scum or the solids removed in wastewater treatment (Tchobanoglous et al., 2013). Sewage comes from human excreta and is a mixture of fats, proteins, carbohydrates, lignin, amino acids, sugars, cellulose, humic material, fatty acids, non-essential trace metals and organic micropollutants, which can be decomposed and produce offensive odors (Environmental Protection Agency, 1994; Kinney et al., 2006; Rogers, 1996; Singh and Agrawal, 2008; Weemaes and Verstraete, 1998).

In 1991, the term *biosolid* was adopted to apply to all sedimentary sludge in which mesophilic or thermophilic digestion diminishes odors, organic matter, and pathogenic risk under anaerobic conditions, which also generates bioenergy (Bright and Healey, 2003; Cain, 2010; Carrère et al., 2010; Snowden-Swan et al., 2016; Zahan et al., 2016). Biosolids have been recognized as a useful soil amendment and source of nitrogen, phosphorus, organic matter, and other nutrients, which can enhance the physical properties of soil as well as plant growth (Kinney et al., 2006); moreover, biosolids contain a good deal of energy at approx. 11,400 BTU per dry pound (Mu et al., 2016). Among Canada, the US, and European countries, 53 percent of biosolids are used in agriculture directly or after composting, totaling more than 2.39 million dry tons per year yield (Kinney et al., 2006; Stasinakis, 2012; Wang et al., 2007). However, biosolids are not completely secure for land application due to offensive odors and toxic elements, e.g., heavy metals and persistent organic pollutants, found in the sewage sludge (Hale et al., 2001; Krach et al., 2008; Lewis et al., 2002; Wei et al., 2003). A total of 87 synthetic organic chemicals were found in biosolids, including chemicals like polychlorinated biphenyls (PCBs), pharmaceuticals such as triclosan (antimicrobial disinfectant), tonalide (a musk fragrance), diphenhydramine (antihistamine), carbamazepine (an antiepileptic drug), and heavy metals like As, Cd, Cr, Pb, Hg, Ni, and Se (Barrios et al., 2015; Cain, 2010; Egan, 2013; Mulla et al., 2016; Venkatesan and Halden, 2014). Biosolids also carry high densities of enteric viruses, helminth eggs, and *Salmonella spp.*, which pose risks to the human health (Barrios et al., 2015; Gerba et al., 2011; Oron et al., 2014). Given that a stabilization process should be performed with high-quality standards to prevent the human health risks, the principal contributions in the present study are:

- An anaerobic process to stabilize sewage sludge using pre-treatments under different conditions like the addition of premontane forest soil under differing nutritional conditions and pasteurization.
- A thermogravimetric technique to monitor sludge stabilization.

In the pursuit of testing the hypothesis, we found out that the interactions between the biological, thermal, and salinity treatments influence the change of total and volatile solids, namely a gravimetric evolution of a biostabilization process.

MATERIALS AND METHODS

Biostabilization process

Residual sludge was collected from the secondary settler tank of a municipal wastewater treatment plant located in Quito (Ecuador). The residual sludge (A) was thickened by sedimentation (B) with a sedimentation column (2 m in length and 30 cm in diameter). Afterwards, the resulting sludge was filtered with cellulose pads (C) and then pre-treated under different conditions (D). The following pre-treatments were made before the anaerobic digestion: (T_1) 0.6 kg of soil from a premontane secondary forest was added per each kg of sludge; (T_2) again, 0.6 kg of soil from a premontane forest was added per each kg of sludge, and also, a 0.1% nutrient solution of 10% N, 40% P and 10% K (Merck, USA) was added twice daily for 30 days; (T_3) the sludge was pasteurized in an oven (Wiseven 165, USA) at 70°C for 30 minutes; (T_4) a 0.1% nutrient solution was added every morning, and a 1.0% NaCl solution (Merck, USA) was added at night for 30 days. After pre-treatment, anaerobic digestion was performed in a sealed 1000 mL Erlenmeyer flask (E). The flasks were thermally insulated and were slowly mixed twice a day for 30 days. At the end of the process, the biosolid was sundried (Figure 1).

Analysis

The temperature and pH were measured three times a day using a soil multiparameter tester (HANNA HI 99121, USA). Total solids (ST) and volatile solids content (SV) were determined at 105°C and 550°C, respectively, using ASTM methods. Arsenic, cadmium, mercury, and lead were quantified by atomic absorption according to APHA standard methods. The biological activity of the sludge was measured using OxiTop® Biological Oxygen Demand (BOD) respirometry system (WTW, Germany) at 25°C. Helminth eggs were determined using EPA method 9132, and total coliform was counted using the violet bile red Petrifilm plates (3M, USA) with tetrazolium as the indicator at 37°C for 48 hours; the sludge samples were diluted at 1:100 with sterile peptone water because sludge is characterized by a dark color.

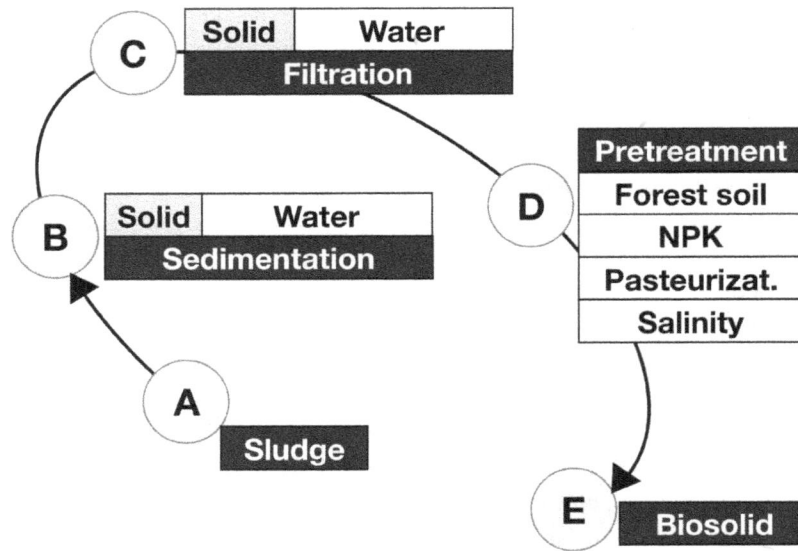

Figure 1. Biostabilization process

RESULTS AND DISCUSSION

Effect on pH and temperature

Different pre-treatment conditions using premontane forest soil, NPK, NaCl, and pasteurization changed pH from 6.0–6.5 to 7.0–7.5 during 30 days of biostabilization, as indicated in Figure 2. According to the results, pH changes due to the microbial decarboxylation of organic anions, heterocyclic compounds, and volatile fatty acids, suggesting that stability was reached (Wu, Ma, and Martinez, 2000; Yuan and Zhu, 2016). Normally, the first days in the anaerobic digestion are the rate-limiting step because of the hydrolysis of insoluble organic matter into a soluble form, which influences the further stabilization kinetics (Eastman and Ferguson, 1981; Parkin et al., 1986). The addition of forest soil, NPK, and NaCl boosted the release of microbial exoenzymes to cleave macromolecules like triglycerides, diglycerides, and fatty acids (Harris and McCabe, 2015). Thus, the more carbon is solubilized, the faster the stabilization process. Therefore, pre-treatments focused on hydrolysis maximize the degradation rate until mineralization of organic matter occurred. A drop in pH shows an increase of volatile fatty acids as acetic acid or propionic acid, which could be a problem in the stabilization process (Ahring et al., 1995; Amani et al., 2010; Berktay and Nas, 2007; Carrère et al., 2010; Harris and McCabe, 2015). In our case, the acid phase lasted for about 15 days, which is an indicator of hydrolysis without interruption of the

high fatty acid concentrations or non-biodegradable (refractory) components. At the end of the process (Fig. 2), the pH was 7.0–7.5 due to the microbial ammonification when ammonium compounds are nitrified to nitrate (Cofie et al., 2016).

In Figure 3, the temperature remained between 30–37°C in all treatments while the control reaction remained at approx. 20°C. One pretreatment involved pasteurization at 70°C for 30 minutes, which is used to improve the organics solubilization to provide a carbon source for microorganisms (Weemaes and Verstraete, 1998). In this way, high temperatures expose macromolecules to achieve their breakdown and further ease the degradation by thermophilic microorganisms (Carrère et al., 2010; Harris and McCabe, 2015). Since the mesophilic digestion does not kill pathogens efficiently, the thermal hydrolysis at 70°C was utilized as another kind of disinfection technique (Oleszkiewicz and Mavinic, 2002). The thermophilic digestion is up to three-fold higher than the mesophilic one, which enhances the hydrolysis conversion rates with an impact of higher volumetric biogas production at a lower hydraulic retention time (HRT) (De Vrieze et al., 2016). However, in economic terms, the mesophilic temperatures are more stable and have lower energy costs than the thermophilic treatments (Braguglia et al., 2015). Moreover, a mesophilic process is less inhibited by ammonium and long-chain fatty acids (LCFA) (Fernández-Rodríguez et al., 2015). According to Figure 4, the process started almost at 45°C and had an acidic pH dur-

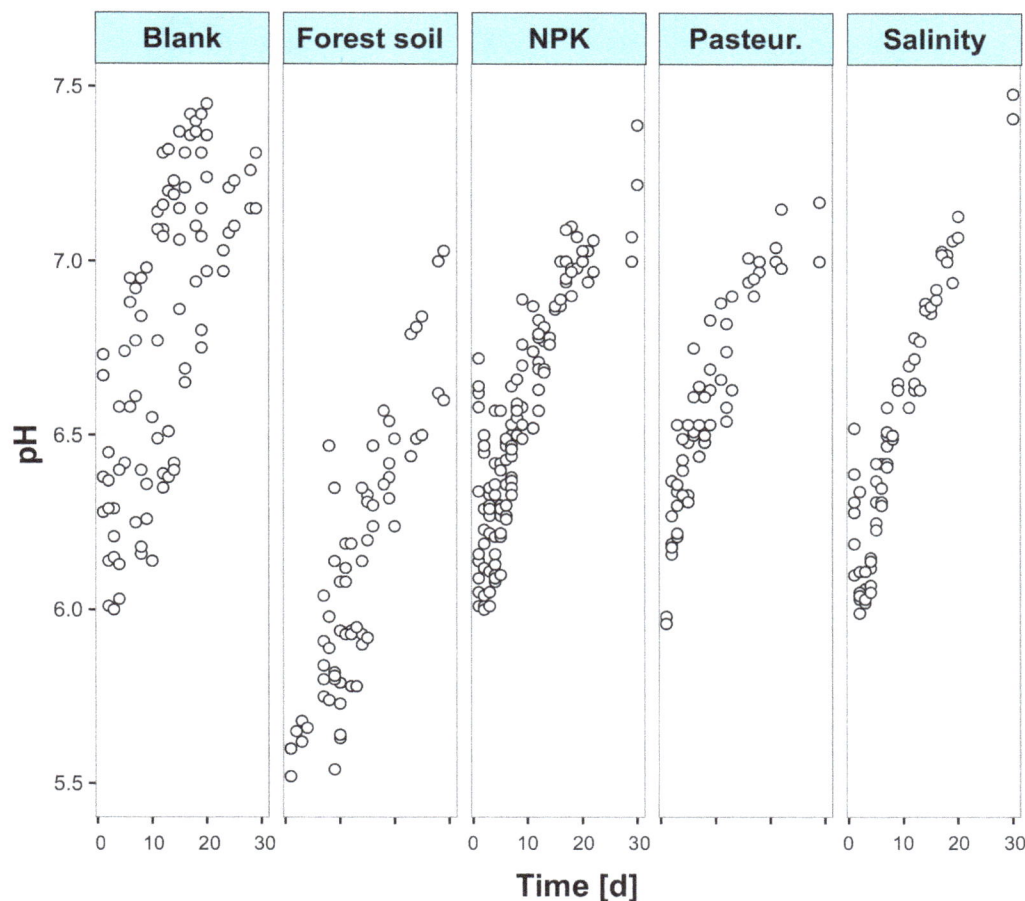

Figure 2. Effect of different biostabilization pre-treatments on pH during 30 days

ing the first few days, which favored the growth of fungi and actinomycetes. However, afterwards, a neutral pH and a temperature close to ambient temperature (20°C) were obtained. Together, pH and temperature progressively interacted due to high microbial activity during a solid residence time (SRT) of 30 days at 20–45°C. Thus, SRT and the mesophilic range in our biostabilization process performed according to the EPA regulations for biosolids under class B (Mustafa et al., 2014).

Gravimetric evolution

One method to evaluate biostabilization is to measure the change in total solids (TS) and volatile solids (VS), which is herein recognized as gravimetric evolution. Over a period of 30 days, the content of TS and VS changed dramatically, and showed tendencies that demonstrated biostabilization (Fig. 4). The content of organic matter is virtually represented by VS, and its reduction is due to the microbial mineralization and the conversion of organic matter into humic substances (Gómez et al., 2005; Otero et al., 2002). The measurement of volatile solids destruction is an indicator of the mineralization degree of organic carbon to mineral forms like CO_2 or CH_4 (Ahring et al., 1995; Bernal et al., 1998). According to VS destruction, the USEPA uses the value of 38% reduction in the threshold for considering the sludge to be stabilized (Environmental Protection Agency, 1994; Oleszkiewicz and Mavinic, 2002). In Figure 4, the VS/TS relationship is presented. Due to the low mineralization rate of the organic nitrogen found in sludge, enrichment with minerals like nitrogen, phosphorous, potassium and bacterial inoculant (forest soil) enhances the microbial activity and further organic matter degradation (Cofie et al., 2016). Our results confirm that nutrients, forest soil, thermal treatment, and even salinity are the enhancers of VS reduction (Bhattacharya et al., 1996; Ruffino et al., 2015).

In Figure 5, the gravimetric evolution of VS/TS is shown in every treatment. The control treatment (blank) did not overcome the stability range

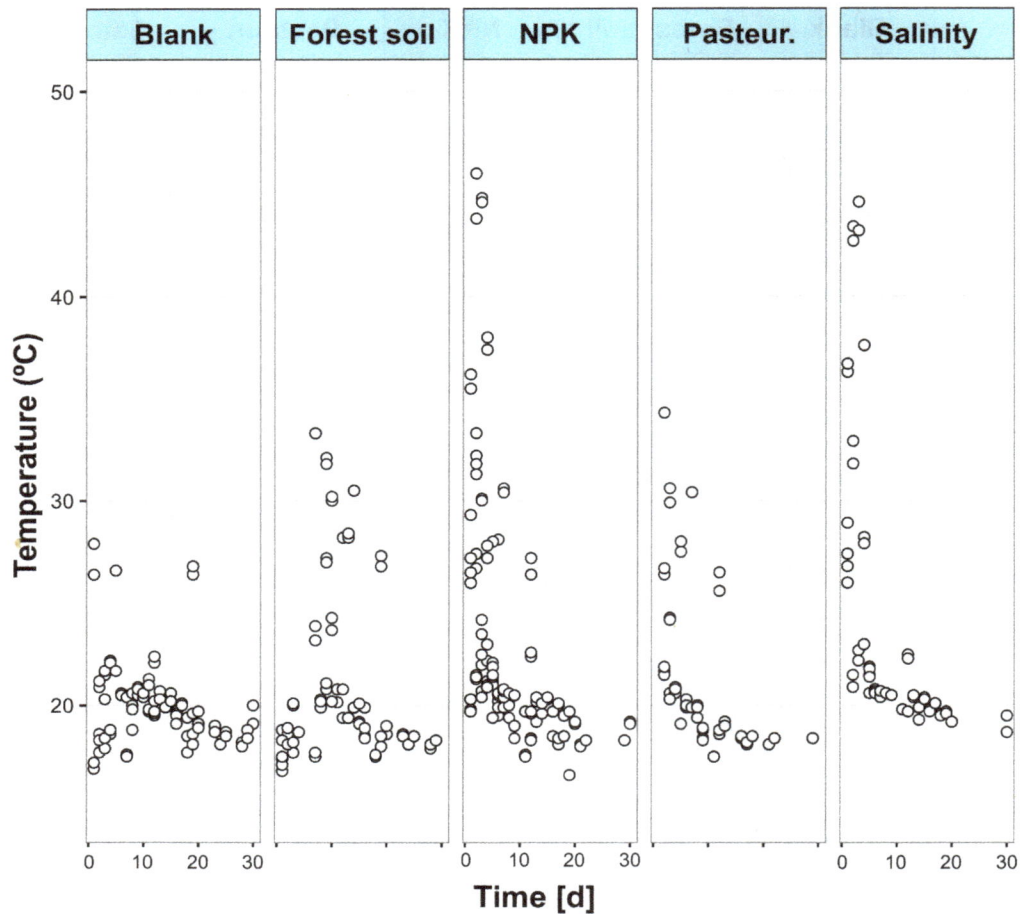

Figure 3. Effect of different biostabilization pre-treatments on temperature during 30 days

above 11.4% VS reduction. The forest soil, as a pre-digestion treatment, reduced 55.7% VS/TS because the more diverse the microbes, the faster digestion occurs. The pre-treatment that used forest soil and NPK nutrients, altogether reduced by 87.3% the VS/TS content. An optimal pre-treatment was also using pasteurization, in which an 87.2% VS/TS reduction was obtained. The pre-treatment with NaCl destroyed the VS content just by 36.7%; in this case, NaCl inhibited the microbial growth.

According to the results shown in Figure 5, a first order kinetic was obtained in two pre-treatments (Bernal et al., 1998). Forest soil enhanced a first order kinetic (k = 0.02930, R^2 = 0.9312, $p < 0.05$), as well as NPK (k = 0.064, R^2 = 0.9706, $p < 0.05$). On the other hand, a second order kinetic was observed in the pre-treatments using pasteurization (k = 0.2390, R^2 = 0.9833, $p < 0.05$) and salinity (k = 0.0262, R^2 = 0.9486, $p < 0.05$). Thus, faster kinetics reduces the size of the reactor and decreases hydraulic retention time (HRT) (Carrère et al., 2010).

According to the literature, the mesophilic digestion requires over a 20-day retention time, and it is not as efficient as the thermophilic process (Rulkens, 2008; Song et al., 2004). However, in the present work, the change of VS/TS in pre-treatments using forest soil, NPK, and pasteurization was up to 87%, which shows that mesophilic digestion can be optimal as the thermophilic counterpart.

Heavy metals

Table 1 shows the reduction in the heavy metal content through the biostabilization process in comparison with the EPA permissible limits. Ac-

Table 1. Chemical characterization before and after the biostabilization process

Parameter	Sludge (mg/kg)	Biosolid (mg/kg)	EPA (mg/kg)
As	0.170	0.010	41
Cd	0.110	0.025	85
Hg	0.002	0.001	57
Pb	2.080	0.400	840

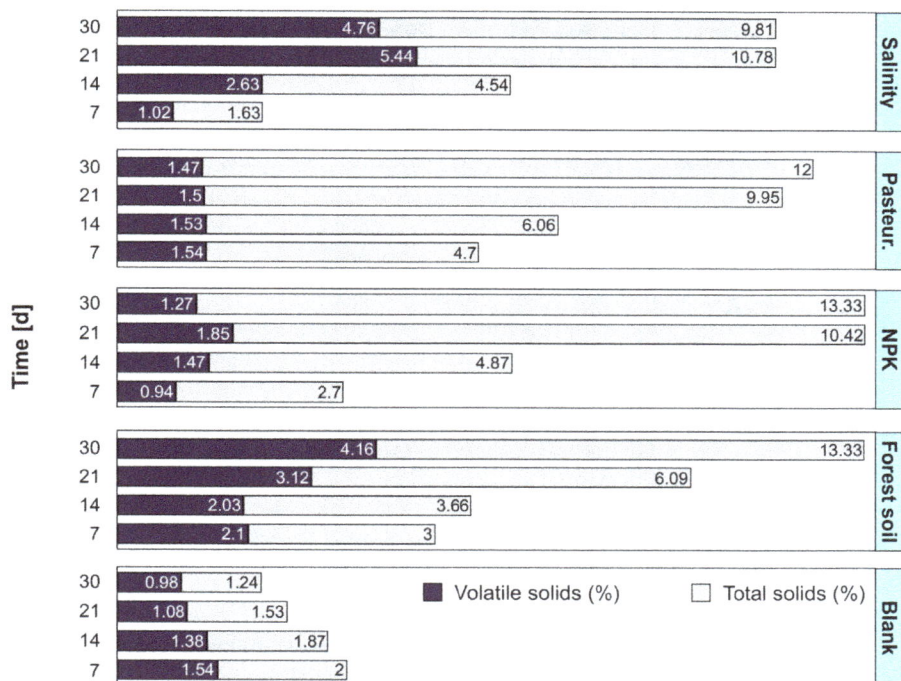

Figure 4. Change in TS and VS (%) during 30 days

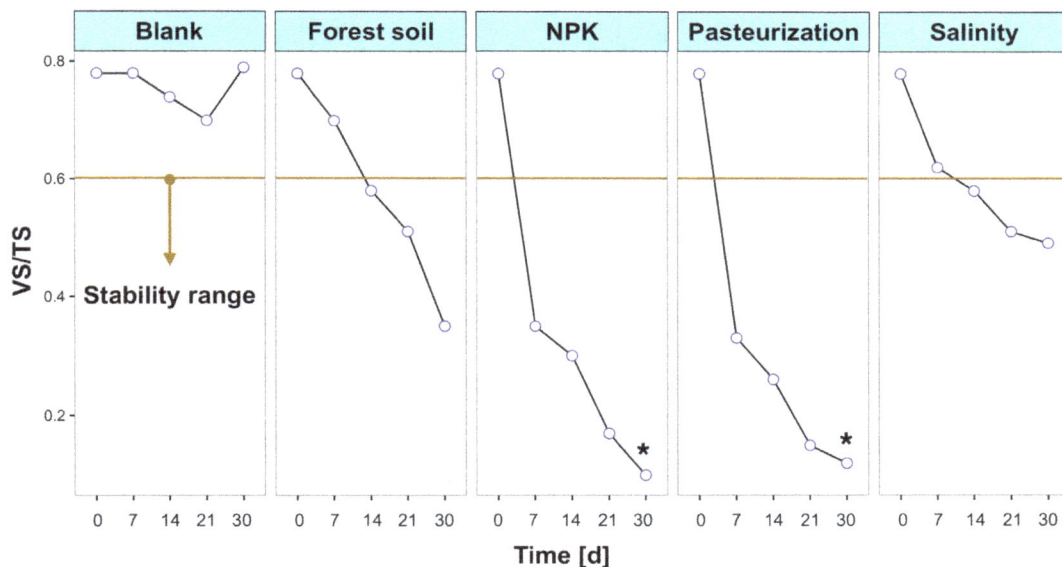

Figure 5. Gravimetric evolution (VS/TS) at different pre-treatments during 30 days

cording to the results, the order of metal content at the end of biostabilization was Pb > Cd > As > Hg. The concentrations of Pb, Cd, As, and Hg decreased by 94.1%, 77.3%, 50%, and 80.8%, respectively. The reduction of the total concentrations of heavy metals in biosolids is significant due to dilution during the mixing process with forest soil; therefore, the concentrations are under the EPA regulation limits. However, the bioavailability and toxicity of heavy metals in the sludge depend on their chemical forms and pH, which in consequence will precipitate with carbonate minerals, complexes and organic ligands at basic pH (Dong et al., 2013).

Pathogen removal

We measured the remaining digestion of samples in mg/L over nine days after the 30-day of biostabilization process to test any microbial

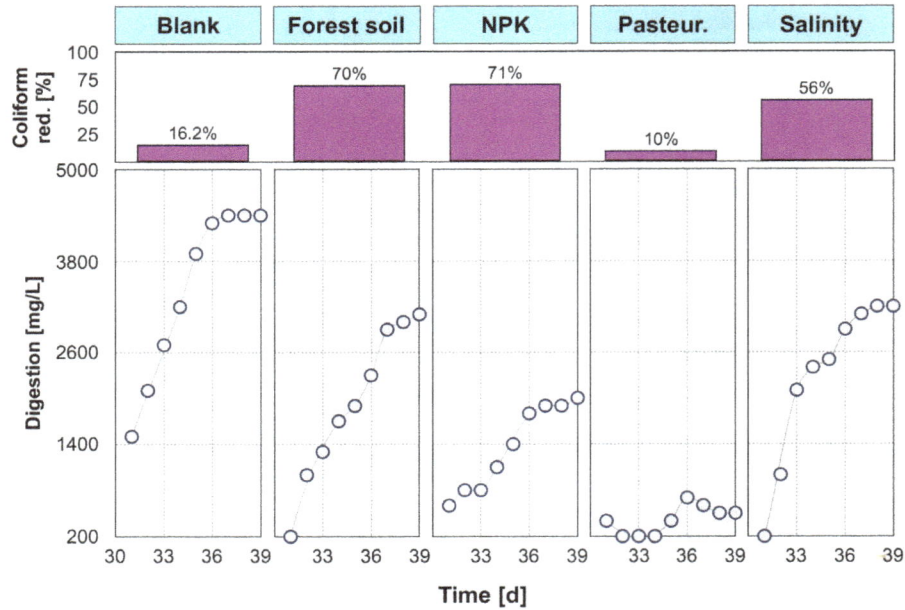

Figure 6. Digestion of sludge samples after the biostabilization process

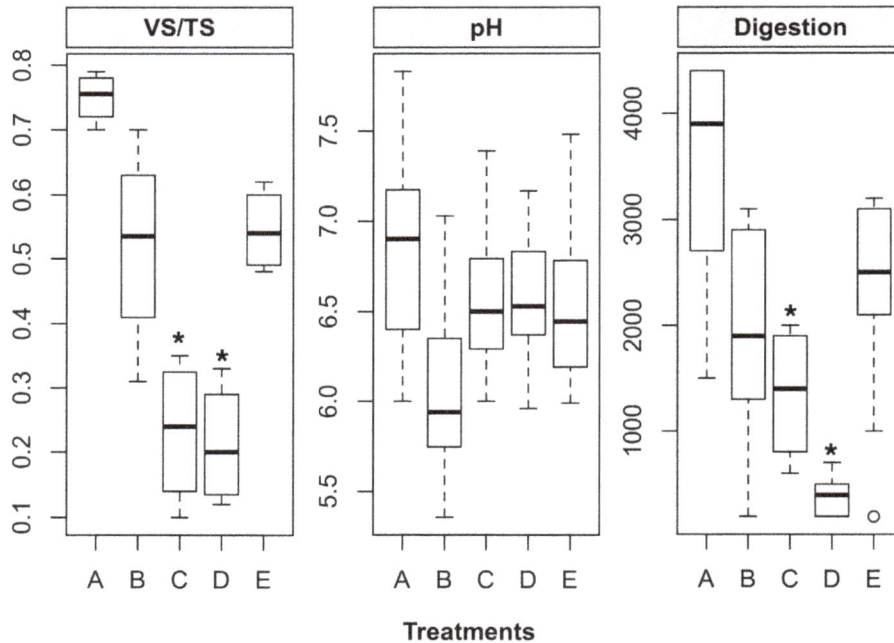

Figure 7. Results of the biostabilization process; (A) blank sample, (B) forest soil, (C) NPK, (D) pasteurization and (E) salinity treatment

activity (Figure 6). After the VS destruction, the CO_2 production rate should decrease, along with the pathogen removal (Wu et al., 2000). Both digestion and coliform reduction were examined to test which treatment was the most efficient way to stabilize sludge through pathogen reduction. The colony-forming units (CFU) per mL were 31700, 26550, 9650, 9150, 28500, and 13700 for blank, forest soil, NPK, pasteurization, and salin-

ity treatments, respectively. According to the results, the order of coliform reduction at the end of biostabilization was NPK > forest soil > salinity > blank > pasteurization, and the removal of coliforms was reduced by 71%, 70%, 56%, 16.2%, and 10%, respectively. In the present case, the best pre-treatment utilized NPK and forest soil.

According to the results, our biosolid is classified as class B because the US EPA recommends a VS reduction by 38% and a mean coliform density of less than 2 million CFU per gram of biosolids (Environmental Protection Agency, 1994). Although a biosolid might be recognized as freshly processed class B, sewage sludges may pose a significant risk of infection due to the pathogen regrowth, and this is why such biosolids should be managed with great care to public access (Gattie and Lewis, 2003). In order to adhere to the human health standards, different treatment methods should demonstrate the ability to neutralize the pathogen viability.

Figure 7 presents the biostabilization results in box plots. The pre-treatments using forest soil as bio-accelerator, NPK and pasteurization stabilized sewage sludge up to 87% in VS/TS reduction. All treatments, except the forest soil one, got a pH between 6.5 and 7. No significant benefit was obtained by the NaCl treatment.

CONCLUSIONS

The legal constraints to using sewage sludge in agriculture and further land applications motivate the search for cheaper stabilization processes and therefore avoiding incineration as the final option for sewage disposal. In the present study, the use of the proper combination of nutrition and temperature enhanced the performance of sludge biostabilization. However, to improve the digestion efficiency, a pasteurization process should be conducted at a higher temperature in a shorter time. According to the results obtained in this work, the further research outlook is focused on the high reduction of pathogen count.

Acknowledgment

The authors thank lab specialists Javier Álava and Pavlova Sigcha from UDLA for their assistance in all experiments.

REFERENCES

1. Ahring, B.K., Sandberg, M., and Angelidaki, I. 1995. Volatile fatty acids as indicators of process imbalance in anaerobic digestors. Applied Microbiology and Biotechnology, 43(3), 559–565.

2. Amani, T., Nosrati, M., and Sreekrishnan, T.R. 2010. Anaerobic digestion from the viewpoint of microbiological, chemical, and operational aspects – a review. Environmental Reviews, 18(NA), 255–278.

3. Barrios, J.A., Becerril, E., De León, C., Barrera-Díaz, C. and Jiménez, B. 2015. Electrooxidation treatment for removal of emerging pollutants in wastewater sludge. Fuel, 149, 26–33.

4. Berktay, A., and Nas, B. 2007. Biogas Production and Utilization Potential of Wastewater Treatment Sludge. Energy Sources, Part A: Recovery, Utilization, and Environmental Effects, 30(2), 179–188. https://doi.org/10.1080/00908310600712489

5. Bernal, M.P., Sanchez-Monedero, M.A., Paredes, C., and Roig, A. 1998. Carbon mineralization from organic wastes at different composting stages during their incubation with soil. Agriculture, Ecosystems and Environment, 69(3), 175–189.

6. Bhattacharya, S.K., Madura, R.L., Walling, D.A., and Farrell, J.B. 1996. Volatile solids reduction in two-phase and conventional anaerobic sludge digestion. Water Research, 30(5), 1041–1048.

7. Braguglia, C.M., Gianico, A., Gallipoli, A., and Mininni, G. 2015. The impact of sludge pre-treatments on mesophilic and thermophilic anaerobic digestion efficiency: Role of the organic load. Chemical Engineering Journal, 270, 362–371.

8. Bright, D.A., and Healey, N. 2003. Contaminant risks from biosolids land application. Environmental Pollution, 126(1), 39–49. https://doi.org/10.1016/S0269–7491(03)00148–9

9. Cain, G. D. (2010). Sanitizing Sewage Sludge: The Intersection of Parasitology, Civil Engineering, and Public Health. Journal of Parasitology, 96(6), 1037–1040. https://doi.org/10.1645/GE-2631.1

10. Carrère, H., Dumas, C., Battimelli, A., Batstone, D.J., Delgenès, J.P., Steyer, J.P., and Ferrer, I. 2010. Pretreatment methods to improve sludge anaerobic degradability: A review. Journal of Hazardous Materials, 183(1–3), 1–15. https://doi.org/10.1016/j.jhazmat.2010.06.129

11. Cofie, O., Nikiema, J., Impraim, R., Adamtey, N., Paul, J., and Kone, D. 2016. Co-composting of solid waste and fecal sludge for nutrient and organic matter recovery: Retrieved from https://books.google.com.ec/books?id=QrukDQAAQBAJ

12. De Vrieze, J., Smet, D., Klok, J., Colsen, J., Angenent, L.T., and Vlaeminck, S.E. 2016. Thermophilic sludge digestion improves energy balance and nutrient recovery potential in full-scale municipal wastewater treatment plants. Bioresource Technology, 218, 1237–1245.

13. Dong, B., Liu, X., Dai, L., and Dai, X. 2013. Changes of heavy metal speciation during high-solid anaerobic digestion of sewage sludge. Bioresource Technology, 131, 152–158.

14. Eastman, J.A., and Ferguson, J.F. 1981. Solubilization of particulate organic carbon during the acid phase of anaerobic digestion. Journal (Water Pollution Control Federation), 352–366.

15. Egan, M. 2013. Biosolids management strategies: an evaluation of energy production as an alternative to land application. Environmental Science and Pollution Research, 20(7), 4299–4310. https://doi.org/10.1007/s11356–013–1621–1

16. Environmental Protection Agency 1994. A Plain English Guide to the EPA Part 503 Biosolids Rule. Retrieved from https://www.epa.gov/biosolids/

17. Fernández-Rodríguez, J., Pérez, M., and Romero, L.I. 2015. Temperature-phased anaerobic digestion of Industrial Organic Fraction of Municipal Solid Waste: A batch study. Chemical Engineering Journal, 270, 597–604. https://doi.org/10.1016/j.cej.2015.02.060

18. Fytili, D., and Zabaniotou, A. 2008. Utilization of sewage sludge in EU application of old and new methods – A review. Renewable and Sustainable Energy Reviews, 12(1), 116–140. https://doi.org/10.1016/j.rser.2006.05.014

19. Gattie, D.K., and Lewis, D.L. 2003. A High-Level Disinfection Standard for Land-Applied Sewage Sludges (Biosolids). Environmental Health Perspectives, 112(2), 126–131. https://doi.org/10.1289/ehp.6207

20. Gerba, C.P., Ross, A., Takizawa, K., and Pepper, I.L. 2011. Efficiency of ASTM Method D4994–89 for Recovery of Enteric Viruses from Biosolids. Food and Environmental Virology, 3(1), 43–45. https://doi.org/10.1007/s12560–011–9054–9

21. Gómez, X., Cuetos, M.J., García, A.I., and Morán, A. 2005. Evaluation of digestate stability from anaerobic process by thermogravimetric analysis. Thermochimica Acta, 426(1–2), 179–184. https://doi.org/10.1016/j.tca.2004.07.019

22. Hale, R.C., La Guardia, M.J., Harvey, E.P., Gaylor, M.O., Mainor, T.M., and Duff, W.H. 2001. Flame retardants: Persistent pollutants in land-applied sludges. Nature, 412(6843), 140–141. https://doi.org/10.1038/35084130

23. Harris, P.W., and McCabe, B.K. 2015. Review of pre-treatments used in anaerobic digestion and their potential application in high-fat cattle slaughterhouse wastewater. Applied Energy, 155, 560–575. https://doi.org/10.1016/j.apenergy.2015.06.026

24. Jenicek, P., Bartacek, J., Kutil, J., Zabranska, J., and Dohanyos, M. 2012. Potentials and limits of anaerobic digestion of sewage sludge: Energy self-sufficient municipal wastewater treatment plant? Water Science and Technology, 66(6), 1277. https://doi.org/10.2166/wst.2012.317

25. Kinney, C.A., Furlong, E.T., Zaugg, S.D., Burkhardt, M.R., Werner, S.L., Cahill, J.D., and Jorgensen, G.R. 2006. Survey of Organic Wastewater Contaminants in Biosolids Destined for Land Application. Environmental Science and Technology, 40(23), 7207–7215. https://doi.org/10.1021/es0603406

26. Krach, K.R., Burns, B.R., Li, B., Shuler, A., Cole, C., and Xie, Y. 2008. Odor Control for Land Application of Lime Stabilized Biosolids. Water, Air, and Soil Pollution: Focus, 8(3–4), 369–378. https://doi.org/10.1007/s11267–007–9147–5

27. Krüger, O., Grabner, A., and Adam, C. 2014. Complete Survey of German Sewage Sludge Ash. Environmental Science and Technology, 48(20), 11811–11818. https://doi.org/10.1021/es502766x

28. Lewis, D.L., Garrison, A.W., Wommack, K.E., Whittemore, A., Steudler, P., and Melillo, J. 1999. Influence of environmental changes on degradation of chiral pollutants in soils. Nature, 401(6756), 898–901. https://doi.org/10.1038/44801

29. Lewis, D.L., Gattie, D.K., Novak, M.E., Sanchez, S., and Pumphrey, C. 2002. Interactions of pathogens and irritant chemicals in land-applied sewage sludges (biosolids). BMC Public Health, 2(1), 11.

30. Li, X., Brown, D.G., and Zhang, W. 2007. Stabilization of biosolids with nanoscale zero-valent iron (nZVI). Journal of Nanoparticle Research, 9(2), 233–243. https://doi.org/10.1007/s11051–006–9187–1

31. Mu, D., Addy, M., Anderson, E., Chen, P., and Ruan, R. 2016. A life cycle assessment and economic analysis of the Scum-to-Biodiesel technology in wastewater treatment plants. Bioresource Technology, 204, 89–97. https://doi.org/10.1016/j.biortech.2015.12.063

32. Mulla, S.I., Wang, H., Sun, Q., Hu, A., and Yu, C.-P. 2016. Characterization of triclosan metabolism in Sphingomonas sp. strain YL-JM2C. Scientific Reports, 6(1). https://doi.org/10.1038/srep21965

33. Mustafa, N., Elbeshbishy, E., Nakhla, G., and Zhu, J. 2014. Anaerobic digestion of municipal wastewater sludges using anaerobic fluidized bed bioreactor. Bioresource Technology, 172, 461–466. https://doi.org/10.1016/j.biortech.2014.09.081

34. Oleszkiewicz, J.A., and Mavinic, D.S. 2002. Wastewater biosolids: an overview of processing, treatment, and management. Journal of Environmental Engineering and Science, 1(2), 75–88. https://doi.org/10.1139/s02–010

35. Oron, G., Adel, M., Agmon, V., Friedler, E., Halperin, R., Leshem, E., and Weinberg, D. 2014. Greywater use in Israel and worldwide: Standards and prospects. Water Research, 58, 92–101. https://doi.org/10.1016/j.watres.2014.03.032

36. Otero, M., Calvo, L.F., Estrada, B., García, A.I., and Moran, A. 2002. Thermogravimetry as a technique for establishing the stabilization progress of sludge from wastewater treatment plants. Thermochimica Acta, 389(1), 121–132.

37. Parkin Gene F., and Owen William F. 1986. Fundamentals of Anaerobic Digestion of Wastewater Sludges. Journal of Environmental Engineer-

ing, 112(5), 867–920. https://doi.org/10.1061/(ASCE)0733–9372(1986)112:5(867)

38. Rogers, H.R. 1996. Sources, behaviour and fate of organic contaminants during sewage treatment and in sewage sludges. Science of the Total Environment, 185(1), 3–26.

39. Ruffino, B., Campo, G., Genon, G., Lorenzi, E., Novarino, D., Scibilia, G., and Zanetti, M. 2015. Improvement of anaerobic digestion of sewage sludge in a wastewater treatment plant by means of mechanical and thermal pre-treatments: Performance, energy and economical assessment. Bioresource Technology, 175, 298–308. https://doi.org/10.1016/j.biortech.2014.10.071

40. Rulkens, W. 2008. Sewage Sludge as a Biomass Resource for the Production of Energy: Overview and Assessment of the Various Options †. Energy and Fuels, 22(1), 9–15. https://doi.org/10.1021/ef700267m

41. Singh, R.P., and Agrawal, M. 2008. Potential benefits and risks of land application of sewage sludge. Waste Management, 28(2), 347–358. https://doi.org/10.1016/j.wasman.2006.12.010

42. Snowden-Swan, L.J., Hallen, R.T., Zhu, Y., Billing, J.M., Jones, S.B., Hart, T.R., et al. 2016. Hydrothermal Liquefaction and Upgrading of Municipal Wastewater Treatment Plant Sludge: A Preliminary Techno-Economic Analysis. Pacific Northwest National Laboratory (PNNL), Richland, WA (US). Retrieved from http://www.pnnl.gov/main/publications/external/technical_reports/PNNL-25464Rev1.pdf

43. Song, Y.-C., Kwon, S.-J., and Woo, J.-H. 2004. Mesophilic and thermophilic temperature co-phase anaerobic digestion compared with single-stage mesophilic- and thermophilic digestion of sewage sludge. Water Research, 38(7), 1653–1662. https://doi.org/10.1016/j.watres.2003.12.019

44. Stasinakis, A.S. 2012. Review on the fate of emerging contaminants during sludge anaerobic digestion. Bioresource Technology, 121, 432–440. https://doi.org/10.1016/j.biortech.2012.06.074

45. Tchobanoglous, G., Burton, F.L., and Stensel, H.D. 2013. Wastewater Engineering: Treatment and Resource Recovery. McGraw-Hill Education. Retrieved from https://books.google.com.ec/books?id=BL3wjgEACAAJ

46. Venkatesan, A.K., and Halden, R.U. 2014. Wastewater Treatment Plants as Chemical Observatories to Forecast Ecological and Human Health Risks of Manmade Chemicals. Scientific Reports, 4. https://doi.org/10.1038/srep03731

47. Wang, H., Brown, S.L., Magesan, G.N., Slade, A.H., Quintern, M., Clinton, P.W., and Payn, T.W. (2008a). Technological options for the management of biosolids. Environmental Science and Pollution Research – International, 15(4), 308–317. https://doi.org/10.1007/s11356–008–0012–5

48. Wang, H., Brown, S.L., Magesan, G.N., Slade, A.H., Quintern, M., Clinton, P.W., and Payn, T.W. 2008b. Technological options for the management of biosolids. Environmental Science and Pollution Research – International, 15(4), 308–317. https://doi.org/10.1007/s11356–008–0012–5

49. Wang, L.K., Hung, Y.-T., and Shammas, N.K. 2007. Biosolids Treatment Processes. Totowa (NJ): Humana Press.

50. Weemaes, M.P.J., and Verstraete, W.H. 1998. Evaluation of current wet sludge disintegration techniques. Journal of Chemical Technology and Biotechnology, 73(2), 83–92.

51. Wei, Y., Van Houten, R.T., Borger, A.R., Eikelboom, D.H., and Fan, Y. 2003. Minimization of excess sludge production for biological wastewater treatment. Water Research, 37(18), 4453–4467. https://doi.org/10.1016/S0043–1354(03)00441-X

52. Wu, L., Ma, L., and Martinez, G. 2000. Comparison of methods for evaluating stability and maturity of biosolids compost. Journal of Environmental Quality, 29(2), 424–429.

53. Ye, F., Liu, X., and Li, Y. 2014. Extracellular polymeric substances and dewaterability of waste activated sludge during anaerobic digestion. Water Science and Technology, 70(9), 1555.

54. Yuan, H., and Zhu, N. 2016. Progress in inhibition mechanisms and process control of intermediates and by-products in sewage sludge anaerobic digestion. Renewable and Sustainable Energy Reviews, 58, 429–438. https://doi.org/10.1016/j.rser.2015.12.261

55. Zahan, Z., Othman, M.Z., and Rajendram, W. 2016. Anaerobic Codigestion of Municipal Wastewater Treatment Plant Sludge with Food Waste: A Case Study. BioMed Research International, 2016, 1–13. https://doi.org/10.1155/2016/8462928

6

Treatment Efficiency and Characteristics of Biomass in a Full-Scale Wastewater Treatment Plant with Aerobic Granular Sludge

Agnieszka Cydzik-Kwiatkowska[1*], Michał Podlasek[1], Dawid Nosek[1], Beata Jaskulska[1]

[1] University of Warmia and Mazury in Olsztyn, Department of Environmental Biotechnology, Słoneczna 45G, 10-709 Olsztyn, Poland

[*] Corresponding author's e-mail: agnieszka.cydzik@uwm.edu.pl

ABSTRACT

Recently, studies have been carried out on an implementation of aerobic granular sludge (AGS) technology in full-scale wastewater treatment plants. The aim of the work was to evaluate the effectiveness of organic, phosphorus and nitrogen compounds removal from municipal wastewater and to characterize the biomass in a wastewater treatment plant upgraded from the activated sludge to AGS technology. In the upgraded facility, granulation was obtained quickly and it was observed that the granule morphology depended of the temperature. In the granular biomass harvested at moderate temperatures in the reactor (15°C), the granules with diameters in the range from 125 to 500 μm constituted the largest share (about 60%), while the second-largest biomass fraction comprised the granules with diameters over 1 mm (25%). The analysis of granule diameters carried out in winter (the temperature in the reactor equaled 8°C) showed a decrease in the share of the largest granules and predominance of the granules with diameters in the range from 90 to 355 μm (about 75%). Upgrading the municipal wastewater treatment plant from activated sludge to aerobic granular sludge significantly improved the settling properties of the biomass and efficiency of wastewater treatment. The average efficiency chemical oxygen demand (COD) and phosphorus removal increased by about 10% and 20%, respectively, while ammonium nitrogen was completely oxidized, regardless of the season. After modernization, the concentration of nitrates in the effluent increased significantly to about 3–6 mg/L. The results of the study show that it is possible to effectively upgrade the existing facilities to aerobic granular sludge technology; it was also proven that this technology is an excellent alternative to a conventional activated sludge.

Keywords: next-generation sequencing, microbial structure, full-scale wastewater treatment plant, wet sieving, wastewater temperature, aerobic granules

INTRODUCTION

Microorganisms in the form of biofilm, activated sludge flocs or granules are commonly used in the aerobic and anaerobic wastewater treatment systems. The use of aerobic granules for wastewater treatment is a relatively new technological solution, intensively implemented at the technical scale, both in in Poland [Świątczak, Cydzik-Kwiatkowska 2017, Podlasek et al. 2017] as well as worldwide [van der Roest et al. 2011, Giesen et al. 2013].

Wastewater treatment in batch reactors based on aerobic granular sludge does not require de-signing the multi-reactor treatment line and the reactor working cycle is much shorter. The presence of anaerobic zones in the granules allows the removal of nitrogen and phosphorus without the need to provide anaerobic conditions in the reaction chamber, while the very good settling properties of the granules allow to reduce the length of a settling phase [Sławiński 2015]. The absence of a conventional recirculation pump and stirrers ensures a significant reduction in the energy consumption, compared to the conventional wastewater treatment plants [de Kreuk et al. 2005a].

The world's first pilot-scale system with granular sludge treating municipal wastewater was launched in 2003 in the Netherlands. It consisted of two biological reactors with a height and diameter of 6 m and 0.6 m, respectively, working in parallel [de Kreuk 2006]. The world's first full-scale municipal wastewater treatment plant (WWTP) based on the aerobic granular sludge technology was built in Gansbaai (South Africa). The treatment plant was designed with a capacity of 4,000 m^3/d and consisted of three reactors operated in parallel [Li et al. 2014]. In Poland, experience indicates that upgrading existing activated sludge WWTPs to granular sludge technology substantially improved COD, phosphorus and nitrogen removal, as well as reducing the volume of the biological reactors by 30% [Świątczak, Cydzik-Kwiatkowska 2017]

The study of microorganisms in a WWTP provides information on the species structure of the community, indicating the dominant microbiological groups responsible for effective wastewater treatment. The discoveries made in recent years have enabled the introduction of next-generation sequencing (NGS), which is regarded as the most suitable method for studying complex microbial consortia [Kotowska, Zakrzewska-Czerwińska 2010].

Wastewater treatment is characterized by the trophic relationships typical to the detritus chain. Microorganisms oxidize organic matter, transform food substrates, and produce polysaccharides and other polymers [Tarczewska 1997]. The basic genera found in the activated sludge systems are *Zooglea*, *Flavobacterium*, *Alcaligenes*, *Bacillus*, *Achromobacter*, *Corynobacterium*, *Comomonas*, *Brevibacterium* and *Acinetobacter* [Bitton 2005]. Aerobic granules have a layered structure; in the outer layer, which is directly in contact with wastewater, nitrifying microorganisms dominate, while denitrifiers and bacteria accumulating polyphosphates are located in the center of the granules. A high proportion of microorganisms capable of producing extracellular polymeric substances (EPS) in the biomass is favorable for the stability of the granule structure.

To date, despite the growing interest in the implementation of aerobic granular sludge technology, limited data on the treatment efficiency and biomass structure in such full-scale WWTPs has been presented in the literature. The aim of the work was to assess the effectiveness of pollutant removal and the characteristics of biomass in a full-scale WWTP upgraded from activated sludge to aerobic granular sludge technology.

MATERIALS AND METHODS

Technological studies

The effectiveness of municipal wastewater treatment after upgrading to aerobic granular sludge technology was investigated from September 2014 to June 2016 at WWTP in Jeziorany (Poland). Municipal wastewater (about 500 m^3/d) and the wastewater from a meat processing plant (about 510 $m^3/month$) were supplied to the treatment plant. The characteristics of wastewater in the investigated period were as follows: total suspended solids (TSS) 0.340 g/L, total Kjeldahl nitrogen (TKN) 0.110 g/L, total phosphorus (TP) 0.021 g/L, 1.431 g COD/L. The technological line included a horizontal grate, a vertical vortex grate, a stepped mechanical grate, a retention tank and a batch granular sludge reactor (GSBR). The excess sludge was directed to the thickener and then to the aerobic stabilization tank. From the GSBR, the wastewater flowed to the secondary settling tank. The sludge from the secondary settling tank was recycled to the aerobic stabilization tank, and the treated wastewater was discharged to the river.

After the upgrading of WWTP to the aerobic granular sludge technology, the GSBR was operated at a hydraulic sewage retention time of 31 h, the organic load of at the beginning of the GSBR cycle of 1.1 kg $COD/(m^3 \cdot d)$ and sludge age of 28 d. The average biomass concentration was 4.8 g MLSS/L.

In the influent and effluent, COD, ammonium nitrogen, nitrates and nitrites were analyzed according to APHA (1992). In the GSBR, the concentration of mixed liquor suspended solids (MLSS) and the sludge volumetric index (SVI) after 30 min of settling were measured [APHA 1992]. The morphology of the granules was evaluated using wet sieving in an AS 200 screening unit (Retsch Sieves with mesh sizes of 2 mm, 1 mm, 710 μm, 500 μm, 355 μm, 125 μm and 90 μm were used for the analysis. During the analysis, the sludge was washed with tap water (12°C) from the spray nozzle over the uppermost sieve; the sieving lasted for 10 min and the vibration amplitude was 50 mm. The granule photographs were taken with a fluorescence microscope Eclipse 50i (Nikon).

Molecular research

The biomass samples were collected in October 2014, August 2015 and October 2015. DNA isolation was performed from 200 µl of centrifuged sludge (8000 rpm, 2 min) using a FastDNA® SPIN Kit for Soil (MP Biomedicals). The quality and quantity of the DNA was assessed in a NanoDrop Lite spectrophotometer (Thermo Scientific) at 260 and 280 nm. The isolated DNA was sent to the Research and Testing Laboratory (USA) for high-throughput sequencing. A primer set 357wF/785R (357wF 5'CCTACGGGNG-GCWGCAG 3'; 785R 5 ,GACTACHVGGG-TATCTAATCC 3') was used to amplify the 16S rDNA gene [Wu et al. 2016]. The data processing was conducted according to the methodology from the Research and Testing Laboratory (http://www.researchandtesting.com/docs/Data_Analysis_Methodology.pdf).

Statistics

The results were analyzed with Statistica 12.5 software (StatSoft). In order to compare the two samples, a t-test for independent samples was used after normality and homogeneity of variance were confirmed with the Shapiro-Wilk test and Levene's test. With all tests, $p < 0.05$ was considered significant.

RESULTS

During the GSBR working cycle, the contents of the reactor were constantly aerated. In the first 38 minutes of the cycle, the oxygen concentration increased to a maximum value of 4.5 mg/L (Fig. 1). After a further 37 minutes, the oxygen concentration dropped to the smallest value of 0.96 mg/L, and then gradually increased for about 2 hours to about 2.5 mg/L at the end of the aeration phase.

Before the upgrading of the sewage treatment plant, the SVI was in the range of 130–170 mL/g MLSS, whereas after the introduction of the aerobic granular sludge, the SVI fell below 50 mL/g MLSS. The microscopic analysis showed that after granulation, dense, spherical microbial consortia were present in the biomass (Fig. 2).

Wet sieving of granular sludge from WWTP was carried out in November 2015 and in February 2016. During the first analysis, the granules with a diameter ranging from 125 µm to 500 µm comprised the largest share of the granular biomass (about 60%) (Fig. 3a). The second largest component of biomass included the granules with a diameter of more than 1 mm (25%). In the sieve analysis conducted in February 2016, the share of the two largest fractions was lower than in November 2015 (Fig. 3b) and smaller granules predominated in the biomass.

The change from activated sludge to granular sludge significantly increased the efficiency of pollutant removal. The average efficiency of COD removal increased by about 10% to 95%, while the average efficiency of phosphorus removal increased by about 20% to about 30%. The average concentration of ammonium nitrogen in the effluent was high (40 mg/L), which corresponded to a nitrification efficiency of 48.5 ± 6.7%. After upgrading, ammonium nitrogen was completely removed (Fig. 4a), and the average nitrification efficiency increased to 99.8 ± 0.57%.

The concentration of nitrates in the effluent after upgrading (from October 2014) was maintained at an average of 3.7 ± 1.5 mg/L (Fig. 4b), which was significantly higher than before modernization (0.4 ± 0.3 mg/L). In the winter periods, higher concentrations of nitrate nitrogen (V) in

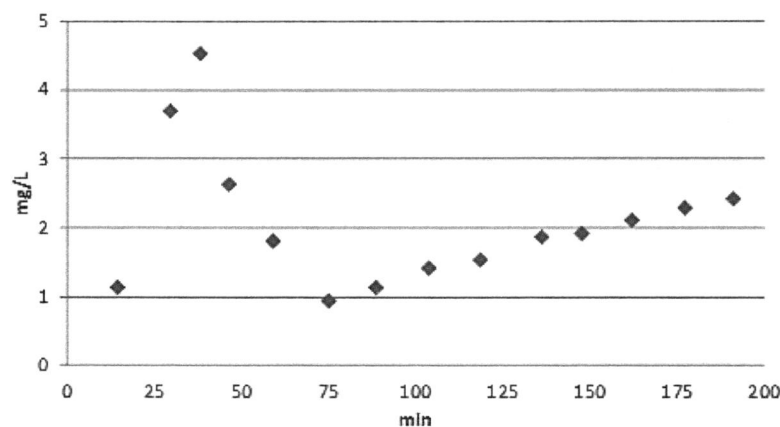

Fig. 1. Oxygen concentration changes in a working cycle of granular sequencing batch reactor

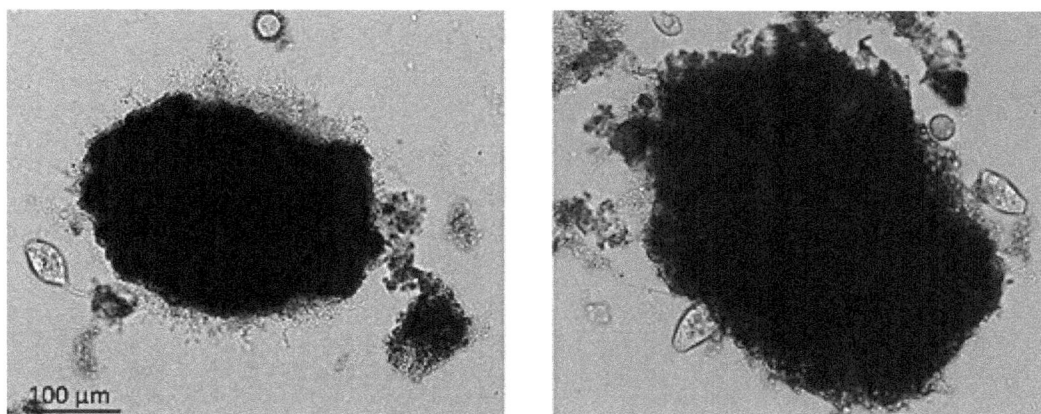

Fig. 2. Aerobic granules from wastewater treatment plant in Jeziorany

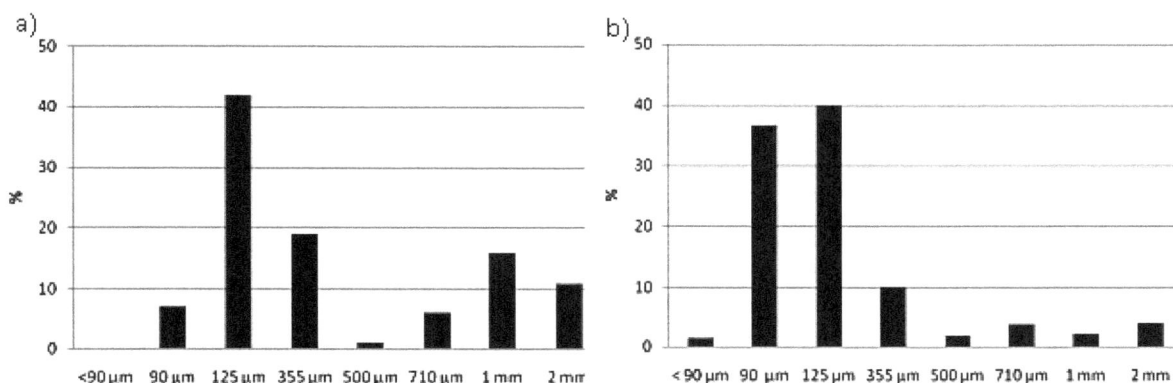

Fig. 3. Results of wet sieving analysis a) November 2015, b) February 2016; sieves with hole sizes of 2 mm, 1 mm, 710 μm, 500 μm, 335 μm, 125 μm and 90 μm were used

the outflow were recorded, which indicated a decrease in denitrification efficiency.

The samples for molecular tests were collected in 1st, 1368th and 1643rd cycle of reactor operation, when the temperature in the aeration chamber was 16.0, 22.5 and 16.0°C, respectively. Archaea and Bacteria were present in the biomass, with Bacteria accounting for over 98% of the sequences. Out of the 43 identified classes of microorganisms, the most numerous were Actinobacteria (36.9%, 25.2%, 22.3% in 1st, 1368th and 1643rd cycle of reactor operation, respectively), Alphaproteobacteria (13.8%, 15.4%, 13.3%), Betaproteobacteria (7.8%, 15.6%, 16.1%), and Gammaproteobacteria (11.2%, 14.3%, 10.0%). In total, 79 orders were identified in the analyzed biomass, the most numerous orders were Actinomycetales (11.8%, 26.9%, 1.7%), Xanthomondales (5.0%, 10.2%, 5.0%), Rhizobiales (5.9% 6.7% 3.30%), Burkholderiales (7.5%, 8.4%, 8.5%), Rhodocyclales (0.5%, 5.3%, 6.0%) and Lactobacillales (1.4%, 4.5%, 2.0%).

In Table 1, the genera of microorganisms with abundances in biomass higher than 1% are presented. The most numerous microorganisms belonged to the genera *Acidothermus* (3.85%, 3.70%, 4.82%), *Tetrasphaera* (4.22%, 3.63%, 5.44%), *Trichococcus* (4.27% 1.33% 1.82%), *Clostridium* (2.57%, 1.38%, 0.63%), *Mesorhizobium* (2.14%, 1.53%, 0.25%) and *Acidovorax* (3.25%, 2.51%, 1.40%).

DISCUSSION

The study was conducted in a municipal wastewater treatment plant in Jeziorany – one of the first full-scale WWTPs with aerobic granules in Poland.

The SVI for aerobic granular sludge should not exceed 50 mL/g MLSS [Toh et al. 2003]. In this study, after WWTP upgrading, the SVI of the biomass was below 50 mL/g MLSS and the settling properties were significantly improved.

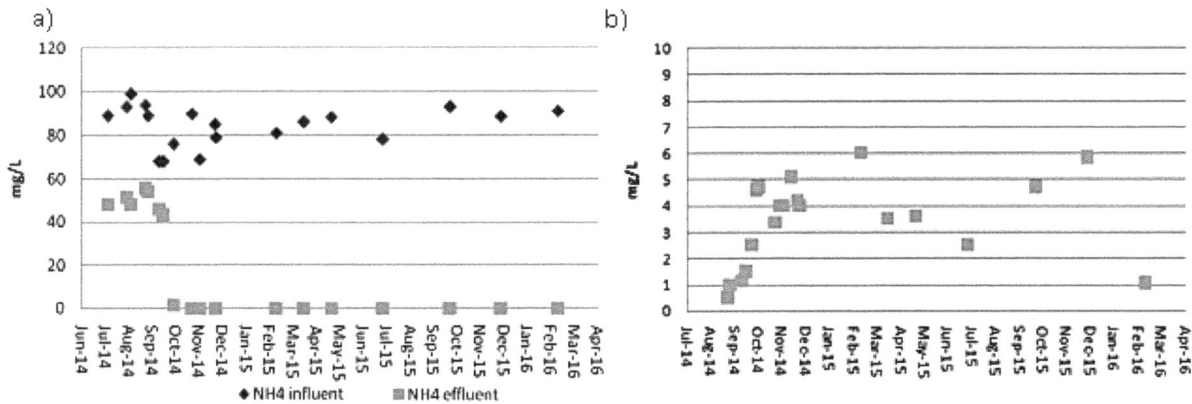

Fig. 4. Concentration of a) ammonium nitrogen in the influent and in the effluent, b) nitrates in the effluent

For comparison, the SVI values at the Yancang WWTP decreased after the introduction of granular sludge technology from 75.5 to 43 mL/g MLSS after 14 days of operation. Over the next 310 days, the average value of SVI stabilized at 48 mL/g MLSS [Li et al. 2014]. The concentration of the granular biomass in full-scale reactors can reach up to 9 g MLSS/L [Pronk et al. 2015]. In the present study, the concentration of the MLSS in the batch reactor averaged 4.8 g MLSS/L, while the concentration of total suspended solids in the effluent was below 25 mg/L.

Upgrading the investigated WWTP improved the efficiency of organic compounds removal; before modernization, the efficiency of COD removal did not exceed 90%, while after modernization it reached 95%. The obtained efficiency was higher than in other aerobic granular sludge WWTPs. In the Yancang WWTP, treating a mixture of domestic and industrial wastewater (dyeing, textile and food industry) (3:7 v/v) with a concentration of 200–600 mg O_2/L, the efficiency of COD removal by aerobic granules was 85% [Li et al. 2014]. Pronk and others (2015) studied the effectiveness of municipal wastewater treatment (506 mg COD/L, 6.7 mg TP/L, 49 mg TKN/L) at the WWTP in Garmerwolde. The reactors with granular sludge were operated at the sludge age of 20–38 d and the MLSS concentration in the aeration chamber of 6.5–8.5 g/L. Under these operating conditions, the authors observed the effectiveness of COD removal of about 87%. In the tested WWTP, after upgrading, full ammonium nitrogen removal was observed. Similar results were observed at the Yancang and Garmerwolde WWTPs, where the ammonium removal efficiency was 96% and 97%, respectively [Li et al. 2014, Pronk et al 2015]. In addition, the Yancang plant had a high efficiency of total nitrogen removal of 60% [Li et al 2014].

According to the literature, the size of the granules ranges from 0.2 mm to 9.0 mm. An increase in the granule diameter may cause problems with proper operation of the batch reactor [Liu, Tay 2004, Cydzik-Kwiatkowska et al. 2009]. During the first analysis of granule morphology in this study, it was observed that 25% of the biomass consisted of granules with a diameter above 1 mm, while the granules with a diameter of 125–500 μm consituted 60%. At the second measurement, the share of granules with diameters higher than 1 mm was much lower. The differences between the sizes of granules at both measurements resulted from the changes in the operating temperature of the GSBR. The first analysis was carried out at a wastewater temperature of around 15°C, and the second was carried out at a temperature of 8°C. According to de Kreuk et al. (2005b), the start-up of the treatment plant with aerobic granular sludge at 8°C resulted in the production of irregularly shaped granules, which caused the biomass washout from the system and unstable operation. Starting-up the granular reactors at 20°C, and then reducing the temperature to 15°C and 8°C, did not significantly affect the stability of the granules and biomass was effectively retained in the technological system. Our own experiments indicate that the start-up of an aerobic granular sludge system during the summer period ensures stable granulation and effective organics, nitrogen and phosphorus removal even during winter.

In the aerobic granules from the full-scale WWTP in the present study, the bacteria belonging to Proteobacteria and Actinobacteria predominated. Alphaproteobacteria, Betaproteobacteria, Gammaproteobacteria and Actinobacteria accounted for between 7.8 and 36.9% of the microor-

Tab.1. Genera of microorganisms present in aerobic granules (% of all sequences)

Kingdom ; phylum ; class ; order ; family ; genus	October 2014	August 2015	October 2015
Bacteria ; Actinobacteria ; Actinobacteria ; Acidimicrobiales ; Acidimicrobiaceae ; Aciditerrimonas	0.98	0.90	0.01
Bacteria ; Actinobacteria ; Actinobacteria ; Acidothermales ; Acidothermaceae ; Acidothermus	3.85	3.70	4.82
Bacteria ; Actinobacteria ; Actinobacteria ; Actinomycetales ; Actinomycetaceae ; Actinomyces	1.09	0.23	0.78
Bacteria ; Actinobacteria ; Actinobacteria ; Actinomycetales ; Intrasporangiaceae ; Tetrasphaera	4.22	3.63	5.44
Bacteria ; Actinobacteria ; Actinobacteria ; Actinomycetales ; Microbacteriaceae ; Frigoribacterium	2.5	0.76	0.12
Bacteria ; Actinobacteria ; Actinobacteria ; Actinomycetales ; Mycobacteriaceae ; Mycobacterium	0.91	2.65	1.33
Bacteria ; Actinobacteria ; Actinobacteria ; Actinomycetales ; Dietziaceae ; Dietzia	0.05	2.84	0.06
Bacteria ; Actinobacteria ; Actinobacteria ; Actinomycetales ; Nocardioidaceae ; Nocardioides	1.19	1.42	0.53
Bacteria ; Actinobacteria ; Actinobacteria ; Unclassified ; Unclassified ; Candidatus Microthrix	3.70	0.96	0.00
Bacteria ; Chloroflexi ; Caldilineae ; Caldilineales ; Caldilineaceae ; Caldilinea	1.18	1.98	0.58
Bacteria ; Firmicutes ; Bacilli ;Lactobacillales; Carnobacteriaceae ; Trichococcus	4.27	1.33	1.82
Bacteria ; Firmicutes ; Clostridia ; Clostridiales ; Clostridiaceae ; Clostridium	2.57	1.38	0.63
Bacteria ; Proteobacteria ; Alphaproteobacteria ; Caulobacterales ; Caulobacteraceae ; Phenylobacterium	1.38	0.61	0.00
Bacteria ; Proteobacteria ; Alphaproteobacteria ; Rhizobiales ; Hyphomicrobiaceae ; Devosia	0.68	1.49	0.55
Bacteria ; Proteobacteria ; Alphaproteobacteria ; Rhizobiales ; Phyllobacteriaceae ; Mesorhizobium	2.14	1.53	0.25
Bacteria ; Proteobacteria ; Alphaproteobacteria ; Rhizobiales ; Rhizobiaceae ; Rhizobium	0.53	0.71	1.07
Bacteria ; Proteobacteria ; Alphaproteobacteria ; Rhodobacterales ; Rhodobacteraceae ; Gemmobacter	1.48	0.38	0.00
Bacteria ; Proteobacteria ; Alphaproteobacteria ; Rhodobacterales ; Rhodobacteraceae ; Paracoccus	1.54	0.75	0.34
Bacteria ; Proteobacteria ; Alphaproteobacteria ; Rhodospirillales ; Acetobacteraceae ; Roseomonas	1.17	0.81	0.19
Bacteria ; Proteobacteria ; Alphaproteobacteria ; Rhodospirillales ; Rhodospirillaceae ; Dongia	0.04	0.68	1.19
Bacteria ; Proteobacteria ; Alphaproteobacteria ; Sphingomonadales ; Sphingomonadaceae ; Sphingobium	1.12	0.51	0.27
Bacteria ; Proteobacteria ; Alphaproteobacteria ; Sphingomonadales ; Sphingomonadaceae ; Sphingopyxis	0.03	0.80	4.40
Bacteria ; Proteobacteria ; Betaproteobacteria ; Burkholderiales ; Comamonadaceae ; Acidovorax	3.25	2.51	1.40
Bacteria ; Proteobacteria ; Betaproteobacteria ; Burkholderiales ; Comamonadaceae ; Rhodoferax	0.15	2.02	1.51
Bacteria ; Proteobacteria ; Betaproteobacteria ; Burkholderiales ; Comamonadaceae ; Simplicispira	1.13	0.07	0.00
Bacteria ; Proteobacteria ; Betaproteobacteria ; Burkholderiales ; Unclassified ; Leptothrix	0.37	1.34	1.34
Bacteria ; Proteobacteria ; Betaproteobacteria ; Rhodocyclales ; Rhodocyclaceae ; Dechloromonas	0.16	0.86	1.73
Bacteria ; Proteobacteria ; Betaproteobacteria ; Rhodocyclales ; Rhodocyclaceae ; Thauera	0.04	1.35	0.63
Bacteria ; Proteobacteria ; Betaproteobacteria ; Rhodocyclales ; Rhodocyclaceae ; Zoogloea	0.01	1.7	0.90
Bacteria ; Proteobacteria ; Gammaproteobacteria ; Oceanospirillales ; Halomonadaceae ; Halomonas	0.00	0.00	1.90
Bacteria ; Proteobacteria ; Gammaproteobacteria ; Pseudomonadales ; Pseudomonadaceae ; Pseudomonas	1.10	1.15	1.46
Bacteria ; Proteobacteria ; Gammaproteobacteria ; Xanthomonadales ; Rhodanobacteraceae ; Dokdonella	0.75	1.62	0.81
Bacteria ; Proteobacteria ; Gammaproteobacteria ; Xanthomonadales ; Xanthomonadaceae ; Aquimonas	0.04	1.31	0.41
Bacteria ; Proteobacteria ; Gammaproteobacteria ; Xanthomonadales ; Rhodanobacteraceae ; Rhodanobacter	0.06	1.60	1.34
Bacteria ; Proteobacteria ; Gammaproteobacteria ; Xanthomonadales ; Xanthomonadaceae ; Thermomonas	1.70	2.31	1.27
Unrecognized	1.53	2.23	1.68

ganisms in aerobic granules, which indicated that they were necessary for the proper formation of the granule structure and played an important role in the transformation of nutrients in wastewater.

The species structure of the bacterial community changed with maturation of the granular biomass. In the initial stages of granulation, filamentous microorganisms serve as a frame upon which single cells of bacteria are deposited [Cydzik-Kwiatkowska 2014]. In the tested samples, the number of the filamentous bacteria of the genus *Microthrix* was high in the initial working cycles (3.70%), when the reactor was dominated by the activated sludge. As mature aerobic granules were formed, the number of *Microthrix* sp. decreased until they completely disappeared in the biomass after about 1,600 GSBR working cycles.

In the granular biomass, numerous microorganisms involved in the transformation of carbon and nitrogen compounds were identified. *Tetrasphaera* sp., which are responsible for denitrification and accumulation of polyphosphates, constituted up to 5.44% of the species identified in mature granules. The abundance of *Tetrasphaera* was high throughout the study which suggested that their presence was not determined by the type of biomass in the biological reactor (activated sludge vs. granular sludge). Muszyński and Załęska-Radziwiłł (2015) observed that the phosphate-cumulating microorganisms comprised from 18 to 36% of all bacteria and most of them belonged to *Tetrasphaera* sp. Their highest abundance was observed in WWTPs with high participation of wastewater from food industry (BOD_5 at a level of thousands of mg/L). In the present study, the wastewater from a meat processing plant was regularly discharged to the treatment plant that probably favored growth of *Tetrasphaera* sp. On the other hand, *Clostridium* sp., which are able to decompose organic substances, constituted 2.57% of the identified species at the beginning of the study but their abundance decreased with granule maturation to 0.63% at the end of the study.

Other denitrifying bacteria were *Thauera* sp., whose percentage in the biomass reached 1.35%. This group of microorganisms has the ability to metabolize organic compounds under the aerobic and anoxic conditions and is also commonly found in technological systems with activated sludge [Liu et al. 2006]. The changes of biomass morphology from loose activated sludge flocs to densely packed spherical granules diminished the abundance of aerobic denitrifiers belonging to *Mesorhizobium* sp. from 2.14 to 0.25%.

EPS have a significant contribution to the phenomenon of microbial adhesion as well as to the mechanical strength of granules. Extracellular polymers favor fast adaptation of microorganisms to the environmental conditions and support the formation of three-dimensional, heterogeneous bacterial structures in wastewater treatment systems. In aerobic granules, the bacteria of *Rhizobium* genus, which have denitrification genes and are able to produce polysaccharides that are part of EPS, were identified. Another EPS-producer capable of dissimilatory reduction of nitrate nitrogen (III) [Adav et al. 2008] that was identified in the biomass was *Pseudomonas* sp.

Observations indicate that the temperature of wastewater treatment affects the removal of nitrogen by aerobic granules [de Kreuk et al. 2005b], whereas it does not significantly affect the efficiency of phosphorus removal [Bao et al. 2009]. The present study indicates that the temperature also affects the microbial structure of granules: the highest species diversity (data not presented) was observed in the granular biomass in summer. During this period, the highest percentages of microorganisms from the genera *Thermomonas*, *Dietzia*, *Rhodobacter*, *Dokdonella* and *Dyella* were observed.

CONCLUSIONS

The results of this study indicate that the aerobic granular sludge is an excellent alternative to the conventional activated sludge. Upgrading a WWTP to the aerobic granular sludge technology improved the quality of the treated wastewater: the COD removal efficiency was about 10% higher and ammonium nitrogen was completely oxidized. The use of aerobic granules for the treatment of municipal wastewater in a moderate climate ensured the maintenance of the required quality of the effluent throughout the year, although the temperature affected the morphology of the aerobic granules. In the aerobic granules, the most numerous group of microorganisms were the bacteria from the Actinobacteria class. At higher temperatures in the granules, the percentage of microorganisms from the genus *Rhodobacter*, *Dietzia* and *Dyella* increased, whereas the bacteria from genera like *Sphingopyxis*, *Dechloromonas* or *Halomonas* were most abundant in the mature aerobic granules collected from the biological reactor at the end of the study. Granulation resulted in the disappearance of the filamentous microorganisms of the *Microthrix* genus from the biomass.

Acknowledgements

This study was supported by the Polish National Science Center (grant number 2016/21/B/NZ9/03627) and by a Ministry of Science and Higher Education in Poland (Statutory Research, 18.610.006–300).

REFERENCES

1. Adav S.S., Lee D.J., Show K.Y., Tay J.H. 2008. Aerobic granular sludge: recent advances. Biotechnology Advances, 26, 411–423.

2. APHA Standard Methods for the Examination of Water and Wastewater (1992). 18th edn. APHA, AWWA and WEF, Washington.

3. Bao R., Yu S., Shi W., Zhang X., Wang Y. 2009. Aerobic granules formation and nutrients removal characteristics in sequencing batch airlift reactor (SBAR) at low temperature. Journal of Hazardous Materials, 168, 1334–1340.

4. Bitton G. 2005. Wastewater Microbiology. A John Wiley & Sons, Inc. Publication, 225–235.

5. Cydzik-Kwiatkowska A. 2014. Zastosowanie oraz perspektywy rozwoju technologii granulacji tlenowej w oczyszczaniu ścieków. Inżynieria Ekologiczna, 38, 156–166.

6. Cydzik-Kwiatkowska A., Białowiec A., Wojnowska-Baryła I., Smoczyński L. 2009. Characteristic of granulated activated sludge fed with glycerin fraction from biodiesel production. Archives of Environmental Protection, 35, 41–52.

7. Cydzik-Kwiatkowska A., Zielińska M. Technologia osadu granulowanego w oczyszczaniu ścieków, in: Trendy w biotechnologii środowiskowej, Ed. Wojnowska-Baryła I., Wydawnictwo Uniwersytetu Warmińsko-Mazurskiego, Olsztyn 2011.

8. de Kreuk M.K. 2006. Aerobic granular sludge: scaling up a new technology. Delft University of Technology, Delft (PhD thesis).

9. de Kreuk M.K., McSwain B.S., Bathe S., Tay J., Schwarzenbeck S.T.L., Wilderer P.A. 2005a. Discussion outcomes. Ede. In: Aerobic granular sludge, water and environmental management series. Munich: IWA Publishing, 165–169.

10. de Kreuk M.K., Pronk M., van Loosdrecht M.C. 2005b. Formation of aerobic granules and conversion processes in an aerobic granular sludge reactor at moderate and low temperatures. Water Research, 39(18), 4476–4484.

11. Giesen A., de Bruin L.M.M., Niermans R.P., van der Roest H.F. 2013. Advancements in the application of aerobic granular biomass technology for sustainable treatment of wastewater. Water Practice Technology, 8(1), 47–54.

12. Kotowska M., Zakrzewska-Czerwińska J. 2010. Kurs szybkiego czytania DNA – nowoczesne techniki sekwencjonowania. Biotechnologia, 4, 24–38.

13. Liu B., Zang F., Feng X., Liu Y., Yan X., Zhang X., Wang L., Zhao L. 2006. Thauera and Azoarcus as functionally important genera in a denitriyfing quinoline-removal bioreactor as revealed by microbial community structure comparsion. FEMS Microbiology Ecology, 55, 274–286.

14. Li J., Ding L.B., Cai A., Huang G.X., Horn H. 2014. Aerobic Sludge Granulation in a Full-Scale Sequencing Batch Reactor. Hindawi Publishing Corporation, Article ID 268789.

15. Liu Y., Tay J.H. 2004. State of the art of biogranulation technology for wastewater treatment. Biotechnology Advances, 22, 533–563.

16. Muszyński A., Załęska-Radziwiłł M. 2015. Polyphosphate accumulating organisms in treatment plants with different wastewater composition. Architecture Civil Engineering Environment, 4, 99–106.

17. Podlasek M., Gudecki M., Cydzik-Kwiatkowska A. 2017. Technologia granul tlenowych alternatywą dla oczyszczalni ścieków. Przegląd komunalny, 2, 48–49.

18. Pronk M., de Kreuk M.K., de Bruin B., Kamminga P., Kleerebezem R., van Loosdrecht M.C.M. 2015. Full scale performance of the aerobic granular sludge process for sewage treatment. Water Research, 84, 207–217.

19. Sławiński J. 2015. Tlenowy granulowany osad czynny. Gaz, Woda i Technika Sanitarna, 9, 342–344.

20. Świątczak P., Cydzik-Kwiatkowska A. 2017. Performance and microbial characteristics of biomass in a full-scale aerobic granular sludge wastewater treatment plant. Environmental Science and Pollution Research, doi: 10.1007/s11356–017–0615–9.

21. Tarczewska T.M. 1997. Biotyczne i abiotyczne uwarunkowania pęcznienia osadu czynnego. Ochrona środowiska, 65, 29–32.

22. Tay S.T.L., Ivanov V., Yi S., Zhuang W.Q., Tay J.H. 2002. Presence of anaerobic Bacteroides in aerobically grown microbial granules. Microbial Ecology, 44, 278–285.

23. Toh S.K., Tay J.H., Moy B.Y.P., Tay S.T.L. 2003. Size-effect on the physical characteristics of the aerobic granule in a SBR. Applied Microbiol and Biotechnology, 60, 687–695.

24. van der Roest H.F., de Bruin L.M.M., Gademan G., Coelho F. 2011. Towards sustainable waste water treatment with Dutch Nereda® technology. Water Practice and Technology, 6(3).

25. Wu B., Li Y., Lim W., Lee S.L., Guo Q., Fane A.G., Liu Y. 2016. Single-stage versus two-stage anaerobic fluidized bed bioreactors in treating municipal wastewater: Performance, foulant characteristics, and microbial community. Chemosphere, 171, 158–167.

7

Trickling Filter for High Efficiency Treatment of Dairy Sewage

Radosław Żyłka[1], Wojciech Dąbrowski[1*], Elena Gogina[2], Olga Yancen[2]

[1] Bialystok University of Technology, Faculty of Building and Environmental Engineering, ul. Wiejska 45E, 15-351 Białystok, Poland
[2] Moscow State University of Civil Engineering, Yaroslawskoyoe Shosse, Moscow, Russia
[*] Corresponding author's e-mail: dabrow@pb.edu.pl

ABSTRACT

The article presents the results of the research on the possibility of using a trickling filter for high efficiency treatment of dairy sewage. Nowadays, to the best of the authors' knowledge, it is possible to change the activated sludge system to the trickling filter technology, especially in small dairy plants with lower raw sewage parameters in comparison to larger plants. In the research, dairy wastewater after dissolved air flotation (DAF) process was treated with a laboratory scale Gunt CE701e research model which allows to control the basic parameters of the treatment with a trickling filter (TF). The conducted study included determining the changing sewage parameters during the DAF process, as well as the trickling filter (TF) efficiency. Such parameters as Biological Oxygen Demand (BOD), Chemical Oxygen Demand (COD), total Organic Carbon (TOC), total Kjeldahl nitrogen (TKN) and total phosphorous (TP) were checked. The research results confirmed the possibility of high efficiency treatment of dairy sewage with DAF and trickling filter technologies. The average efficiency of DAF treatment was 59.3% for BOD, 49.0% for COD and 80.0% for TP, while the average treatment efficiency of TF was 87.3%, 78.3% and 27.9% without recirculation and 95.2%, 85.5% and 42.0% with 100% recirculation applied, respectively. The load of TF during the operation without recirculation was on average 0.22 $kgBOD_5 \cdot m^{-3} \cdot d^{-1}$ and 0.25 with 100% recirculation. Applying recirculation allowed to reach the BOD, COD and total phosphorus standards for the sewage discharged to a receiver from Bielmlek dairy WWTP. On the other hand, the concentration of total nitrogen exceeded the permitted standards in this facility.

Keywords: dairy sewage, trickling filter, dissolved air flotation, efficiency

INTRODUCTION

TFs have been used to provide biological sewage treatment for more than 100 years. A TF is a non-submerged fixed-film biological reactor using rock or plastic media over which sewage is distributed continuously [Tchobanoglous, 1997; Godoy-Olmos et al., 2016]. They were commonly used in Poland in the 1970s and 1980s for treating both municipal and food industry sewage. Classic TFs (sprinkling system) generally operated with Imhoff tanks. Additionally, rotary biological contactors (RBC) were widely applied in small towns and such facilities as hotels, holiday resorts etc. Discontinuing their usage in Poland resulted from introducing the demands concerning intensive removal of biogenic compounds in 1991 [Regulations, 1991; Regulations, 2014].

It practically caused the disappearance of this technology for the sake othe f activated sludge. In Russia, for example, a limited use of trickling filter stemmed from very high standards for removing the organic substance for specific sewage receivers. [Gogina & Yantsen, 2015]. Unlike in Poland, wastewater treatment plants using TF or RBCs still function successfully, for instance in Great Britain, where RBCs operate with vertical or horizontal flow constructed wetlands [Obarska-Pempkowiak et al., 2010]. The experiments involving RBCs for landfill leachate treatment confirm that an effective usage of constructed wetlands technology is possible not only for the treatment of municipal sewage [Passeggi et al., 2012, Habte & Eckstadt, 2013]. This technology, contrary to the activated sludge, is much simpler and bears lower exploitation cost. Operating TFs or RBCs do not generate the waste typical for the activated sludge method; they are also characterized by lower power requirements [EPA, 2000; Daigger & Boltz, 2011; Post & Medlock, 2002, Henrich & Marggraff, 2013]. TFs can be classified as raughing, carbon oxidation, carbon oxidation and nitrification and tertiary nitrification [Tchobanoglous, 1997, Daigger & Boltz 2011, Henrich, 2014]. The classification of TFs into low, medium and highly loaded ones is based on the value of $kgBOD_5 \cdot m^{-3} \cdot d^{-1}$ indicator. Highly loaded ones operate with 0.65–3.2 kg $BOD_5 \cdot m^{-3} \cdot d^{-1}$ and low loaded ones with 0.07 to 0.22kg $BOD_5 \cdot m^{-3} \cdot d^{-1}$ [Tchobanoglous, 1997]. In the TFs classification, surface hydraulic load parameter, expressed as $m^3 \cdot m^{-2} \cdot d^{-1}$, is also used [Dymaczewski et al., 1997]

MATERIAL AND METHODS

Dairy WWTP characteristic

The research was carried out in a wastewater treatment plant of dairy cooperative Bielmlek in Bielsk Podlaski, Poland. The dairy plant specializes in the production of cheese, butter, powdered milk and powdered whey. The designed hydraulic capacity of their dairy WWTP is 750 $m^3 \cdot d^{-1}$, while the maximum is 1200 $m^3 \cdot d^{-1}$. Mechanical and biological blocs are capable of treating daily up to 1106 $kgO_2 \cdot d^{-1}$, 2531$kgO_2 \cdot d^{-1}$, 348$kg \cdot d^{-1}$, 90$kgN \cdot d^{-1}$ and 13,5$kgP \cdot d^{-1}$ of BOD_5, COD, TSS, TN and TP respectively. The production profile of the site determines a high concentration of or-

ganic pollutants in sewage, as well as high fluctuations in the quality and quantity of wastewater [Dąbrowski & Żyłka, 2015; Danalewich et al., 1998]. According to the data obtained in 2016, the average flow capacity was 550 $m^3 \cdot d^{-1}$ (range 220–2020), the personal equivalent was on average 13500 and daily dewatered sludge production was 0.22 tons of dry mass. The average concentration of pollutants in the inflow in 2016 was 1521 (350–2650) $mgO_2 \cdot dm^{-3}$, 2055 (456–4410) $mgO_2 \cdot dm^{-3}$, 90.1 (7.2–265.0) $mgN \cdot dm^{-3}$ and 19.5 (4.7–36.0) $mgP \cdot dm^{-3}$ for BOD_5, COD, TN and TP respectively. The primary sewage treatment is based on screening, sand removal and dissolved air flotation DAF (Fig. 1). The biological treatment is carried out in two Sequencing Batch Reactors (SBRs). Excessive sludge is mechanically dewatered and used as fertilizer. The treated wastewater is discharged to the Biała river. Figure 2 presents a scheme of Bielmlek dairy WWTP (sewage and sludge treatment). The required treatment efficiency for dairy WWTPs is determined by Water Permission [IPPC, 2015]. It specifies the maximum concentration of pollutants in the outflow as 25, 125, 30 and 1 $mg \cdot dm^{-3}$ for BOD_5, COD, TN and TP, respectively.

Laboratory scale installation

In laboratory tests, the pretreated sewage from Bielmlek dairy WWTP after DAF flotation was used. A testing system designed by the GUNT company (figure 3) was applied mainly for carbon removal and nitrification. It consisted of a TF filled with fittings made of HDPE ma-

Figure 1. DAF flotation in Bielmlek WWTP

terial. The total volume of the TF is 90 liters, while its diameter measures 340 mm. The supplying tank has the volume of 200 liters. The carrier material (HDPE) has the specific surface of 180 m²·m⁻³. The rotary distributor allows an even supply of the bed (sprinkling), while the aeration vents located below a fixed bed supply the biofilm process with oxygen. The CE 701 model is also equipped with a collecting tank and recirculation pumps. A secondary clarifier and aeration compressor are used only during the startup of the biofilm process.

Sampling, measuring methods and startup of biofilm process

The efficiency tests of DAF flotation in Bielmlek dairy WWTP were carried out in October and November 2017 (ten series). The laboratory scale part of the research with the TF was conducted in November 2017 (12 series with 0.44 m³·m⁻²·d⁻¹ hydraulic load, with and without recirculation). During the operation, 100% recirculation was applied to achieve the results allowing for reaching the parameters required for the quality of treated sewage in Bielmlek WWTP [IPPC, 2015]. The air temperature during the laboratory scale experiment was stable and varied from 16°C to 18°C. The basic physical and chemical analyses were performed, including: biochemical oxygen demand BOD_5, chemical oxygen demand COD, total organic carbon TOC, total Kjeldahl nitrogen TKN, ammonia nitrogen $N-NH_4^+$, nitrate nitrogen $N-NO_3^-$, nitrite nitrogen $N-NO_2^-$ and total phosphorus TP. The parameters of dairy sewage supplying the installation were used to calculate the load of the TF. Deter-

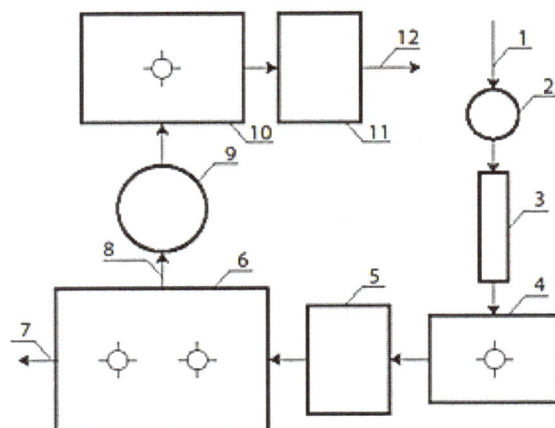

Figure 2. Flow diagram of Bielmlek dairy WWTP: 1. Raw dairy sewage, 2. Screen, 3. Grit chamber, 4. Buffer tank, 5. DAF flotator, 6. SBRs, 7. Treated sewage discharged to receiver (Biała river), 8. Excess sludge, 9. Thickener, 10. Aerobic sewage sludge stabilization chamber, 11. Dewatering press, 12. Sludge for disposal

minations were conducted in a certified laboratory, in accordance with the procedures set out in the Regulation of the Environmental Protection Minister from 18th November 2014 and in line with the American Public Health Association [APHA, 2005].

In accordance the instruction, the start-up of the TF model lasted eleven weeks. At the beginning, the TF was supplied with activated sludge using aeration to obtain biofilm. Then, supplying the research installation with dairy sewage began. The examination of the biofilm structure (its content) in a BUT laboratory confirmed the proper functioning of the bed. Figure 4 presents microorganisms in the biofilm obtained during the startup.

Figure 3. Research installation with main elements during operation (Nov 2017)

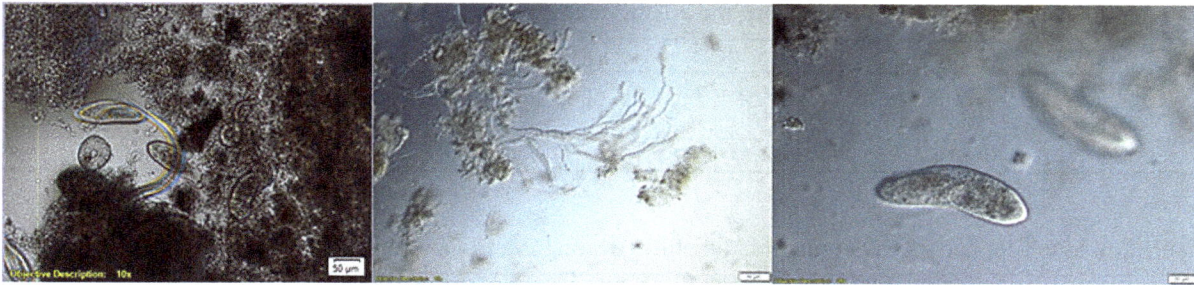

Figure 4. Biocenosis of biofilm, research installation after the startup [Butarewicz]

RESULTS AND DISCUSSION

Efficiency of DAF treatment

The DAF device proved to be very effective in removing the pollution load from the dairy wastewater. During the research period, the mean value of percentage removal was 59.3%, 49.0%, 39.8%, and 80% for BOD_5, COD, TN and TP, respectively (Fig. 5). It corresponds to the literature data. Passeggi *et al.* (2012) obtained 20% efficiency in terms of COD removal, while Rusten *et al.* (1993) examined the influence of different coagulants concentration and obtained 60% efficiency. Organic pollutants expressed as BOD and COD in the influent ranged from 850 $mgO_2 \cdot dm^{-3}$ to 1650 $mgO_2 \cdot dm^{-3}$ and from 1215 $mgO_2 \cdot dm^{-3}$ to 2465 $mgO_2 \cdot dm^{-3}$ respectively, while in the effluent, they ranged from 410 $mgO_2 \cdot dm^{-3}$ to 660 $mgO_2 \cdot dm^{-3}$, and from 780 $mgO_2 \cdot dm^{-3}$ to 930 $mgO_2 \cdot dm^{-3}$, respectively. Nutrients, expressed as total nitrogen and total phosphorus in the influent ranged from 52 $mgN \cdot dm^{-3}$ to 152 $mgN \cdot dm^{-3}$ and from 13,6 $mgP \cdot dm^{-3n}$ to 25.4 $mgP \cdot dm^{-3}$ respectively, while in the effluent they ranged from 24 $mgN \cdot dm^{-3}$ to 59 $mgN \cdot dm^{-3}$ and from 1.0 $mgP \cdot dm^{-3}$

to 6,8 $mgP \cdot dm^{-3}$ respectively (table 1). Babatola *et al.* (2011) studied the efficiency of DAF device for dairy sewage and received the removal rates equal to 66.09%, 65.89% and 94.49% for BOD, COD and TN respectively. The DAF system is also very effective in removing fats and oils from wastewater. The research carried by Al-Shamrani *et al.* (2002) showed that properly adjusted mixing and flocculation times can lead to significant oil separation, achieving more than 99% removal.

Efficiency of TF treatment

Table 2 presents the dairy wastewater characteristic after the DAF process (average and SD) which supplied the TF during the research period. It also presents the results of treatment during the operation with and without recirculation. The average chemical composition of the dairy wastewater feeding the TF varied from 460 to 540 $mgO_2 \cdot dm^{-3}$ for BOD_5, 740–810 $mgO_2 \cdot dm^{-3}$ for COD, 190–240 $mg \cdot dm^{-3}$ for TOC, 5–9.4 $mgN\text{-}NH_4^+ \cdot dm^{-3}$ for ammonia nitrogen and 1.4–3 $mgP \cdot dm^{-3}$ for total phosphorus. The content of sewage before the flotation process (table 1) did not differ from the typical sewage content in broad production range (milk, cheese, powder milk). The sewage content after the DAF process is very individual and depends mainly on the amount of chemicals used in this process [Dąbrowski, 2011, Dąbrowski et al. 2016.] It was observed that during the experiment without treated sewage recirculation, sewage parameters significantly exceeded the permitted emission standards for Bielmlek WWTP [IPPC, 2015]. Only the phosphorous concentration was below the permitted value.

The load indicator during treatment without recirculation was calculated using the average values of BOD_5, hydraulic flow and TF volume. It reached 0.22 $kgBOD_5 \cdot m^{-3} \cdot d^{-1}$ and 0.25 with 100% recirculation. The treatment efficiency of the TF with recirculation was 95.2% for BOD and 85.5%

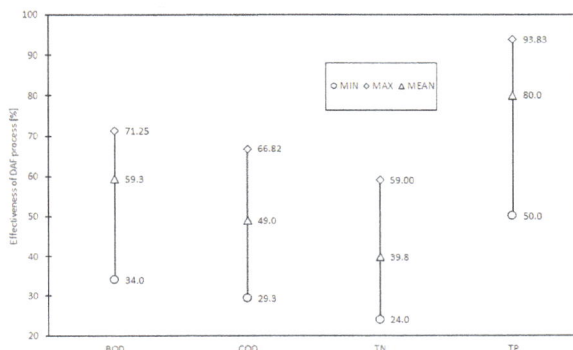

Figure 5. Pollutants removing effectiveness of the DAF process in Bielmlek dairy WWTP during the research period

Table 1. Concentration of pollutants before and after the DAF process. Mean values ± standard deviation.

BOD$_5$		COD		N-tot.		P-tot.	
mgO$_2$·dm^{-3}		mgO$_2$·dm^{-3}		mgN·dm^{-3}		mgP·dm^{-3}	
IN	OUT	IN	OUT	INL	OUT	IN	OUT
1336±281.1	518±70.7	1735±374.3	843±57.8	84±29.9	40±12.4	20.3±4.5	3.8±1.5

Table 2. Wastewater parameters during trickling filter treating. Mean value ± standard deviation

Item	Unit	Inlet	Outlet – no recirculation	Outlet – 100% recirculation
BOD$_5$	mgO$_2$·dm^{-3}	505.5±26.71	63.7±7.73	24.2±2.10
COD	mgO$_2$·dm^{-3}	782±22.39	169.3±12.53	113.6±5.83
TOC	mg·dm^{-3}	212.5±18.06	32.6±4.53	11.3±1.64
TKN	mgN·dm^{-3}	48±4.83	35.8±2.15	25.2±2.86
N-NH$_4^+$	mgN-NH$_4^+$·dm^{-3}	6.89±1.53	1.48±0.51	0.76±0.21
N-NO$_3^-$	mgN-NO$_3^-$·dm^{-3}	1.54±0.61	13.03±1.89	14.04±2.44
TP	mgP·dm^{-3}	2.32±0.60	1.66±0.43	1.33±0.40

for COD. In the sewage treated with recirculation, the average BOD was 24.2 mgO$_2$·dm^{-3} and COD 113.6 mgO$_2$·dm^{-3}. Such parameters met the demands set for Bielmlek WWTP (BOD$_5$, COD, P Total). Recirculation of the treated sewage caused a decrease in the BOD$_5$ value in the influent from 505.5 to 283.7 mgO$_2$·dm^{-3}. In the case of COD, recirculation caused its decrease in the TF influent from 782.0 to 475.6 mgO$_2$·dm^{-3}. Reaching a high level of removing organic substance during the TF treatment was possible due to a very low bed load, as well as high temperature during the tests. In the research conducted by Rodziewicz et al. (2014) on an anaerobic disc batch reactor, the value of COD decreased from 530.5 to 74.0 mgO$_2$·dm^{-3} during the synthetic sewage primary treatment with the use of four stage RBC. The TFs operating at low load are dedicated to conduct the nitrification process [Daigger & Boltz, 2011]. Increased effectiveness in removing the

organic substance measured by BOD$_5$ and COD values can be achieved by, for example, using modified media for biofilm growing [Łobos-Mojsa et al., 2016]. In the authors' own research, a very low concentration of ammonia nitrogen in the dairy sewage after the DAF process was observed. The concentration of ammonia nitrogen was on average 6.89 mgN-NH$_4^+$·dm^{-3} (table 2), while the concentration of TKN was on average 48.0 mgN·dm^{-3}. Finally, 88.8% decrease of ammonia nitrogen and nearly 47.0% of TKN (with recirculation) were achieved. The concentration of nitrates raised from 1.54 mgN-NO$_3^-$·dm^{-3} to 13.03mgN-NO$_3^-$·dm^{-3} and 14.04mgN-NO$_3^-$·dm^{-3} in the experiments with and without recirculation, respectively. The calculated value of total nitrogen in the treated sewage, on the basis of the average concentration of Kjeldahl nitrogen, nitrates and nitrites, equaled 48.8 mgN·dm^{-3} for the treatment without recirculation and 39.0 mgN·dm^{-3} for

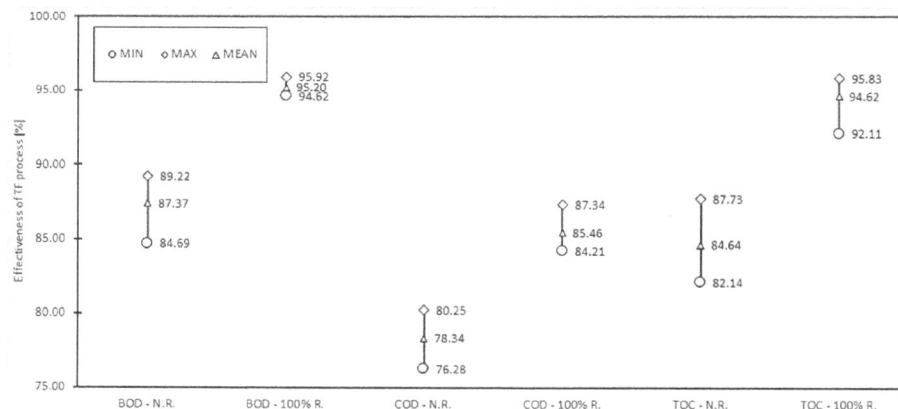

Figure 6. Effectiveness of the TF process. Parameters BOD, COD and TOC; N.R. – no recirculation applied (N.R.), 100% R. – 100% recirculation applied

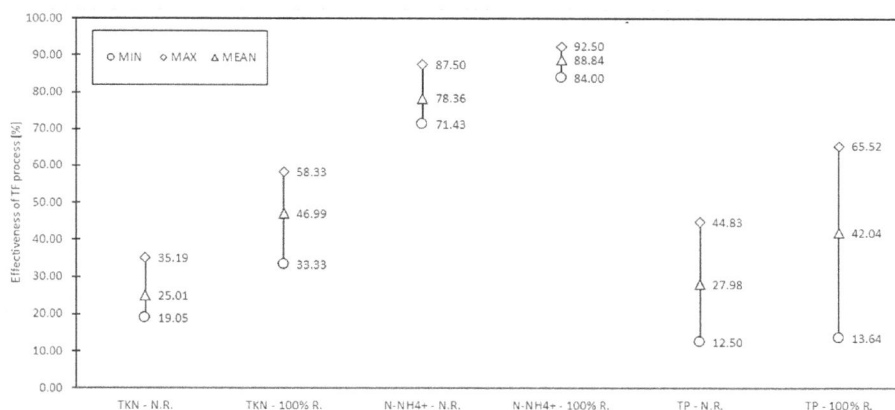

Figure 7. Effectiveness of the TF process. Parameters TKN, N-NH$_4^+$ and TP; N.R. – no recirculation applied, 100% R. – 100% recirculation applied

the treatment with recirculation. The concentration of nitrites was below 0.2 mg N-NO$_2$·dm^{-3} in raw sewage and after treatment with and without recirculation. The conditions of conducting the process ensured high level of nitrification typical for TFs [Post & Medlock, 2002]. Intensification of the nitrogen removal through the process of biological denitrification requires applying multi-layered beds with modified filling or increasing recirculation level [Kanda et al., 2016; Łobos-Moysa et al., 2016].

CONCLUSIONS

Currently, the DAF process is commonly used in the food industry WWTPs, especially with meat and dairy production wastewater. Its application can decrease the load of organic matter, phosphorus and nitrogen in raw wastewater. It substantially decreases the load of the biological treatment part of WWTPs. During the research concerning treating dairy wastewater after DAF process with TF technology, it was proven that it is possible to achieve high efficiency of treatment. The parameters after treatment with low load TF allowed to discharge sewage to the receiver (BOD, COD and TP). The average efficiency of treatment was up to 95.9% for BOD and up to 87.3% for COD. The issue of high efficiency phosphorus removal was solved during the DAF treatment. During the authors' own research, the concentration of TP decreased from 20.3 to 3.8 mgP·dm^{-3} with DAF and reached 1.33 mgP·dm^{-3} after TF treatment, on average. The efficiency of the chemical phosphorus removal depends mainly on the amount of chemicals used during its precipitation. It confirms the

possibility of changing the activated sludge into TF technology, especially in small dairy plants. The primary advantage of this solution is a lack of excess sludge and lower consumption of energy. Further research will be conducted to determine the influence of temperature on the possibility of high efficiency treatment and the possibility of intensifying the removal of total nitrogen.

Acknowledgements

The study was conducted as a research project MB/WBIIS/17/2017 at the Faculty of Civil and Environmental Engineering of Bialystok University of Technology and financed by Ministry of Science and Higher Education of Poland. Scientific and technical cooperation agreement between BUT and Bielmlek allowed for conducting the research.

REFERENCES

1. Al-Shamrani A. A., James A., Xiao H. 2002. Destabilisation of oil–water emulsions and separation bydissolved air flotation. Water Research, 36: 1503–1512.

2. American Public Health Association (APHA). 2005. Standard Methods for Examination of Water and Wastewater. 21st edition. American Public Health Association. Washington

3. Babatola J. O., Oladepo K. T., Lukman S., Olarinoye N. O., Oke I. A. 2011. Failure Analysis of a Dissolved Air Flotation Treatment Plant in a Dairy Industry. Journal of Failure Analysis and Prevention. 11: 110–122.

4. Daigger G. T., Boltz J. P. 2011. Trickling Filter and Trickling Filter-Suspended Growth Process Design

and Operation: A State of Art Review. Water Environmental Research. 83(5), 388–404.

5. Danalewich J. R., Papagiannis T. G., Belyea R. L., Tumbleson, M. E., Raskina L. 1998. Characterization of dairy waste streams, current treatment practices, and potential for biological nutrient removal. Water Research. 32(12), 3555–3568.

6. Dąbrowski W. 2011. Determination of pollutants concentration changes during dairy

7. wastewater treatment in Mlekovita Wysokie Mazowieckie. Ecological Engineering. 24, 236–243 (In Polish).

8. Dąbrowski W., Żyłka R., Rynkiewicz M. 2016. Evaluation of energy consumption in agro-industrial wastewater treatment plant. Journal of Ecological Engineering. 17: 73–78.

9. Dąbrowski W., Żyłka R. 2015. Evaluation of energy consumption in dairy WWTP Bielmlek Bielsk Podlaski. Ecological Engineering. 43, 68 – 74 (In Polish).

10. Godoy-Olmos S., Martinez –Lloren S., Tomas-Vidal A., Jover-Cerda M. 2016. Influence of filter medium type, temperature and ammonia production on nitrifying trickling filters performance. Journal of Environmental Chemical Engineering. 4, 328–340.

11. Gogina E., Yantsen O. 2015. Research of biofilter feed properties. International Journal of Applied Engineering Research. 10(24), 44070–44074.

12. Habte Lemji H., Eckstädt H. 2013. A pilot scale trickling filter with pebble gravel as media and its performance to remove chemical oxygen demand from synthetic brewery wastewater. Journal of Zhejiang University – SCIENCE B (Biomedicine & Biotechnology). 14(10), 924–933,

13. Henrich C. D., Marggraff M. 2013. Energy-efficient Wastewater Reuse – The Renaissance of Trickling Filter Technology. Proc. 9th International Conference on Water Reuse, 27–31.

14. Henrich C. D. 2014. German research underway on trickling filter practices. World Water – Wastewater Treatment. 9–10/2014, 37–38.

15. Integrated Pollution Prevention and Control (IPPC) Permission for Bielmlek Dairy Cooperative WWTP. 2015. (In Polish).

16. Kanda R., Kishimoto N., Hinobayashi J., Hashimoto T. 2016. Effects of recirculation rate of nitrified liquor and temperature on biological nitrification–denitrification process using a trickling filter. Water and Environment Journal. 30, 190–196.

17. Łobos-Moysa E., Bod M., Śliwa A. 2016. Influence of modified porous aggregates on the efficiency of treatment by trickling filter systems. Proceedings of ECOpole. 10(2), 693–698. (In Polish)

18. Tchobanoglous, G. 1997. Wastewater Engineering: Treatment, Disposal and Reuse, 2nd edition, Mc Graw Hill, New York

19. Obarska-Pempkowiak H., Gajewska M., Wojciechowska E. 2010. Wetland treatment of water and wastewater. PWN Publishing, Warsaw (In Polish).

20. Dymaczewski Z., Oleszkiewicz J. A., Sozański M. 1997. Wastewater treatment plant exploiter's guide. PZiTS, Poznań (In Polish).

21. Passeggi M., López I., Borzacconi L. 2012. Modified UASB reactor for dairy industry wastewater: performance indicators and comparison with the traditional approach. Journal of Cleaner Production. 26, 90–94.

22. Post T., Medlock J. 2002. Wastewater Technology Fact Sheet Tricking Filters, EPA

23. Regulations of the Minister of Environment from 5th of November 1991 on conditions to be met for disposal of treated sewage into water and soil (Dz. U. 116, no. 503). (In Polish).

24. Regulations of the Minister of Environment from 18th of November 2014 on conditions to be met for disposal of treated sewage into water and soil and concerning substances harmful to the environment (Dz.U. 2014. no. 1800). (In Polish).

25. Rodziewicz J., Janczukowicz W., Mielcarek A., Filipkowska U., Kłodowska I., Ostrowska K., Duchniewicz S. (2014). Anaerobic rotating disc batch reactor nutrient removal process enhanced by volatile fatty acid adsorption. Environmental Technology. 36(8), 953–958.

26. Rusten B., Lundar A., Eide O., Ødegaard H. 1993. Chemical Pretreatment of Dairy Wastewater. Water Science and Technology. 28(2): 67–76.

27. U.S Environmental Protection Agency (EPA). 2000. Wastewater Technology Fact Sheet – Trickling filters. Office of Water, Washington D.C.

8

Using Ion Exchange Process in Removal of Selected Organic Pollution from Aqueous Solutions

Jadwiga Kaleta[1], Dorota Papciak[1], Alicja Puszkarewicz[1]

[1] Rzeszów University of Technology, The Faculty of Civil and Environment Engineering and Architecture, Department of Water Purification and Protection, ul. Poznańska 2, 35-084 Rzeszów, Poland

* Corresponding author's e-mail: jkaleta@prz.edu.pl

ABSTRACT

Surfactant and phenol were removed using AMBERLITE IRA 900 Cl ion-exchange resin, which is a strong alkali. In the process, the tests were carried out under non-flow conditions, the effect of contact time and ionite dose on the surfactant and phenol exchange was determined. The tests under the through-flow conditions were realized in three consecutive cycles, preceded by regeneration and rinsing. The obtained results served for determination of ion-exchange capabilities of the studied ionite. The usable ion-exchange capabilities of the resin obtained after the second and third ionite operation cycle were lower by about 10% (surfactant) and 14.29–17.86% (phenol) than those after the first cycle. It shows that the process of sorption occurred simultaneously with the ion-exchange process.

Keywords: detergent, phenol, ion-exchange

INTRODUCTION

Detergents, as well as phenol and its compounds belong to the group of organic compounds which are dangerous to human health. Detergents constitute a commonly used name for surface-active agents (SAAs), which mix with the surface water and groundwater through the municipal and/or industrial wastewater. The compounds hinder the self-purification processes of waters and accumulate in living organisms [Sarkar and Das 2001, Kida and Koszelnik 2015].

SAAs belong to the group of very durable and slowly biodegradable compounds. In the water intended for human consumption, they are present as a result of insufficient water purification. The compounds facilitate absorption of other toxic substances (such as pesticides and polycyclic aromatic hydrocarbons) from the digestive tract to the blood through the decrease of surface tension. They have a harmful effect on skin, causing its dryness, cracking, irritation, and various kinds of eczema. They destroy mucosa of the digestive tract, which causes gastritis [Kaleta 2008].

An effective method of detergents removal is adsorption. In the process of granular activated carbon adsorption, 31.7–67.4% of anionic and non-ion SAAs were removed [Cserhati et al. 2002]. The adsorption capacity of granular activated carbon produced in Poland towards anionic SAAs was between 11 and 17 mg/g [Kaleta 2008]. Meanwhile, the activated carbon dust removed 34–38% of anionic SAAs, and 68–98% of cationic SAAs [Basar et al. 2004]. Other sorbents used for SAAs sorption were: activated oxygen and mineral clays.

The ultrafiltration process was found to be very effective in anionic surfactant removal in a wide range of concentrations, the retention coefficient amounted to 82–90% [Kowalska 2008].

The anionic surface-active agents were effective in the ion exchange process of gel and macroporous anion-exchange resins [Kowalska 2009].

The source of phenol in the natural waters can be products of degradation of organic substances, as well as industrial wastewater, i.a. from coking plants and chemical industry. Phenols are toxic substances, which negatively influence the water environment. At concentration over 5 mg/dm³, they are toxic to the aquatic fauna and flora; they also hinder the water self-purification processes [Piekutin 2011, Puszkarewicz et al. 2015].

Phenol is a strong toxin. It has transfixing effect on the nervous and cardiovascular systems. It damages the respiratory tract, causes denaturation of proteins, and results in skin irritation. The water chlorination process creates chlorophenols, toxic compounds that cause the water intended for consumption to have unpleasant smell and changed taste [Kaleta 2006].

The most effective method of water purification of phenols and their products is the adsorption of activated carbons [Oghenejoboh et al. 2016]. The adsorption capacity of activated carbons produced by Gryfskand Sp. z o.o. company in Hajnówka, Poland, is between 35 and 50 mg/g [Kapica et al. 2002, Kaleta 2006]. The spherical activated carbons, received in the process of agglomeration of granular hard coking coal with vegetables oils (rapeseed and flaxseed) are characterised with a higher adsorption value (between 70 and 150 mg/g) [Lorenc–Grabowska and Gryglewicz 2002].

An attempt has been made to replace activated carbons with other adsorbents, such as: granular coke, bituminous coal, peat, calcium carbonate, bog iron ore, natural and modified zeolites, bentonites, and modified clays. These adsorbents, despite their lesser adsorption capacity, may replace activated carbons under certain conditions [Nayak and Singh 2007, Kaleta et al. 20013, Dudzik and Werle 2015].

Phenolic compounds were effectively removed by Amberlite XAD resin. Its sorption capacity was higher for chlorophenol (2.27 mmol/g) than for phenol (1.50 mmol/g). Resin regeneration was conducted through extraction (elution) with methanol as a solvent. After subsequent desorption processes, the adsorption capacity did not change [Abburi 2003/B].

An effective method of removal of organic pollution from water is MIEX process (Magnetic Ion Exchange Process), in which selective ion-exchange resins of strongly alkaline function groups, working in a chloride cycle were used. The resins, produced in the form of "magnetic" microions, under appropriate hydraulic conditions underwent agglomeration and separation from the purified water. The MIEX method was utilised along with the coagulation process in order to maximize the effect of water purification process [Mołczan and Biłyk 2006, Kabsch-Korbutowicz 2013].

RESEARCH SUBJECT

The research focused on the removal of an anionic surface-active agent and phenol. The pollution is being limited in water for human consumption (Table 1).

According to the World Health Organisation (WHO) recommendations, the maximum concentration of anionic surface-active agents in the water intended for consumption is 0.2 mg/dm³. In the water intended for human consumption, WHO also limits by-products of chlorination, among them 2,4,6 – trichlorophenol, the maximum concentration of which is 0.01 mg/dm³ [Guidelines 2017].

The research was conducted on model solutions, called also initial solutions, which were prepared on the basis of distilled water. The means of control determination were indicators of the pollution types under research.

Model solution of the detergent (D) was prepared by adding distilled water to the Aerosol OT-100 anionic surface-active agent (chemical formula $C_2H_{37}NaO_7S$) in an amount that results in detergent concentration in the solution equal to 20.00 g/m³. Determination of the detergent was carried out using the method involving methylene blue. The method consisted in creating blue organic compound as a result of reaction of methylene blue with anion-exchange surface-active agent. The created compound was extracted with chloroform in alkaline environment, and the intensity of the blue hue, measured photometrically at the wavelength equal $\lambda=652$ nm, was proportional to the concentration of detergents.

The phenolic pollution was investigated on the example of the most often occurring compound, which is phenol (C_6H_5OH). Phenol was dissolved in distilled water, transforming into model solution (F) with the concentration of 20.00 g/m³. Determination of the phenol concentration was carried out using an indirect method through the absorption measurement (with the wavelength of $\lambda=254$ nm) by means of a Schimadzu UV-1601

Table 1. The admissible concentration of surfactant and phenol in natural waters, mg/dm³ (Council Regulation No. 203, item 1718 of 27 November 2002)

Type of pollution	Category A1	Category A2	Category A3
Surfactant	0,2	0,2	0,5
Phenols (phenolic index)	0,001	0,005	0,1

Visible Spectrophotometer. Quartz cuvettes with 1 cm absorption layer were used. Before carrying out the actual measurements, a model curve was created, and the correlation between phenol content F and absorbency A was determined by the following formula (1):

$$F = tg\ 35 \cdot A\ tg\ 35 = 0.700,$$
$$\text{that is } F = 0.700 \cdot A \tag{1}$$

A strongly alkaline AMBERLITE IRA900 Cl resin was used in the ion exchange process (Table 2). It is a macroporous type I resin, consisting of quaternary amine groups. It allows to remove all anions from the water, including the poorly decomposed ones, e.g. silica. The macroporous structure of the resin resembles a sponge. That feature, combined with a strong alkalinity, allows to remove large organic particles.

RESEARCH METHODOLOGY

Tests conducted under non-flow (static) conditions

Before the tests were conducted, the ion exchanger had been soaked in distilled water for 24 hours. An appropriate amount of the swelled ion exchanger was placed in a bottle of 10% NaCl. The bottle contents were shaken for 30 min; then the ion exchanger was drained and rinsed with distilled water until the disappearance of Cl- ions.

The regenerated and rinsed resin was used for further tests. The kinesis of the ion exchange process was determined using the following method: 3 g of the ion exchanger were added to 8 bottles, each of them containing 1 dm³ of an appropriate model solution. The samples were shaken for 10–180 min, and then underwent 30 minutes of decantation. Afterwards, the solution above the

ion exchanger was removed, infiltrated, and the control determination was carried out.

In order to determine the influence of the ion exchanger dose on the ion exchange process, 1 dm³ of a model solution and ion exchanger doses increasing from 0.5 to 7.0 g/dm³ were added to 8 bottles. The contact time was 30 min. After 30 min of sedimentation, control determination of decanted and infiltrated solutions was carried out. The results obtained in this series of tests were used to calculate the exchange capacity of the ion exchanger.

The exchange capacity of the ion exchanger in a moist state (Zcw) was calculated by the following formula (2):

$$ZCW = \frac{(Co - Ck) \cdot V}{G} \tag{2}$$

where: Co – initial concentration of the model solution, mg/dm³,
Ck – final concentration of the model solution, mg/dm³,
V – volume of the of the model solution used in tests, dm³,
G – mass of the moist ion exchanger, g,
Zcw – the exchange capacity of the moist ion exchanger, mg/g.

In order to calculate the exchange capacity of the dry ion exchanger, the moisture of the ion exchanger sample used in the research was determined (p). The exchange capacity of the ion exchanger in a dry state (Zcs) was calculated by the following formula (3):

$$Zcs = \frac{100 \cdot Zcw}{100 - p} \tag{3}$$

where: Zcs – exchange capacity of the ion exchanger in a dry state, mg/g,
p – percent of moisture content in the ion exchanger sample.

Determining exchange capacity of the ion exchanger under flow (dynamic) conditions

The useful (working) exchange capacity was determined using laboratory ion exchanger columns made of organic glass with the diameter of 15 mm and height of 600 mm.

In order to conduct the tests, a 20 g sample of the swelled (soaked) ion exchanger was transferred to the ion exchanger column. Then, about 1.5 dm³ of the regenerant solution (10% NaCl) was flushed through the ion exchanger at the

Table 2. Characteristics of AMBERLITE IRA900 Cl

Designation	Unit	Value
Bulk density	g/dm³	640–710
Granulation	m	>500
Effective diameter	m	600–800
Coefficient of homogeneity	-	<1,8
Humidity	%	58–64
Total ion exchange capacity	val/dm³	1,0

speed of 5 m/h, downwards. Afterwards, the ion exchange bed was rinsed with distilled water until the disappearance of the regenerant solution in the outflow.

The actual tests consisted of flushing the appropriate model solution through the ion exchanger at the speed of 10 m/h. After each flow of 1 dm³ of the model solution through the ion exchanger had finished, the samples were collected and control determination was carried out. The ion exchange cycle was stopped when the concentration of the detergent in the outflow increased above 0.6 mg/dm³, and of phenol above 0.2 mg/dm³ (cycle I).

Then, the ion exchanger underwent the regeneration and rinsing again; then it was used for another filtration cycle (cycle II) in the way described above. The whole sequence was repeated one more time (cycle III).

The useful (working) exchange capacity of the moist ion exchanger was calculated with the following formula (4):

$$Zuw = \frac{Vr \cdot \Delta C}{Gj} \qquad (4)$$

where: ΔC – difference between the concentrations of the removed ions (initial concentration minus concentration puncture point), mg/dm³,

Vr – volume of outflow from the column until puncture point, dm³,

Gj – mass of the ion exchanger in the column, g,

Zuw – exchange capacity of the moist ion exchanger, determined in the flow conditions, mg/g.

Considering the percentage content of moisture, the useful exchange capacity of the ion exchanger in a dry state was calculated by the following formula (5):

$$Zus = \frac{100 \cdot Zuw}{100 - p} \qquad (5)$$

where: Zus – exchange capacity determined under dynamic conditions in the dry state, mg/g,

p – moisture content, %.

RESULTS

The ion exchange proceeded very quickly. In the case of the detergent, it decreased by 95.55% after 30 minutes of contact with the ion exchanger. After 30 minutes of contact, the phenol was reduced by 97.50%. A longer contact time did not yield better results in the case of the analysed kinds of pollution (Table 3). The assumed optimal time of contact between the ion exchanger and purified solutions was 30 min.

An increase of the ion exchanger dose up to 4.0 g/dm³ resulted in an improvement of the detergent removal effectiveness to 95.55%. A larger amount of the ion exchanger increased the effectiveness of the ion exchange process only by 0.05 % (Table 4).

In the case of phenol, the ion exchanger sample that yielded satisfactory result of 97.50% reduction, amounted to 3.0 g/dm³. Similarly to the detergent removal process, a larger amount of the ion exchanger caused only a slight effectiveness increase of the ion exchange process (Table 4).

In order to calculate the exchange capacity of the resin under non-flow conditions, a dose of 4.0 g/dm³ was used in the case of the detergent, and a dose of 3.0 g/dm³ in the case of the phenol (Table 5). The moistness of the regenerated and rinsed ion exchanger used for tests under static conditions was 64%.

On the basis of tests conducted under flow (dynamic) conditions, the puncture curve of the

Table 3. Effect of contact time the ion exchange

Contact time, min	Concentration of surfactant, mg/dm³	Decreasse of surfactant, %	Concentration of phenol, mg/dm³	Decreasse of phenol, %
10	0.96	95.20	0.87	95.65
20	0.91	95.45	0.63	96.85
30	0.89	95.55	0.50	97.50
60	0.89	95.55	0.50	97.50
80	0.90	95.50	0.50	97.50
100	0.90	95.50	0.49	97.55
120	0.89	95.55	0.50	97.50
180	0.90	95.50	0.49	97.55

Table 4. Influence of ionite doses on the removal of surfactant and phenol

Ionite doses, g/dm³	Concentration of surfactant, mg/dm³	Decreasse of surfactant, %	Concentration of phenol, mg/dm³	Decreasse of phenol, %
0.5	1.52	92.40	0.94	95.30
1.0	1.50	92.50	0.82	95.90
2.0	1.12	94.40	0.68	96.6
3.0	0.90	95.50	0.50	97.50
4.0	0.89	95.55	0.50	97.50
5.0	0.88	95.60	0.50	97.50
6.0	0.89	95.55	0.49	97.55
7.0	0.88	95.60	0.49	97.55

Table 5. Ion exchange capacity determined under static conditions

Type of pollution	Ion exchange capacity, mg/g	
	Ionite moisture	Ionite dry
Surfactant	4,78	13,28
Phenol	6,50	18,05

ion exchange process were constructed for the detergent and the phenol (Figure 1 and Figure 2). The drawn graphs were used to determine the useful exchange capacity of the used ion exchanger (Table 6).

The anion-exchanger puncture point for the detergent was determined at the level of 0.1 mg/dm³, and for the phenol it amounted to 0.00 mg/dm³.

Moistness of the ion exchanger samples placed in the filtration columns amounted to 65%.

The ion-exchange capacity of anion exchanger under the flow conditions was higher than the ion-exchange capacity determined under the non-flow conditions. The differences in ion exchange capacity values are related to the method of determining the ion exchange balance. In the ion exchange process carried out under non-flow conditions, the equilibrium is determined once. The exchange process carried out under flow conditions due to the batching of the model solution allows to establish new equilibrium states. This results in a better use of the ion exchange capacity. The useful exchange capacity determined for detergent during the second and the third filtration cycles were smaller than the ones received in the 1st cycle of the ion exchange process by 10%. In the case of phenol, the ion-exchange capacity was smaller than in the first filtration cycle by 14.29% in the second cycle, and by 17.86% in the third cycle.

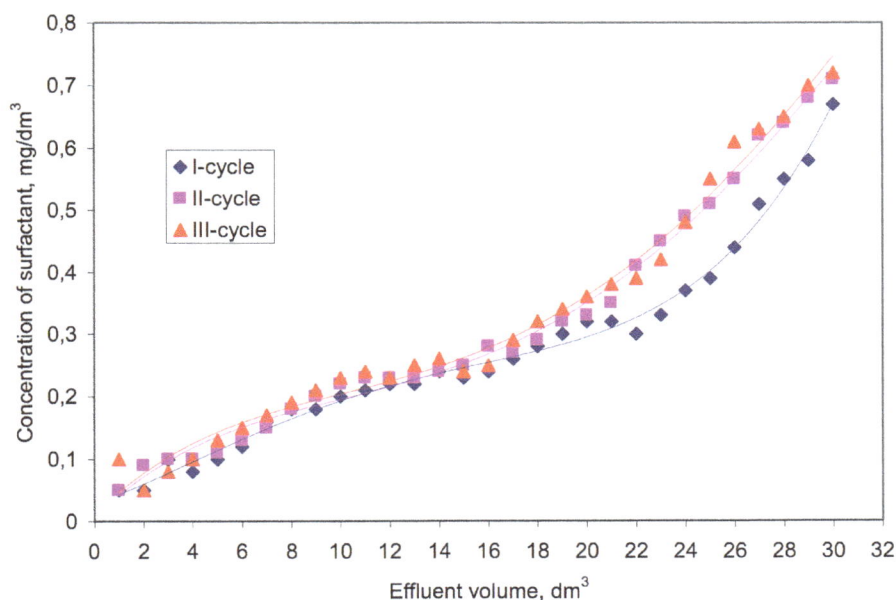

Figure 1. Ion exchange breakthrough curves (isoplanes) – (initial surfactant concentration of 20.00 mg/dm³, filtration rate of 10 m/h)

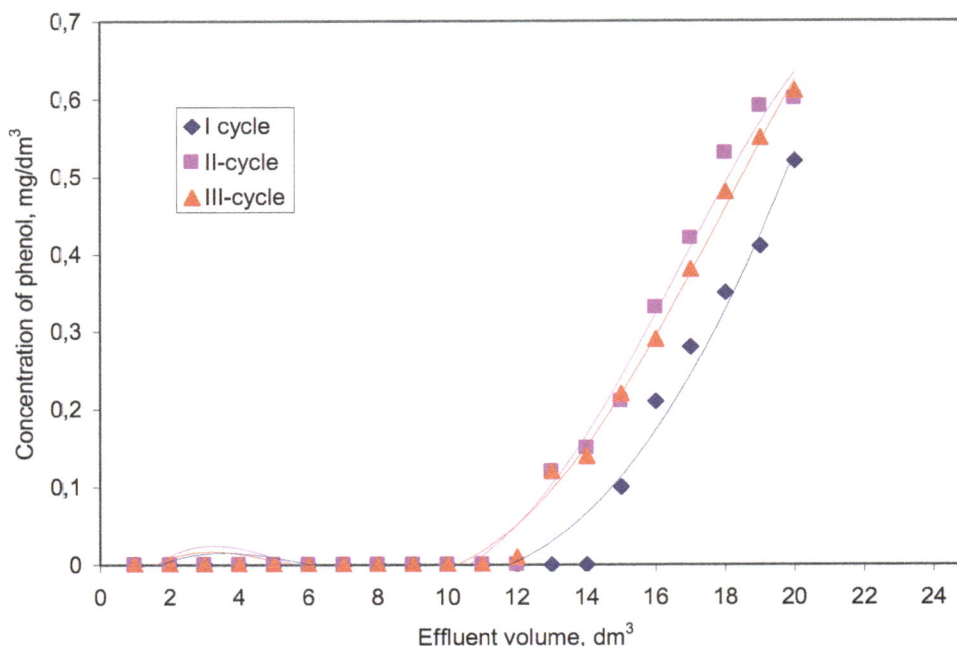

Figure 2. Ion exchange breakthrough curves (isoplanes) – (initial phenol concentration of 20.00 mg/dm³, filtration rate of 10 m/h)

Table 6. Usable ion exchange capacity of *AMBERLITE IRA900 Cl* resin determined under flow conditions

Type of pollution	Usable ion exchange capacity, mg/g					
	I – cycle		II – cycle		III – cycle	
	Ionite moisture	Ionite dry	Ionite moisture	Ionite dry	Ionite moisture	Ionite dry
Surfactant	9,90	26,77	8,91	24,08	8,91	24,08
Phenol	14,00	37,84	12,00	32,43	11,50	31,08

CONCLUSIONS

In the ion exchange process conducted using strongly alkaline, anion-exchange AMBERLITE IRA900 Cl resin yielded good results in the removal of both anionic detergent and phenol.

Anionic detergents, the largest group of the surface-active agents (ca. 80%), decompose into surface-active anions, which is why utilising them for the anion-exchange removal was effective. The useful capacity of ion-exchange resin towards the detergent amounted to 25.60 mg/g for the first run cycle, and 22.74 for the second and the third run cycles.

Phenol is a weak acid which decomposes into a phenolic ion and a proton. The phenolic anion was removed through the ion exchange with anion-exchange resin used in the tests. The useful exchange capacity of the tested ion exchanger towards phenol amounted to 37.14 mg/g in the first run cycle, 31.42 mg/g in the second run cycle, and the least, that is, 28.57 mg/g, in the third run cycle.

The useful ion-exchange capacity of the anion-exchanger was larger in the case of phenol than in the case of the detergent. The detergent, due to its larger particles, entered the channels of the ion exchanger at a slower rate, and even partially obstructed the access to the place of ion exchange.

Besides the ion exchange process of the macroporous AMBERLITE IRA900 resin, an adsorption process took place as well, as proven by the useful ion-exchange capacities achieved after second and third regeneration of the ion exchanger (after second and third cycle of the ion exchanger filter). When designing an ion exchange bed, one should use exchange the capacity values received in the further work cycles of the ion exchanger, since part of the pollution filtered during the first run cycle is not removed during the regeneration process.

REFERENCES

1. Abburi, K. 2003. Adsorption of phenol and p-chlorophenol from their single and bisolute aqueous solutions on Amberlite XAD – 16 resin. Journal of Hazardous Materials, 105, 143–156.

2. Basar, C.A., Karagunduz, A., Cakici, A., Keskinler, B. 2004. Removal of surfactants by powdered activated carbon and microfiltration. Wat. Res., 38, 2117–2124.

3. Cserhati, T., Forgacs, E., Oros, G. 2002. Biological activity and environmental impact of anionic surfactants. Environmet International, 28, 337–348.

4. Dudzik, M., Werle, S. 2015. Phenol sorption from water solution onto conventional and unconventional sorbents . Inżynieria i Ochrona Środowiska, 18(1) 67–81.

5. Guidelines for Drinking – water Quality 2017. World Health Organization, Genewa.

6. Kabsch-Korbutowicz, M. 2013 Application of Ion Exchange to Natural Organic Matter Removal from Water. Ochrona Srodowiska, Vol. 35, No. 1, 11–18.

7. Kaleta, J. 2006. Removal of phenol aqueous solution by adsorption. Canadian Journal of Civil Engineering, 33, 546–551.

8. Kaleta, J. 2008. Removal of surfactant substances from aqueous solution by adsorption. VIII Międzynarodowa Konferencja Naukowo-Techniczna "Zaopatrzenie w wodę, jakość i ochrona wód", Gniezno, 505–514.

9. Kaleta, J., Papciak, D., Puszkarewicz, A. 2013. Assessment of Usability of Bentonite Clays for Removing Phenol from Water Solutions, Annual Set the Environment Protection, 15, 2352–2368.

10. Kapica, J., Kaleńczuk, R.J., Morawski, A.W. 2002. Studies of influence specific surface area on adsorption of phenol from water by activated carbon CWZ. Mat. Konf. "Węgiel aktywny w ochronie środowiska" Wyd. Pol. Częstochowskiej, Częstochowa, 178–187.

11. Kida, M., Koszelnik, P. 2015. Environmental fate of selected micropollutants. Journal of Civil Engineering, Environment and Architecture, Rzeszów, 62(1) 279–298.

12. Kowalska, I. 2008. Separation of Surface Active Agents from Water Solutions Using Polymer Ultrafiltration Membranes. Annual Set the Environment Protection, 10, 593–604.

13. Kowalska, I. 2009. Anionic Surfactant Removal by Ion Exchange. Ochrona Środowiska 2009, Vol. 31, No. 1, pp. 25–29.

14. Lorenc-Grabowska, E., Gryglewicz, G. 2002. Adsorption of chloro- and nitrophenols from aqueous solutions on spherical activated carbons. Mat. Konf. "Węgiel aktywny w ochronie środowiska" Wyd. Pol. Częstochowskiej, Częstochowa, 195–205.

15. Mołczan, M., Biłyk, A. 2006. Use on Anion-Exchange MIEX Resin in the Treatment of Natural Water with High Colored Matter Contend. Ochrona Środowiska vol.28, no 2, 23–26.

16. Nayak, P.S., Singh, B.K. 2007. Removal of phenol from aqueous solutions by sorption on low cost clay. Desalination 207, 71–79.

17. Oghenejoboh, K.M, Otuagoma, S.O., Ohimor, E.O. 2016. Application of cassava peels activated carbon in the treatment of oil refinery wastewater – a comparative analysis. Journal of Ecological Engineering, 17, 52–58.

18. Piekutin, J. 2011. Water pollution with oil products. Annual Set the Environment Protection, 13, 1903–1914.

19. Puszkarewicz, A., Kaleta, J., Papciak, D. 2015. Removal of phenol in adsorption proces. Journal of Civil Engineering, Environment and Architecture, 62(3), 35–362.

20. Sarkowicz M., Das, M. 2001. Design and performance of fixed bed adsorption from treatment of malachite green as single solute as well as in bisolute composition. Annual Set the Environment Protection, 3, 129–237.

9

Using Sawdust to Treat Synthetic Municipal Wastewater and its Consequent Transformation into Biogas

Zaidun Naji Abudi[1]

[1] Environmental Engineering Department, Faculty of Engineering, Al-Mustansiryiah University, Baghdad, Iraq, e-mail: Zaidun.naji77@uomustansiriyah.edu.iq

ABSTRACT

Sawdust, as an agricultural waste which is highly efficient, readily available, and relatively inexpensive, has the potential to be an applicable alternative adsorbent for the total organic carbon (TOC) removal from synthetic domestic wastewater. This study aims firstly to investigate the feasibility of sawdust as a new adsorbent and understand its adsorption mechanism for TOC. The impact of particle size, pH, contact time, and temperature has been evaluated as the controlling factors on the adsorption process. The results presented that the removal efficiency rose with the decrease of particle size, pH, and temperature, as well as the increase of the contact time. The maximum adsorption was obtained at particle size of 0.05 mm, pH of 1, contact time of 1.5 h, and temperature of 15°C, respectively. The second aim of this study is to utilize the sawdust that is used in the adsorption process as biomass in batch anaerobic digestion (AD) to produce methane. Spent sawdust was characterized by the methane production which was 5.9 times greater than in the case of raw sawdust. Four operating parameters were checked, Carbon/Nitrogen ratio (C/N), inoculation, particle size, and total solid (TS) content. The batch results indicated that the optimum parameters were: 20%, 30%, 2 mm, and 15%, respectively.

Keywords: Anaerobic digestion; sawdust; adsorption; total organic carbon

INTRODUCTION

The three main problems faced by the world today are water, food, and energy supply. In order to solve these problems, the domestic wastewater is now treated as a resource rather than as a waste (Benetti, 2008). The 1st and 3rd problems can be addressed through the use of treated wastewater for domestic consumption, including landscape and crop irrigation, which is widely accepted and used to save water as well as utilize the fertilizing elements it contains. The domestic wastewater can be used as a source for energy through anaerobic digestion, to solve the 3rd problem, which involves the methane gas (CH_4) production from wastewater organic content by anaerobic conversion (Speece, 2007). The organic fraction in wastewater is the most direct and commonly exploited energy source. This fraction is a diverse mixture of molecules with a varied structure and molecular weight. It contains ~42% dissolved organic carbon, ~27% settleable organic carbon, ~20% supracolloidal organic carbon, and ~11% colloidal organic carbon (Rickert & Hunter, 1971). If these compounds are not removed properly from wastewater through the wastewater treatment plants (WWTPs), and are discharged into water resources (rivers or lakes), they may constitute an environmental problem. This problem has a harmful effect on the water quality and also affects the aquatic life (Huerta-Fontela et al., 2011).

WWTPs remove or minimize the nutrients such as carbon, nitrogen, and phosphorus. Most of the existing treatment processes involve advanced methods (likes activated carbon, photo-oxidation, ozonation, and UV radiation) because the biological treatment alone is not sufficient (Ahn & Logan, 2010; Katsoyiannis & Samara, 2005). Nevertheless, these advanced methods using ozonaiton and UV light are expensive since they require greater technical experience and are characterized by high energy consumption. Additionally, when activated carbon is used, the steps of its activation and regeneration are delicate and too costly; moreover, additional tertiary filtration is frequently required (Shukla et al., 2010; Yin et al., 2007). In order to treat wastewater by removing nutrients such as TOC efficiently, new waste materials, especially agricultural wastes like sawdust, wood chips etc., have the economic advantages of obtaining low-cost feedstocks for wastewater treatment and minimizing waste disposal costs. In addition, it offers environmental advantages like reusing and reducing wastes and energy recovery (Yang et al., 2017).

Although sawdust, like any biomass, can be explored as a biogas source, it is far less used as a feedstock for biogas production (Castoldi et al., 2017). Therefore, in this study, it is used for the removal of TOC from wastewater and then for biogas (CH_4) production by anaerobic digestion process. Anaerobic digestion (AD) is the process of converting biomass into biogas energy. This process was used for the production of renewable energy so as to achieve stable energy alternatives that will meet the world demand while mitigating the climate change through reduction of emissions. AD constitutes an economical, eco-friendly, renewable energy source which can produce bio-fertilizers as a by-product (Kimming et al., 2011; Lizasoain et al., 2016).

The main objectives of the present study were to (i) investigate the effects of the adsorption process parameters (particle size, dose, temperature, reaction time, and wastewater pH) on the TOC removal by sawdust and (ii) evaluate its potential biogas production by AD of sawdust after the adsorption process; the influence of different operation factors (particle size, inoculum percent, C/N ratio, and T.S content) was checked as well.

MATERIALS AND METHODS

Raw materials

Our main material (pine sawdust) was brought from a furniture factory of Huazhong University of Science and Technology (HUST), Wuhan, China. The pine sawdust was naturally dried for eight days, then crushed and sieved to four different particle sizes. The ultimate and proximate analyses of sawdust are given in Table 1 (wt%, air-dried basis). Synthetic wastewater in this study was prepared to simulate the characteristics of Chinese municipal wastewater. Its composition was

as follows (mg/L): Glucose (332), NH_4CL (210), KH_2PO_4 (21.95), $MgSO_4$ (50), $ZnCl_2$ (50), and a 1 ml trace element mixture, consisting of 0.075 g $CaCl_2$, 0.04 g $CuCl_2H_2O$, 0.048 g $NiCl_2.6H2O$, 0.044 g $FeSO_4.7H_2O$ and 0.120 g H_3BO_3. In order to keep the pH of the synthetic wastewater at 6.8–7.2, 100 mg/L $NaHCO_3$ was added. The inoculum used for anaerobic digestion process with a total solid content (TS%, 4.25), volatile solid content (VS%, 69.0) and pH (7.2) was collected from the active mesophilic anaerobic reactor in the same lab that treated pig manure to produce biogas. All chemicals and reagents used in this study were of analytical grade.

Batch adsorption process

In order to study removal of TOC from synthetic domestic wastewater by pine sawdust, adsorption experiments in batch mode were carried out. The effects of different factors, including particle size (0.05–1 mm), pH (1–9), adsorbent dose (0.1–0.9 g/50 ml), contact time (10 min – 4 h), and temperature (15–50 °C) were investigated. The pH of the solution was adjusted with 0.1 M HCl and 0.1 M NaOH. The weighted absorbents were added to 50 mL of wastewater in a 150mL Erlenmeyer flask and then mixed at 200 rpm for the desired contact time at a specific temperature in a water bath shaker. At the end of adsorption process, and after centrifuging solutions, the suspensions were separated; afterwards, the TOC concentrations were measured according to a standard method (APHA). The adsorption capacity and percentage removal of TOC by sawdust were determined by Eqs. (1) and (2), respectively:

$$q_e = \frac{C_o - C_e}{m} \qquad (1)$$

$$\%removal\ TOC = \frac{C_o - C_e}{C_o} \times 100 \qquad (2)$$

where q_e(mg/g) represents the adsorption capacity.
C_o(mg/L) and C_e(mg/L) is the initial and final concentration of TOC, respectively.
V(ml) is the volume of TOC solution,
m(g) is the weight of the dried adsorbents.

Anaerobic digestion process

The anaerobic digestion experiments were conducted in a batch mode using 1 L Duran glass bottles with a working volume of 0.6 L. Two

Table 1. Characteristics of pine sawdust

Ultimate analysis[a]					Proximate analysis[b]			
Carbon	Hydrogen	Nitrogen	Oxygen[c]	Sulfur	Moisture	Volatile	Ash	Fixed carbon
47.15	6.22	0.18	45.20	0.16	6.95	78.01	1.09	13.95

groups of batches were conducted based on the objectives of this study; the first one was fed with raw sawdust and the other was fed with sawdust used in the adsorption process. At the same time, the bottles only fed with the inoculum were used as a control. All the experiments were conducted in duplicate. Four different parameters were studied in these batch tests, i.e. substrates particle size (0.3, 0.8, 1.0, and 2.0 mm), inoculum percent (10, 20, and 30%), carbon to nitrogen (C/N) ratio (20, 25, and 30%), and total solid (TS) content (5, 10, and 15%). Each parameter was studied alone and others were kept fixed. After adding the designed amounts of sawdust and inoculum in the batch digesters, the C/N ratio and TS% were checked. Prior to sealing, each bottle was flushed with nitrogen gas (99.9% purity) for 2–5 min to confirm anaerobic conditions. Then, the bottles were tightly closed with rubber stopper and caps and placed in a water bath shaker at (35°C) and stirred at 100 rpm. During the batch test, the daily biogas produced in the reactor flowed through a bottle filled with 3M NaOH solution for CO_2-fixation and remaining methane (CH_4), which was measured with the water displacement method.

Analytical methods

The water quality parameters, including total organic carbon (TOC), total Kjeldahl nitrogen (TKN), total solids (TS) and volatile solids (VS) were determined according to Standard Methods (Federation & Association, 2005). The pH of the solution was measured using a pHS-25C pH meter made by Shanghai Precision & Scientific Instrument Co., Ltd. The carbon and nitrogen analysis was conducted using a Vario EL (element analyzer) made by Elementar Analysensysteme GmbH. The CH_4 content was analyzed via gas chromatography (SP-2100, China).

RESULTS AND DISCUSSIONS

Adsorption results

Adsorptive removal of TOC was studied for four different particle sizes of pine sawdust equalling r<0.05, 0.2<r<0.3, 0.3<r<0.5, and 0.5<r<1 mm. As shown in Figure 1a, the rate of adsorption was influenced by the size distribution. It is interesting to note that an increase in the TOC adsorption percentage occurs from 13.0% to 46.5% when the adsorbent size decreases from 0.5<r<1mm to r<0.05 mm. This behaviour can be explained due to the accessibility to larger surface area of the adsorbent for adsorption (Gupta et al., 2015).

In order to study the effect of adsorbent dose in solution on the TOC adsorption percentage, different dosage was used (ranging from 0.1 to 0.9 g/50 mL) and the results are shown in Figure 1b. It was observed that the removal efficiency increases along with the amount of adsorbent in the solution. This can be attributed to the increase in the accessibility of adsorption sites for the TOC. Many studies are in agreement with the current work results (Kazemi et al., 2016; Khamparia & Jaspal, 2016; Mor et al., 2016) where they reported that the sorption of pollutant increases as dosages increased. On the other hand, a reverse trend was observed with the adsorption capacity which decreased as the adsorbent dosage was increased (Figure 1b). The reason of this was possibly due to the particle interaction, such as aggregation, resulting from high adsorbent dosage, which lead to a decrease in the total surface area of the adsorbent and increase in the diffusional path length (Ghaedi et al., 2012).

The pH value of the initial wastewater is the most important parameter affecting the adsorption of TOC. The uptake of TOC by pine sawdust was studied as a function of pH ranging from 1 to 9. The results show that the removal percentage of TOC from the synthetic wastewater decreased with the rise in pH of the solution (Figure1c). Maximum 62% TOC adsorption was obtained at pH 1 indicating that the TOC removal is more efficient under acidic conditions and it minimizes to 12% with a rise in pH to 9. These results were consistent with the previous studies, where it also showed that the adsorption percentage increased with pH reduction (Khamparia & Jaspal, 2016; Liu et al., 2016; Mor et al., 2016).

The extent of TOC taken up from synthetic municipal wastewater by pine sawdust was additionally verified as a function of contact time. The contact time was changed from 10 min to 4 h. The results of the contact time effects are shown in Figure 1d. It is clear from Figure 1d that equilibrium was reached after 1.5 h of contact time (removal percent was 36%). Afterwards, as shown in Figure 1d, the TOC removal was decreased to 32%. On the basis of the findings from contact time experiments, 1.5 h was taken as equilibrium time in next adsorption studies.

The impact of temperature on the TOC adsorption by pine sawdust was checked (Figure1e). Five temperatures were used in this part of adsorp-

tion experiment (15–50°C). As can be seen, the TOC removal percent decreases sharply with the rising temperature from 15°C to 50°C. Maximum TOC removal of 37% was observed at 15°C and decreased to 17% at 50°C. Many studies reported increases in the adsorption percentage along with temperature (El-Naas et al., 2010; Kazemi et al., 2016). This indicates the exothermic nature of the adsorption process. Moreover, the decrease of adsorption with increasing temperature can be explained by the weakening of the sorptive forces between the active sites on the sorbent and the adsorptive species (Mor et al., 2016).

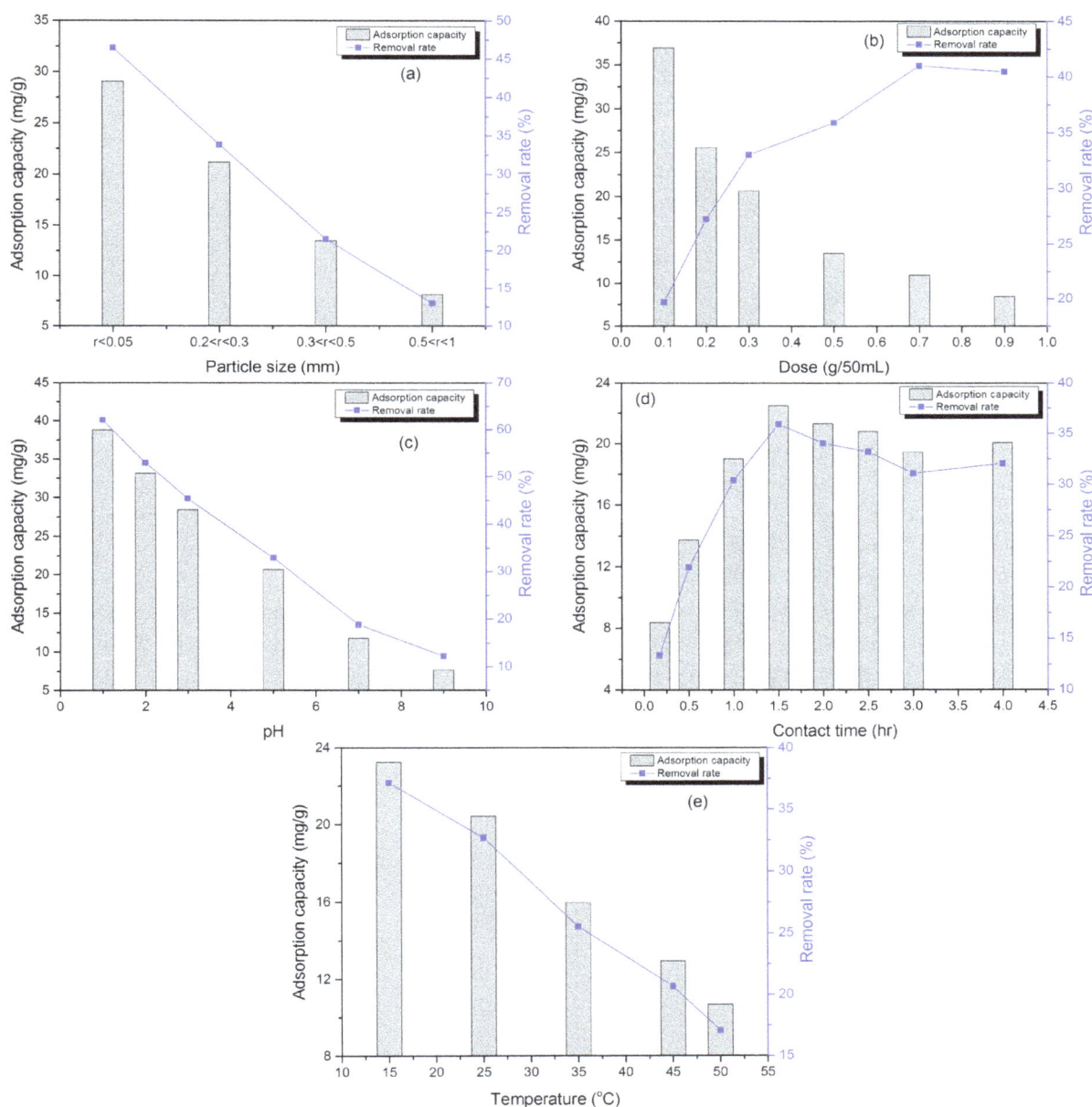

Figure 1. Effect of (a) particle size, (b) adsorbent dosage, (c) pH, (d) contact time, and (e) temperature on adsorption process

Anaerobic digestion results

The second part of this paper was to study the anaerobic digestion of sawdust after using it in the adsorption process. This part was divided into five groups: the first one checked the effect of TOC adsorbed on sawdust on AD of sawdust, while the other four groups investigated the effects of C/N ratio, inoculum percentage, particle size, and TS content on the AD process and biogas production.

In order to study the feasibility of anaerobic digestion of sawdust before and after the adsorption process, batch tests were conducted on both sawdust (raw and spent) and the results were illustrated in Figure 2. It is clear from Figure 2 that the AD of spent sawdust was a very important step of adsorption process because of the great amount of energy that can be recovered from it instead of being wasted. Figure 2 shows that the digestion process of raw sawdust did not take more than 25 days and stopped, while for spent sawdust, the process continued for 45 days. Maximum daily CH_4 produced by spent sawdust was 1250 mL/d, which was 290% more than maximum CH_4 produced from raw sawdust (Figure 2a), this may be due to the large amount of TOC adsorbed on sawdust, which can be digested easily by microorganisms. Figure 2b showed the cumulative CH_4 produced from digestion process. It can be seen that the reused sawdust had a significantly higher CH_4 production (5.9 times) compared to raw sawdust. Therefore, using low-cost biomass has a great theoretical and practical value for treatment of municipal sewage, removal of the hydrocarbon resources from sewage, and finally using absorbent as raw materials to produce biogas energy.

Three different C/N ratios (20%, 25%, and 30%) were carried out to find the optimum ratio. The results were shown in Figure 3. As can be seen in Figure 3, the best C/N ratio was 20% for CH_4 production volume. The batch reactor with C/N of 20% had the maximum daily CH_4 production (1075 mL/d), which was 28% and 55% more than the production of reactors with C/N of 25% and 30%, respectively (Figure 3a). Additionally, cumulative CH_4 produced from a batch with C/N ratio of 20% (8775 mL) was higher than other batches by 14% and 23%, respectively (Figure 3b). The result of this parameter was consistent with the previous studies, where the optimum C/N ratio was 20–30%. In a study conducted by Yen and Brune to investigate the effect of C/N ratio on the anaerobic co-digestion of algal sludge and waste paper, the results showed that the optimized C/N ratio for the co-digestion was 20/1 to 25/1 (Yen & Brune, 2007). Haider et al. (2015) carried out another study to investigate the anaerobic co-digestion of food waste and rice husk. Their results showed that highest specific biogas yield of 584 L/kg VS was obtained from feedstock with C/N ratio of 20 (Haider et al., 2015).

In order to study the effect of inoculum amount on the anaerobic digestion of sawdust, batch experiments with three inoculum amounts (10, 20, and 30%) were conducted. The experiments were continued for 40 days. Figure 4 shows the daily and cumulative CH_4 production of sawdust digestion. As can be seen, the CH_4 production increased along with the inoculum percentage. This can be explained by the fact that when the inoculum amount decreases, it may cause overload, the unfavourable situation where there was too much biomass for the microorganisms to digest.

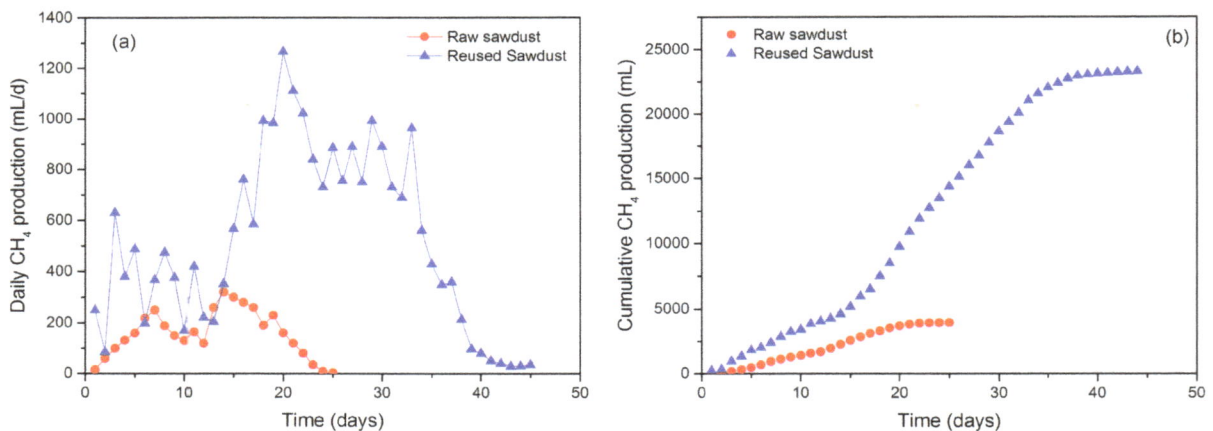

Figure 2. Effect of TOC on AD of sawdust

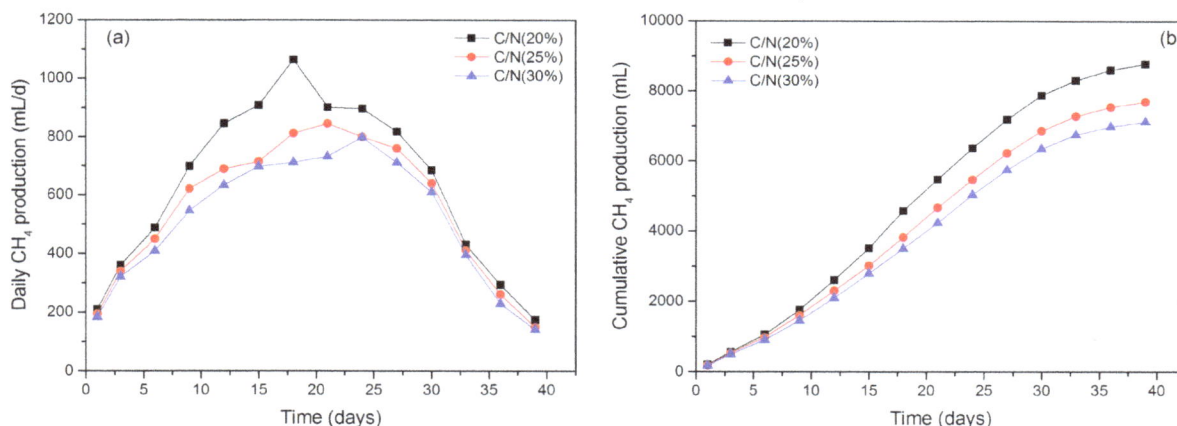

Figure 3. Effect of C/N ratio on AD of sawdust

Thus, higher inoculum could supply some kind of safety factor in the BMP assay since the appropriate inoculum is able to process a higher flow of metabolites (Dechrugsa et al., 2013). Maximum daily CH_4 produced was 1370 mL/d achieved by 30% inoculum amount, which was 52% and 26% more than other inoculum percentages, respectively (Figure 4a). The cumulative CH_4 production after 40 days of AD of sawdust with 10%, 20%, and 30% inoculum percent were 7940, 8275, and 8700 mL, respectively (Figure 4b). This result was comparable to previous studies, where the biogas production increased along with the inoculum amount (Dechrugsa et al., 2013; Haider et al., 2015).

One of the parameters investigated in this study was biomass particle size and its effect on AD of sawdust and biogas produced. As can be seen from Figure 5, the best particle size was 2 mm. Maximum daily CH_4 production (1700 mL/d) was achieved by 2 mm particle size with no significant difference with a 1 mm par-

ticle size (1590 mL/d), but it has a significant difference with other sizes 0.3 and 0.8 mm, 107% and 31%, respectively (Figure 5a). Additionally, the cumulative CH_4 produced during the entire digestion time, the batch with 2 mm particle size had the highest amount of CH_4 generated. This particle size showed a 2.3- and 1.5-fold higher CH_4 volume than 0.3 and 0.8 mm particle sizes, respectively. A small difference (2.5%) was achieved with 1mm particle sizes. These results were in agreement with previous studies (Agyeman & Tao, 2014; De la Rubia et al., 2011; Zhang & Banks, 2013). Zhang and Banks conducted a study to check the impact of different particle sizes on AD of the organic fraction of municipal solid waste. Their results indicated that the digester with the 2 mm mean particle size was characterized by slightly higher methane production than the 4 mm mean particle size (Zhang & Banks, 2013). In another study, Agyeman and Tao found in their work that the maximum methane production rate and specific methane yield were signifi-

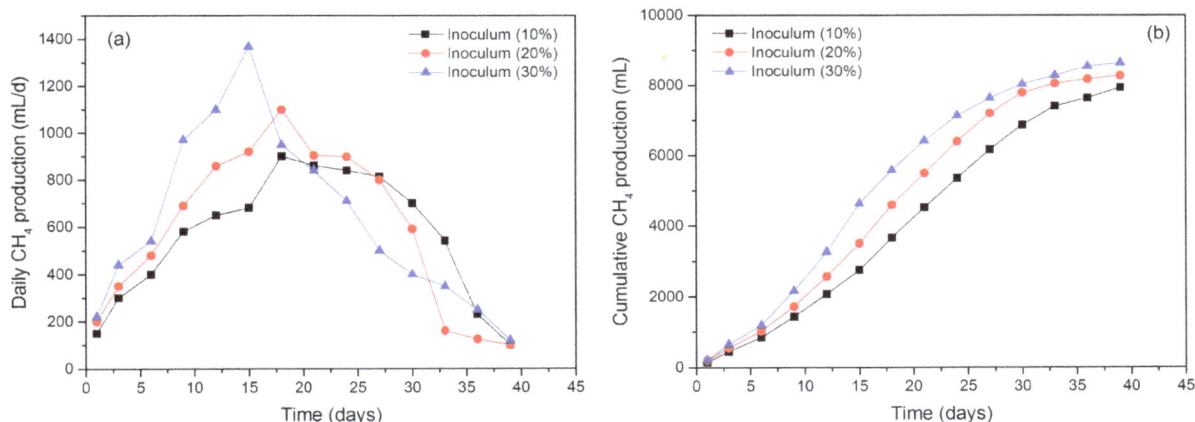

Figure 4. Effect of inoculum percentage on AD of sawdust

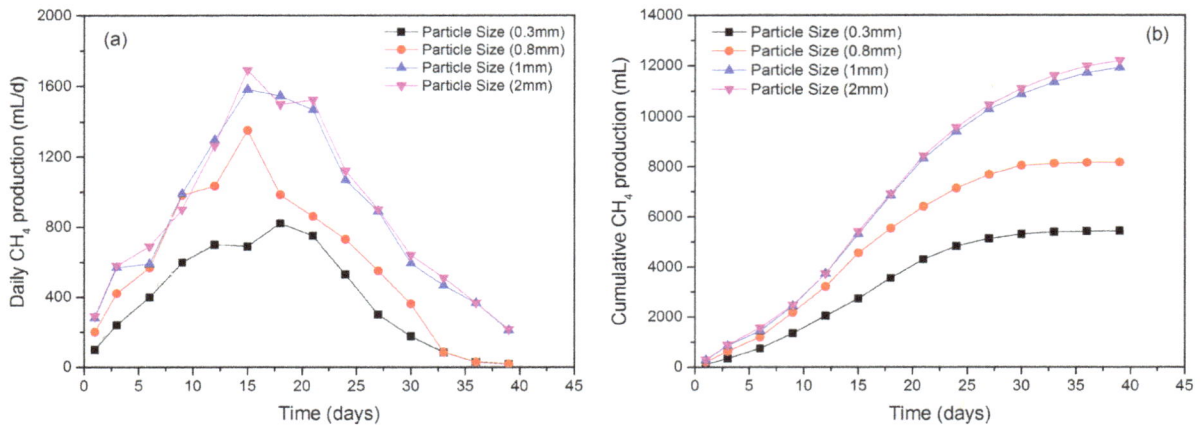

Figure 5. Effect of particle size on AD of sawdust

cantly higher in the digester with 2.5mm (fine) particle size compared to other reactors (Agyeman & Tao, 2014). On the other hand, De la Rubia et al. evaluated three particles size ranges of (1) 0.355–0.55 mm, (2) 0.710–1.0 mm and (3) 1.4–2.0 mm. They showed that the highest methane yield was obtained for the largest particle size analysed (3) (De la Rubia et al., 2011).

The last parameter investigated was total solid (TS) content. Three different contents were checked (5%, 10%, and 15%) and the results were illustrated in Figure 6. As can be seen in Figure 6, the CH_4 production increased along with the TS content. This result was consistent with a previous work (Yi et al., 2014). For daily methane produced, highest methane was achieved by TS content with 15%, which was 1.30 and 1.19-fold higher than other TS contents, 5% and 10%, respectively (Figure 6a). In addition, the cumulative CH_4 production had the same trend with respect to 15% TS content, where 15% TS showed a 28% and 18% more cumulative CH_4 than other

TS contents, respectively (Figure 6b). These results agreed with previous studies, which proposed that the increase of feeding TS contents lower than 20% has a favourable effect on the CH_4 production (Dai et al., 2013; Yi et al., 2014).

CONCLUSIONS

Anaerobic digestion of pine sawdust after adsorption of TOC from synthetic domestic wastewater can increase the recovery of energy products from sawdust. The sawdust used in this study showed a good performance in the adsorption process. The removal efficiency of TOC was increased with the decrease in adsorbent particle size, pH, and temperature. In contrast, the efficiency increased along with the contact time. The main AD conclusions obtained from this study are: the big amount of energy recovered and methane produced from AD of used sawdust demonstrate the importance of this step after adsorption pro-

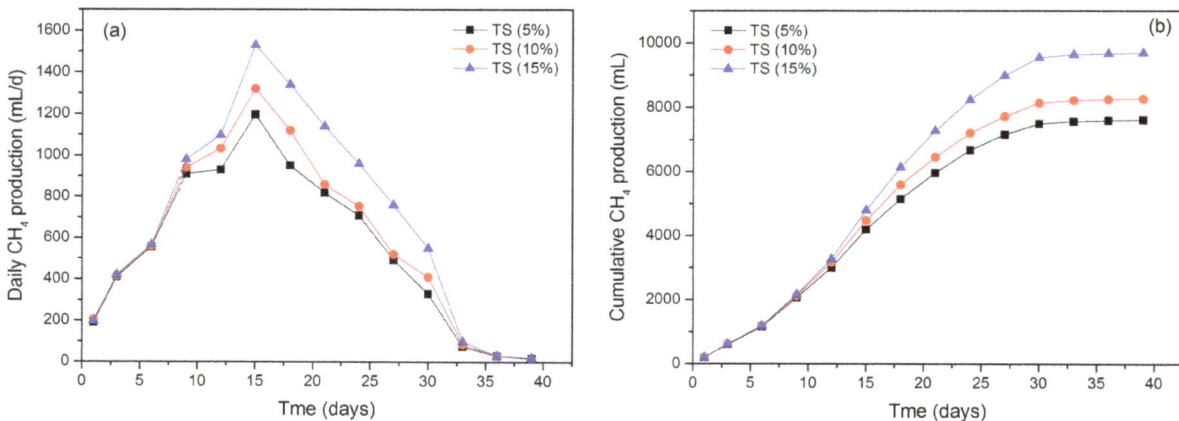

Figure 6. Effect of TS content on AD of sawdust

cess. In addition, the best operating parameters of AD (C/N, inoculation, particle size, and TS) were: 20%, 30%, 2 mm, and 15%, respectively.

Acknowledgements

The author would like to thank Al-Mustansiriyah University Baghdad in Iraq for its support in the present work.

REFERENCES

1. Agyeman, F.O., Tao, W. 2014. Anaerobic co-digestion of food waste and dairy manure: Effects of food waste particle size and organic loading rate. Journal of Environmental Management, 133, 268–274.

2. Ahn, Y., Logan, B.E. 2010. Effectiveness of domestic wastewater treatment using microbial fuel cells at ambient and mesophilic temperatures. Bioresource technology, 101(2), 469–475.

3. Benetti, A.D. 2008. Water reuse: issues, technologies, and applications. Engenharia Sanitaria e Ambiental, 13(3), 247–248.

4. Castoldi, R., Correa, V.G., de Morais, G.R., de Souza, C.G., Bracht, A., Peralta, R.A., Moreira, R.F.P.-M., Peralta, R.M. 2017. Liquid nitrogen pretreatment of eucalyptus sawdust and rice hull for enhanced enzymatic saccharification. Bioresource technology, 224, 648–655.

5. Dai, X., Duan, N., Dong, B., Dai, L. 2013. High-solids anaerobic co-digestion of sewage sludge and food waste in comparison with mono digestions: Stability and performance. Waste Management, 33(2), 308–316.

6. De la Rubia, M.A., Fernández-Cegrí, V., Raposo, F., Borja, R. 2011. Influence of particle size and chemical composition on the performance and kinetics of anaerobic digestion process of sunflower oil cake in batch mode. Biochemical Engineering Journal, 58–59(0), 162–167.

7. Dechrugsa, S., Kantachote, D., Chaiprapat, S. 2013. Effects of inoculum to substrate ratio, substrate mix ratio and inoculum source on batch co-digestion of grass and pig manure. Bioresource Technology, 146, 101–108.

8. El-Naas, M.H., Al-Zuhair, S., Alhaija, M.A. 2010. Reduction of COD in refinery wastewater through adsorption on date-pit activated carbon. Journal of Hazardous Materials, 173(1–3), 750–757.

9. Federation, W.E., Association, A. 2005. Standard methods for the examination of water and wastewater. American Public Health Association (APHA): Washington, DC, USA.

10. Ghaedi, M., Sadeghian, B., Pebdani, A.A., Sahraei, R., Daneshfar, A., Duran, C. 2012. Kinetics, thermodynamics and equilibrium evaluation of direct yellow 12 removal by adsorption onto silver nanoparticles loaded activated carbon. Chemical Engineering Journal, 187, 133–141.

11. Gupta, V., Agarwal, A., Singh, M. 2015. Belpatra (aegel marmelos) bark powder as an adsorbent for the color removal of textile dye "torque blue". Int J Sci Eng Tech, 4(2), 56–60.

12. Haider, M.R., Zeshan, Yousaf, S., Malik, R.N., Visvanathan, C. 2015. Effect of mixing ratio of food waste and rice husk co-digestion and substrate to inoculum ratio on biogas production. Bioresource Technology, 190(0), 451–457.

13. Huerta-Fontela, M., Galceran, M.T., Ventura, F. 2011. Occurrence and removal of pharmaceuticals and hormones through drinking water treatment. Water research, 45(3), 1432–1442.

14. Katsoyiannis, A., Samara, C. 2005. Persistent organic pollutants (POPs) in the conventional activated sludge treatment process: fate and mass balance. Environmental Research, 97(3), 245–257.

15. Kazemi, S.Y., Biparva, P., Ashtiani, E. 2016. Cerastoderma lamarcki shell as a natural, low cost and new adsorbent to removal of dye pollutant from aqueous solutions: Equilibrium and kinetic studies. Ecological Engineering, 88, 82–89.

16. Khamparia, S., Jaspal, D. 2016. Investigation of adsorption of Rhodamine B onto a natural adsorbent Argemone mexicana. Journal of Environmental Management, 183, Part 3, 786–793.

17. Kimming, M., Sundberg, C., Nordberg, Å., Baky, A., Bernesson, S., Norén, O., Hansson, P.-A. 2011. Biomass from agriculture in small-scale combined heat and power plants–a comparative life cycle assessment. Biomass and bioenergy, 35(4), 1572–1581.

18. Liu, Q., Liu, Q., Ma, W., Liu, W., Cai, X., Yao, J. 2016. Comparisons of two chelating adsorbents prepared by different ways for chromium (VI) adsorption from aqueous solution. Colloids and Surfaces A: Physicochemical and Engineering Aspects, 511, 8–16.

19. Lizasoain, J., Rincón, M., Theuretzbacher, F., Enguídanos, R., Nielsen, P.J., Potthast, A., Zweckmair, T., Gronauer, A., Bauer, A. 2016. Biogas production from reed biomass: Effect of pretreatment using different steam explosion conditions. Biomass and Bioenergy, 95, 84–91.

20. Mor, S., Chhoden, K., Ravindra, K. 2016. Application of agro-waste rice husk ash for the removal of phosphate from the wastewater. Journal of Cleaner Production, 129, 673–680.

21. Rickert, D.A., Hunter, J.V. 1971. General nature of soluble and particulate organics in sewage and secondary effluent. Water Research, 5(7), 421–436.

22. Shukla, P., Fatimah, I., Wang, S., Ang, H., Tadé, M.O. 2010. Photocatalytic generation of sulphate and hydroxyl radicals using zinc oxide under low-power UV to oxidise phenolic contaminants in wastewater. Catalysis Today, 157(1), 410–414.

23. Speece, R.E. 2007. Anaerobic biotechnology and odor/corrosion control for municipalities and industries. Nashville (Ten.) : Archae press.

24. Yang, H.I., Lou, K., Rajapaksha, A.U., Ok, Y.S., Anyia, A.O., Chang, S.X. 2017. Adsorption of ammonium in aqueous solutions by pine sawdust and wheat straw biochars. Environmental Science and Pollution Research, 1–10.

25. Yen, H.-W., Brune, D.E. 2007. Anaerobic co-digestion of algal sludge and waste paper to produce methane. Bioresource Technology, 98(1), 130–134.

26. Yi, J., Dong, B., Jin, J., Dai, X. 2014. Effect of increasing total solids contents on anaerobic digestion of food waste under mesophilic conditions: performance and microbial characteristics analysis. PloS one, 9(7), e102548.

27. Yin, C.Y., Aroua, M.K., Daud, W.M.A.W. 2007. Review of modifications of activated carbon for enhancing contaminant uptakes from aqueous solutions. Separation and Purification Technology, 52(3), 403–415.

28. Zhang, Y., Banks, C.J. 2013. Impact of different particle size distributions on anaerobic digestion of the organic fraction of municipal solid waste. Waste Management, 33(2), 297–307.

10

Impact of Sedimentation Supported by Coagulation Process on Effectiveness of Separation of the Solid Phase from Wastewater Stream

Piotr Maciołek[1], Kazimierz Szymański[2*], Rafał Schmidt[3]

[1] Municipal Water Supply and Sewerage Systems Co. Ltd., ul. Wojska Polskiego 14, 75-711 Koszalin, Poland

[2] Faculty of Civil Engineering, Environmental and Geodetic Sciences, Koszalin University of Technology, ul. Śniadeckich 2, 75-453 Koszalin, Poland

[3] Municipal Public Utilities Company Co. Ltd., 1 Maja 1, 76-150 Darłowo, Poland

[*] Corresponding author's e-mail: kazimierzszymanski@wp.pl

ABSTRACT

The objective of this work was to test the impact of coagulant and flocculant at the stage of mechanical wastewater treatment on the wastewater treatment plant operation, performed in the A2O process. In this paper, the principles of correct conduct of coagulation in wastewater treatment have been discussed. It appears from the research performed that significant elimination of BOD_5 such as 20÷30%, total suspended matter up to 90%, COD up to 50% and total nitrogen at 30% level was achieved supported by the coagulation process in the pre-settling tanks. Approximately 50% of phosphorus was eliminated after the mechanical part. Additionally, a significant impact of Superfloc flocculant on the effectiveness of the solid phase separation (activated sludge) in the secondary settling tank was noted under diversified flow conditions.

Keywords: wastewater treatment, coagulation, flocculation, separation

INTRODUCTION

Municipal wastewater treatment creates serious technical and technological problems. This originates from the fact that municipal wastewater has non-uniform composition and high concentration of organic pollutants as well as contents of specific substances. The supernatant water from wastewater treatment that is returned to the wastewater treatment technological line has an impact on the wastewater treatment process. It features high contents of organic compounds and non settling suspensions containing adsorbed metals as well as gases (hydrogen sulphide and methane) and biogenic compounds. Numerous research works performed by many authors indicate that the phosphorus contents in the supernatant water coming from gravitational concentrators may amount up to 500 mg/L, whereas nitrogen up to 300 mg/L [Malej & Boguski, 2000, Malej, 2000, Piaskowski & Ćwikałowska, 2007]. However, application of centrifugal sedimentation increases the concentrations of these pollutants up to 600 mg/L and up to 1000 mg/L, respectively [Malej, 2000, Przywara, 2006]. In order to increase the effectiveness of wastewater treatment, in physicochemical processes in coagulation-flocculation and coagulation-sedimentation-flocculation arrangements are dominant in the modern wastewater treatment systems. In the event of a necessity to particularly protect receiving waters, such as protection of seawater in the vicinity of beaches and watering places as well as protection of water intakes for alimentation of cities/towns or food industry, disinfection process is applied additionally [Duan & Gregory, 2003, Hansen, 2002, Malej, 2008, Przywara, 2006].

The objective of this work was to test the impact of coagulant and flocculant at the stage of mechanical wastewater treatment on the wastewater treatment plant operation, performed in the A2O process.

METHODOLOGY OF RESEARCH

Wastewater is transported to "Jamno" Wastewater Treatment Plant by gravity via a 1800 mm sewer pipe. The wastewater treatment plant operating in the A2O arrangement/system, which ensures highly efficient elimination of carbon, nitrogen and phosphorus compounds, is composed of mechanical and biological parts. The designed wastewater treatment plant throughput was defined at 40 000 m^3/d level, whereas the average volume of wastewater supplied to the plant in 2017 via the sewerage system, was only 24 783 m^3/d.

The mechanical part is composed of a hand-operated screen with 100 mm clearance, three step screens with 3 mm clearance and three sand traps with horizontal flow before which PIX-113 coagulant is added to wastewater in the amount of 35÷90 g PIX/m^3 and two rectangular two-chamber oblong pre-settlings tanks. The coagulation process is supported, before the settling tanks, with anionic polymer added in proportion to wastewater flow rate (0.02÷0.1 g/m^3 PRAESTOL 2530). Such process supports the wastewater purification, particularly when high pollutant load is delivered in effluents from the sludge management area [Cherchi et al., 2009, Dziubek & Kowal, 1988, Dincer & Kargi, 2000, Lemmer, 2000, Malej, Boguski, 2000, Maciołek et al, 2016]. Following primary treatment, wastewater flows by gravity into two multistage A2O biological reactors operating in parallel. The dephosphatization, denitrification, nitrification and carbon biodegradation processes proceed in individual activated sludge chambers. The flow diagram of the biological wastewater treatment plant part is shown in Figure 1.

Beyond the pre-settling tank part of raw wastewater (approx. 20%) flows to the pre-denitrification chamber, whereas the remaining volume flows directly into the anaerobic chamber (KB1, then KB2). The denitrification process is performed in three subsequent chambers with internal recirculation flowing from the last nitrification chamber to the first denitrification chamber. The volume of such internal recirculation depends on the concentration of nitrates (V) in the aerobic chamber (nitrification). Before wastewater is supplied to the secondary settling tank, it is subjected to fine bubble aeration in nitrification chambers (KN). The same coagulant as the one used in the mechanical treatment stage, i.e. iron (III) sulphate (VI) with commercial name PIX-113 is added (from 15 g to 80 g/m^3) before the radial secondary settling tanks to the distribution chamber; it prevents the secondary liberation of phosphorus from the activated sludge bacteria cells and promotes a simultaneous phosphorus precipitation. The sludge external recirculation flows from the settling tank to the pre-denitrification chamber (KPDN) at 100–150% Q level. The purified wastewater is channelled to the receiving water, which is the Dzierżęcinka river, supplying the Jamno coastal lake.

Raw and mechanically purified wastewater (beyond the pre-settling tanks) was taken as daily average samples in 2015 with automatic samplers in proportion to the flow rate, in weekly intervals. The following pollution indicators were determined in the taken averaged samples: COD,

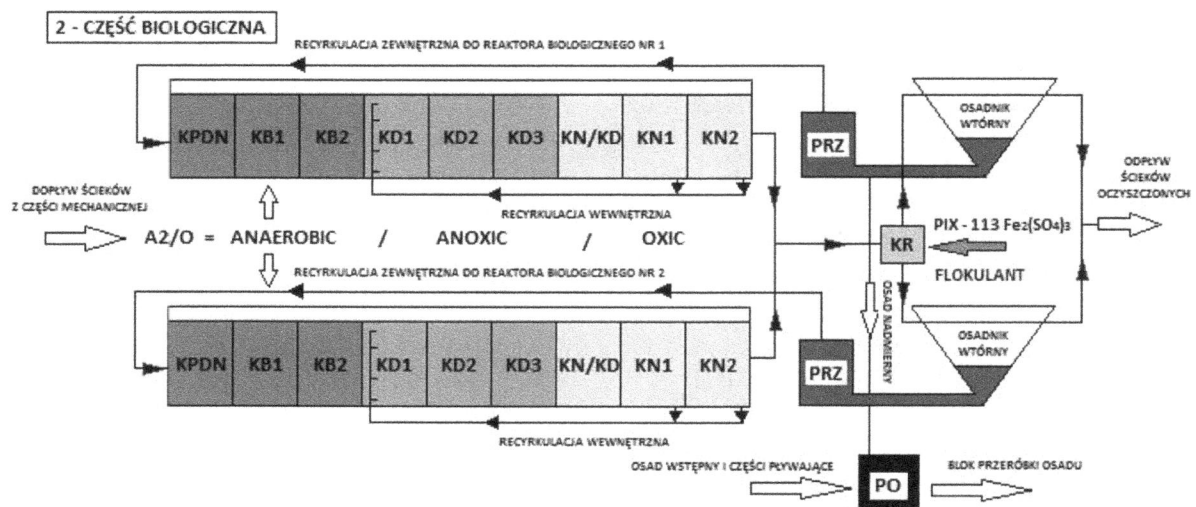

Figure 1. Flow diagram of the biological part in JAMNO Wastewater Treatment Plant

BOD_5, total suspended matter and total phosphorus. The analyses were performed in the wastewater treatment plant laboratory using recommended laboratory methods: COD Merck method 1.14541.0001, BOD_5 – PN-EN 1899, total suspended matter – PN-EN 872, total phosphorus – Merck method 1.14729.0001.

Furthermore, variations of the height of the clarified zone in secondary settling tanks No 1 and No 2, monitored in March 2016, are presented in this paper. Variations of the clarified zone are illustrated in a graph based on the on-line record of Turbimax CUS71D Endress+Hauser probe, under normal wastewater flow conditions (during a single day) and during water hammering (two days). At the same time, SUPERFLOC polymer was added during the experiment into settling tank No 2 to improve the deposit sedimentation rate. Secondary settling tank No 1 operated without flocculant additive, which allowed for comparison and testing of its effectiveness in phase separation.

RESEARCH RESULTS

Supporting of wastewater treatment with coagulant and polymer in the mechanical stage is particularly significant due to high variability of raw wastewater composition, practically during the entire year. Variability of concentration of the analysed indicators in raw wastewater and after sedimentation in the pre-settling tank is illustrated in Figures 2–5. The presence of pre-settling tanks supported by chemical precipitation allows for reduction and stabilisation of pollutant loads entering the biological part. Therefore, by increased effectiveness of elimination of the solid phase from wastewater stream in the pre-settling tanks, the load of biological chambers with pollutants is reduced, resulting in practical benefits. Those benefits originate, first and foremost, from the reduction of excess sludge increase, with a simultaneous reduction of the oxygen consumption and demand. By applying such solution we strive for improvement of the economic and energetic wastewater treatment effectiveness.

The coagulation process conducted before the pre-settling tanks definitely improved the quality of mechanically purified wastewater. Nevertheless, the lack of chemical support of sedimentation process in the pre-settling tanks contributed, in connection with the periodical problem of preliminary sludge disposal (due to low temperature), to deterioration of the solid phase separation results. The height of sludge laying in pre-settling tanks at that period (without chemical support from PRAESTOL 2530 flocculant), measured with hand-operated Burkle Vampire Sampler, was approximately 1 metre below the wastewater table. Reintroduction of the chemical support from PRAESTOL 2530 caused reduction of sludge layer height in the pre-settling tanks down to approximately 3.4 metre below the wastewater table.

The absence of chemical sedimentation support in the analysed period caused a significant deterioration of the mechanical wastewater treatment results. Table 1 shows examples of the analysed parameters values achieved after the pre-settling tank without flocculant application, which are similar to those of raw wastewater.

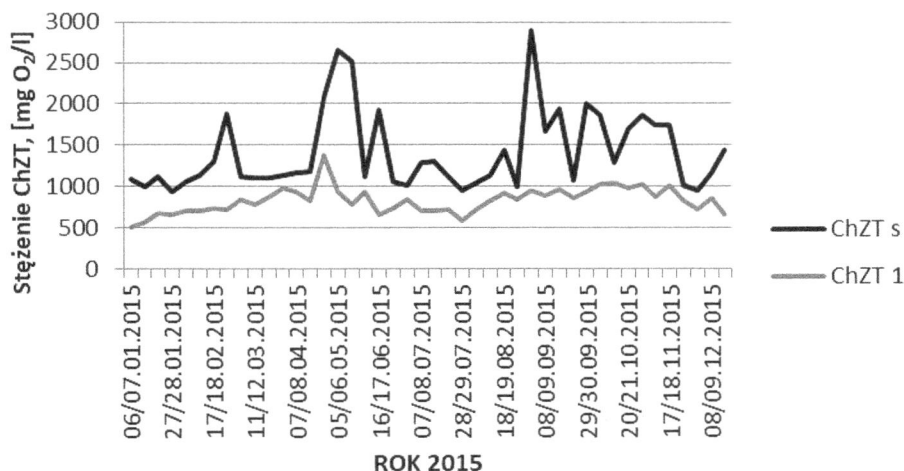

Figure 2. COD values in raw wastewater (ChZT s) and after pre-settling tank No 1 (ChZT 1) in 2015

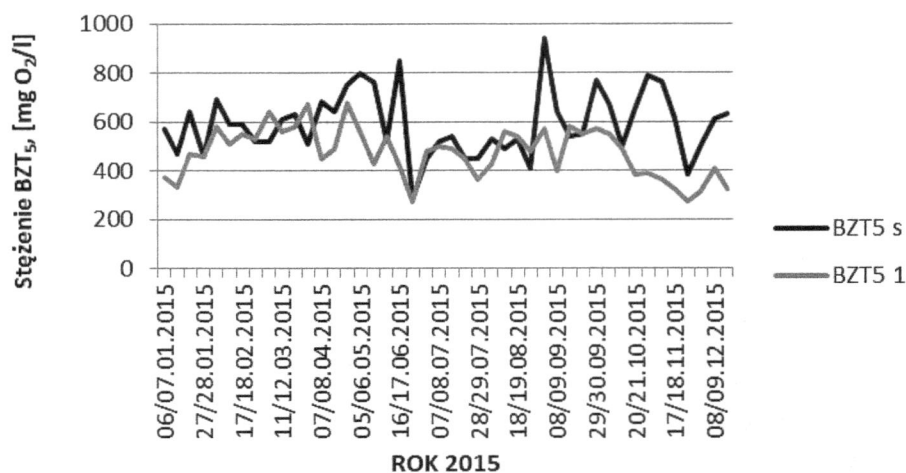

Figure 3. BOD$_5$ values in raw wastewater (BZT$_5$ s) and after pre-settling tank No 1 (BZT$_5$ 1) in 2015

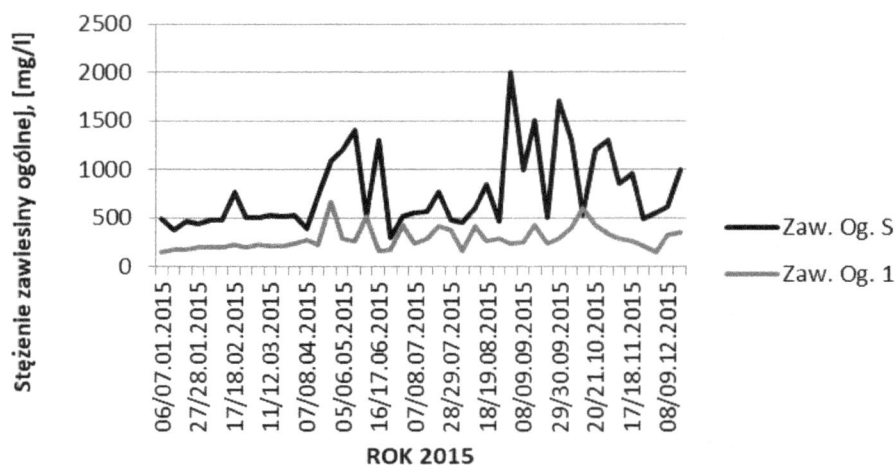

Figure 4. Total suspended matter values in raw wastewater (Zaw. Og. s) and beyond the pre-settling tank No 1 (Zaw. Og. 1) in 2015

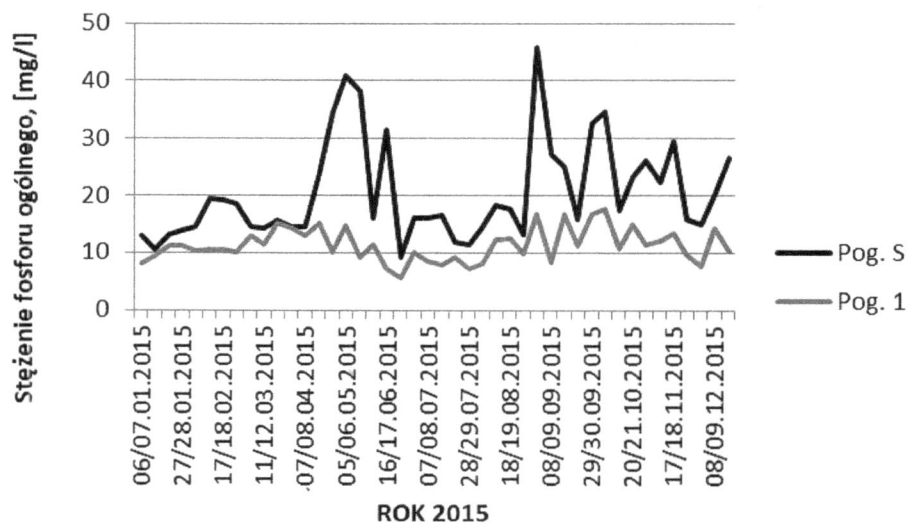

Figure 5. Total phosphorus values in raw wastewater (P og. s) and beyond the pre-settling tank No 1 (P og. 1) in 2015

During the research period, the performance of the secondary settling tanks was also observed. Flocculant is added in Koszalin wastewater treatment plant, depending on needs, also into the secondary settling tanks, in which the activated sludge is separated from the purified wastewater through sedimentation. Under normal flow conditions, such separation proceeds without any disturbance, and the sludge sedimentation rate depends on its concentration. If the sludge concentration is high, the rate of its sedimentation considerably decreases. In Figure No 6, variations of the height of activated sludge in the secondary settling tanks under variable flow conditions, during addition of Superfloc C 18450 polymer, are illustrated. The difference of height between a well-concentrated sludge layer in the pre-settling tank No 2, supported by addition of Superfloc C 18530, and the concentrated sludge layer without the reagent was, during water hammering, maximum 50 cm. Under normal flow conditions, this difference was 20 cm on average (Maciołek, 2016). The difference in sludge layer height was determined through direct on-line measurements of height of the clarified layer using Turbimax CUS71D Endress+Hauser probes, set for "distance from water to the dead zone" measurement mode. The added flocculant (quarternary polyamine) caused, under normal flow conditions (1 day), a quick change of the sludge concentration, which was confirmed by the on-line record. The height of the clarified zone during polymer addition remained similar to that under low flow conditions. During an increased inflow of wastewater, the sludge was intensely mixed and agitated, causing interference in readings. This may prove limited application of Turbimax CUS71D probes set for "distance from water to the dead zone" measurement mode under normal flow conditions.

The "Jamno" Wastewater Treatment Plant also collects wastewater from rural areas. Such wastewater features high hydrogen sulphide content. The PIX – 113 preparation, which has been applied for many years, considerably eliminates hydrogen sulphide from wastewater. Furthermore, it improves the condition of the activated sludge thus improving the operation of the biological part of the plant. During a long-term operation of said wastewater treatment plant, it was noted that cessation or reduction of coagulant addition to raw wastewater led to an excessive increase of the total phosphorus concentration in the supernatant water and effluents from the sludge management returned to the main stream of raw wastewater. An additional advantage of using PIX in preliminary chemical precipitation performed at the mechanical part of the wastewater treatment plant, can be the reduction of biological reactors volume and operating costs, particularly reduction of electrical power consumption cost. This operation had an impact on reduction of the excess sludge volume, whereas the increased volume of the preliminary sludge (30÷60% reduction of the total suspended matter) promoted the process of anaerobic stabilisation of mixed sludge. It appears from laboratory tests and observation of the wastewater treatment plant operation, that pre-settling tanks make an important and essential part before the biological process of biogenic compounds elimination. They protect the biomass from elution of phosphorus eliminating bacteria and allow for intensification of the nitrogen and phosphorus elimination from wastewater. Improperly performed preliminary coagulation process may deteriorate the denitrification and nitrification effectiveness [Malej, 2000a, Malej, 2008, Szyszko, 2015]. The speed of those processes is reduced radically.

CONCLUSIONS

Total phosphorus is eliminated in „Jamno" Wastewater Treatment Plant approximately at the

Table 1. Listing of parameters of mechanically treated wastewater beyond the pre-settling tank no 1 without flocculant added

Date	Tested parameter			
	COD [mg/l O2]	BOD$_5$ [mg/l O2]	Tot. s.m. [mg/l]	Tot. P [mg/l]
2018–02–06	1900	560	1200	25,8
2018–02–20	2340	1380	1300	29,2
2018–02–27	1306	400	550	22,7
2018–03–06	1142	-	430	16,3
Average value	1672	780	870	23,5

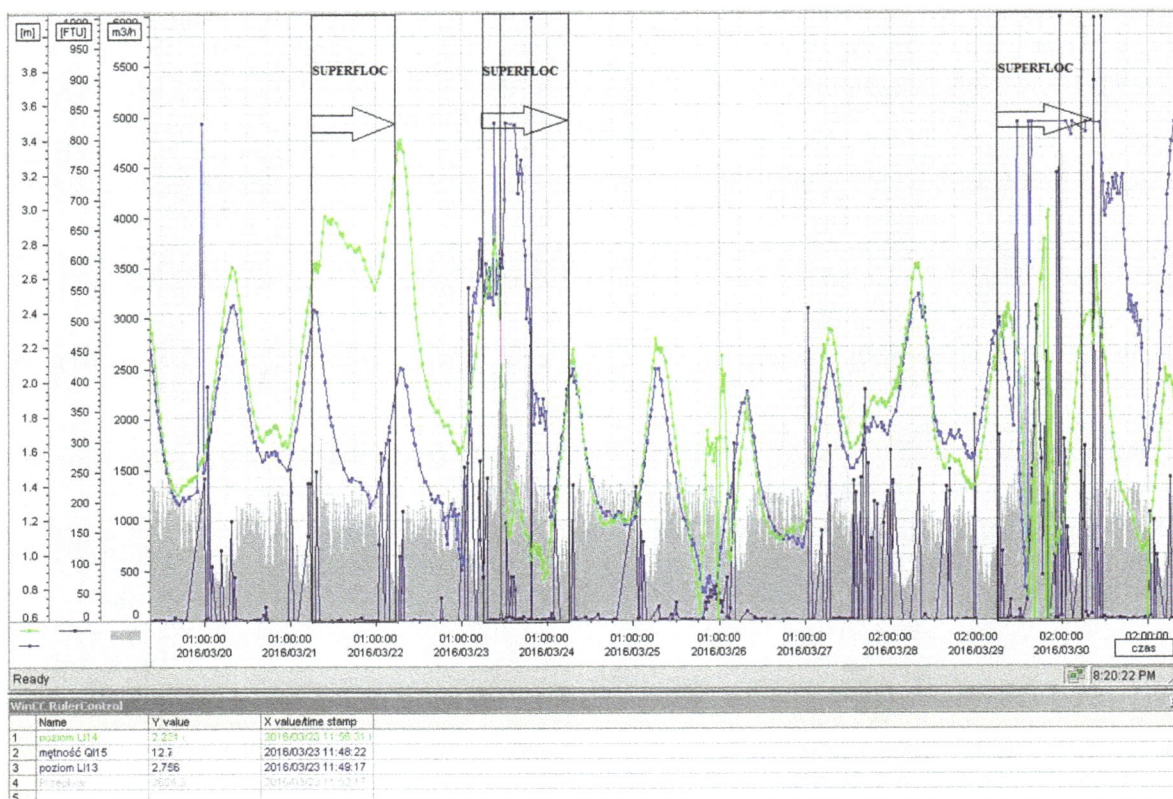

Figure 6. Variations of height of the clarified zone in the secondary sedimentation tanks No 1 (blue) and No 2 (green) with flow rate shown in the background (full graph – grey) and the measurements of turbidity (violet) under normal wastewater flow conditions – 1th day (left area) and during water hammers – 2nd and 3rd day (middle and right areas). SUPERFLOC indicates the area in the course of addition of the polymer improving sludge sedimentation rate

level of 50%, already after the mechanical stage of wastewater treatment supported by preliminary precipitation with iron (III) sulphate (VI) with the trade name of PIX – 113. Furthermore, the coagulant doses of 50 – 100 g/m³ allow for BOD_5 reduction by approximately 50 – 60%, thus reducing the reactors capacity by half.

High variability of raw wastewater composition caused, as indicated by the test results of 2015 presented herein, very high variations in the biological part load during the year, when the coagulant and flocculant were not added into the pre-settling tanks,. Such irregularity may imply a non-uniform increase of the excess sludge, oxygen consumption and demand, concentration of nitrogen forms and the external and internal recirculation degree associated therewith.

Despite the application of a highly efficient biological dephosphatization method, it is still necessary to add PIX before the pre-settling tanks. The PIX-113 coagulant can be added in said wastewater treatment plant simultaneously into the biological reactors and to the distribution

chamber before the secondary settling tanks; this allows for considerable improvement of the phosphorus elimination efficiency.

In the event of intensified rainfall, causing increased hydraulic load in the secondary settling tanks, the use of the polymer ultimately led to the achievement of positive effects, preventing flotation of the sludge outside overflow channels of the secondary settling tanks.

REFERENCES

1. Cherchi C., Onnis-Hayden A., El-Shawabkeh I.N., Gu A.Z. 2009. Implication of using different carbon sources for denitrification in wastewater treatments. Water Environment Research, 81(8), 788–799.

2. Dziubek, A.M., Kowal, A.L. 1988. High-pH coagulation-adsorption: a new technology for water treatment and reuse, Water Science and Technology, 21, 1183–1188.

3. Dincer K., Kargi F. 2000. Effects of operating parameters on performances of nitrification and deni-

trification processes. Bioprocess Engineering, 23, 75–80.

4. Duan, J., Gregory, J. 2003. Coagulation by hydrolysing metal salts, Advances in Colloid and Interface Science, 100 –102, 475–502.

5. Hansen B. 2002. Chemiczne oczyszczanie ścieków – stare i nowe zastosowania. Nowe rozwiązania problemów technicznych w oczyszczalniach ścieków. Materials from a scientific and technical seminar held in Świnoujście and Copenhagen. Szczecin.

6. Klaczyński E. 2013. Oczyszczalnia ścieków – chemiczne usuwanie fosforu, Wodociągi – Kanalizacja, No 2.

7. Lemmer H. 2000. Przyczyny powstawania i zwalczania osadu spęczniałego. ATV monography series. Wydawnictwo Seidel-Przywecki. Szczecin.

8. Maciołek P. 2016. Wpływ polimeru kationowego Superfloc C 18530 na właściwości sedymentacyjne osadu czynnego w zróżnicowanych warunkach przepływu. Technologia Wody, 3(47).

9. Maciołek P., Szymański K., Janowska B. 2016. Usuwanie azotu ze ścieków komunalnych z wyko-

rzystaniem zewnętrznego źródła węgla organicznego. Rocznik Ochrona Środowiska, 18, 885–896.

10. Malej J., Boguski A. 2000. Zmiany ilościowe ładunku zanieczyszczeń w cieczy nadosadowej w procesie zagęszczania osadu czynnego. Rocznik Ochrony Środowiska.

11. Malej J. 2008. Wysoko sprawne oczyszczalnie ścieków a zagrożenia kąpielisk publicznych. Wodociągi – Kanalizacja, No. 11.

12. Malej J. 2000. Właściwości osadów ściekowych oraz wybrane sposoby ich unieszkodliwiania i utylizacji. Rocznik Ochrony Środowiska

13. Piaskowski K., Ćwikałowska M. 2007. Profil zmian stężenia ortofosforanów podczas oczyszczania ścieków i przeróbki osadów ściekowych. Rocznik Ochrony Środowiska.

14. Przywara L. 2006. Warunki i możliwości usuwania fosforanów i fosforu ogólnego ze ścieków przemysłowych. rozprawa doktorska, Politechnika Krakowska.

15. Szyszko M. 2015. Wpływ strącania wstępnego polimerem organicznym na szybkość denitryfikacji, www.eko-dok.pl

11

Methods of Flushing of Sewage Sludge Collected on the Bottom of a Retention Chamber

Robert Malmur[1], Maciej Mrowiec[1]

[1] Czestochowa University of Technology, Faculty of Infrastructure and Environment, Institute of Environmental Engineering, ul. Brzeźnicka 60a, 42-200 Częstochowa, Poland

* Corresponding author's e-mail: rmalmur@is.pcz.czest.pl

ABSTRACT

This paper presents new achievements in the field of designing the gravitation and gravitation-underpressure chamber for bottom flushing in the detention chamber immediately after its emptying from the accumulated sewage. Temporary accumulation of the sewage in the detention chamber of the reservoir causes partial sedimentation of the solids suspended in the liquid. The location of the flushing chamber at the maximal level of filling with sewage in the retention chamber substantially improves the effectiveness of flushing of the bottom of the retention chamber with the stream of sewage that is discharged from the flushing chamber. An opportunity for effective flushing of the sludge that remains on the bottom of a detention chamber represents an important operation which is essential for reliable functioning of retention reservoirs within sewerage systems.

Keywords: drainage system, retention, sewage, sewage reservoirs, transfer reservoirs

INTRODUCTION

Retention reservoirs represent an inherent element of contemporary sewerage systems and, depending on the purpose, may perform various functions at individual stages of sewage transport. Their basic role is to collect and temporarily store specific volumes of sewage in order to reduce the intensity of outflow to the system below the reservoir. Normal operation of the sewerage system does not assume a technological parameter which would manage short but substantial volumes of sewage. Therefore, the economically justified need (or even the necessity) to use retention reservoirs arises [Wolski et al., 2011]. During accumulation of sewage in the retention chamber, each retention reservoir, especially in the stormwater system and the combined sewer system, acts as a settling tank and accumulates sludge transported with sewage to the reservoir on its bottom [Ociepa et al., 2015; Zawieja et al., 2013].

Preventing collection of sewage sludge on the bottom of retention chambers is one of the basic functional problems of sewerage systems.

The use of an effective flushing method to remove the sludge from the retention chambers of the reservoir following their emptying should be planned at each stage of the technological design. Neglecting the activities aimed to prevent collecting of the sludge on the bottom of retention chambers of the reservoir leads not only to collecting excessive amount of sludge but also makes removing the sludge more difficult over the longer period of use. Therefore, this operation is required and more justified if flushing has to be performed automatically without direct or indirect human intervention.

METHODS

The easiest method to flush the sewage sludge from the bottom of a retention chamber involves pressure flushing with the stream of water (sometimes stream of sewage) using the WUKO special-purpose vehicle. However, with many-year perspective of the use of reservoir, this method of emptying is uneconomical and it is critical to perform this process automatically, without direct or indirect human intervention.

If the sewage flows through the entire length of the chamber or in the cases of the SIMPLEX conventional gravitational reservoir, the suggested solution is to employ the system of parallel troughs with semicircular, triangular or trapezoidal cross-section. The multi-chamber retention reservoirs with the gravitational hydraulic system presented in many publications [Deska et al., 2016; Dziopak, 1990] have a flow chamber with insignificant volume and are connected to the retention chambers with an overflow edge. In such cases, a part of the suspensions that are transferred to the retention chamber is settled on the bottom through the sedimentation processes, although the time of sewage storage is around several minutes. The automated hydraulic transport of sewage sludge outside the chamber is impossible in the process of reservoir emptying. After the total emptying of the reservoir and drying the bottom of the retention chamber, the sludge forms a hard crust which cannot be removed using individual cycles of filling and emptying of the reservoir. This is conducive to the formation of individual layers of sludge and increases the difficulty of removal.

Out of the bottom flushing systems used in practice, the most popular solutions are:

a) Capsizable flushing units located at the end of the retention chamber. Flushing of the bottom is performed with the emptied chamber of the reservoir through a fast rotation of the flushing unit and outflow of the whole volume (from 300 to 1500 dm^3) of the fluid to the bottom of the chamber (Fig. 1a). The stream of the fluid flowing from a substantial height has an energy that ensures efficient cleaning of the bottom from the sludge. After emptying from the fluid, the flushing unit returns to the initial position. The filling of the flushing unit can occur, as previously, in two ways. In the first variant, the flushing unit is filled with the sewage flowing to the retention chamber. In the second variant, the flushing unit is filled by means of the pump that supplies pure water or sewage from the reservoir.

b) Flushing chambers, the adequate content of which is separated from the total retention volume of the reservoir (Fig. 1b), operate in two basic versions. In the first version, the chamber is filled gravitationally through the overflow edge whereas in the second, the chamber is filled usually using the negative pressure by reducing the pressure of air closed and cut-off from the atmosphere in the chamber.

c) Systems using the pressure pumps (Fig. 1c) which are mounted whenever the need for flush-

Fig. 1. Examples of flushing of the bottom of the chambers of retention reservoirs a) using the submersible pumps, b) with capsizable flushing unit, c) with separated flushing unit

ing the reservoir arises. Using the pumps controlled by an operator, the reservoir is flushed (both its side walls and the bottom). During the operation, the reservoir is unused, whereas the retention chamber is entirely emptied.

d) Systems using the submersible pumps located at the end of the retention chamber at the wall opposite to the outlet hole (Fig. 1d). Pumps are started after a total emptying of the retention chamber and flush its bottom with the high velocity stream. The flushing fluid can be provided by the sewage stopped in the depression of the reservoir or pure water supplied from the external water supply system.

DESIGN OF THE FLUSHING RESERVOIR

The flushing reservoir is always located on the opposite side with respect to the outflow collector. The reservoir works based on the principle of negative pressure and filling the container occurs through the bottom drain trap hole, while transporting the excess air is carried out via the vent pipe located in the roof (Fig. 2).

The process of reservoir flushing and, therefore, the outflow of water from flushing reservoir occurs when the retention reservoir is entirely emptied, and, using the vent pipe, the air negative pressure over the water surface in the flushing reservoir is equal to the atmospheric pressure.

The end of the vent pipe is installed in a point of the outflow stream where the biggest values of negative pressure occur (Figs. 3 and 4). This solution accelerates the rise of the water surface in the flushing reservoir compared to the retention reservoir.

The vent pipe should represent a separate installation in order for it to act as a lever during the emptying of the retention reservoir. During the filling of the retention and flushing reservoir in a separate vent installation, the air is compressed, thus effectively preventing from its filling with water.

The working principle of the air flow system is presented in Figures 5 and 6.

a)

b)

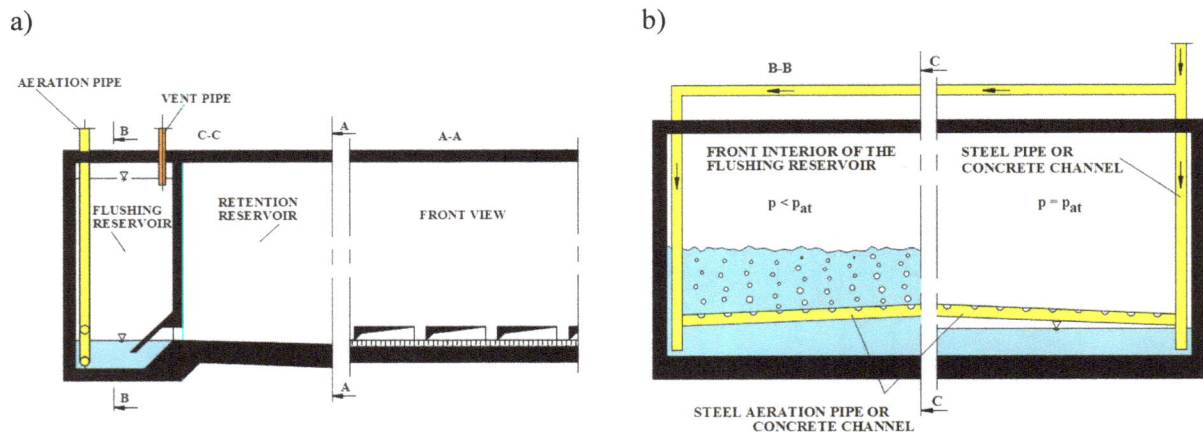

Fig. 2. Design of the flushing reservoir

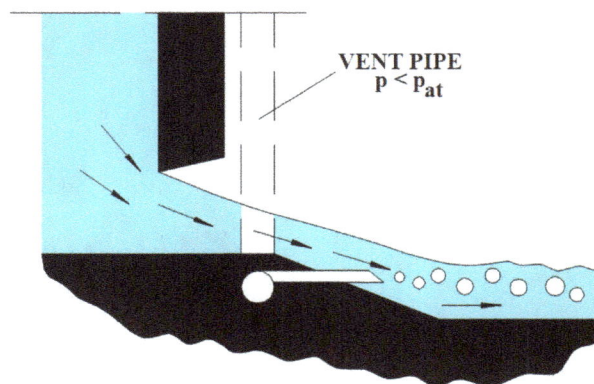

Fig. 3. Installation of the tip of the vent pipe at free flow

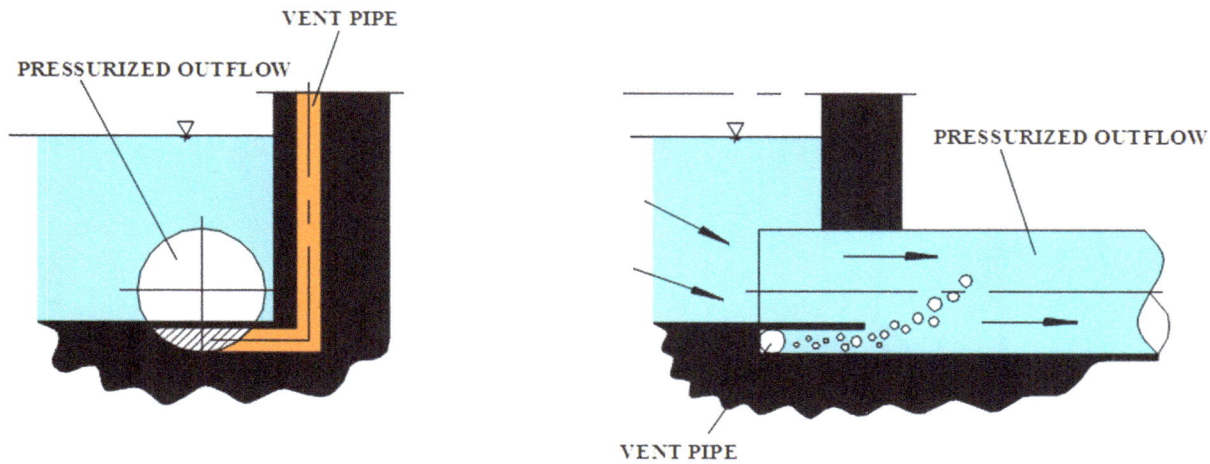

Fig. 4. Installation of the tip of the vent pipe with pressurized emptying of the conventional retention reservoir

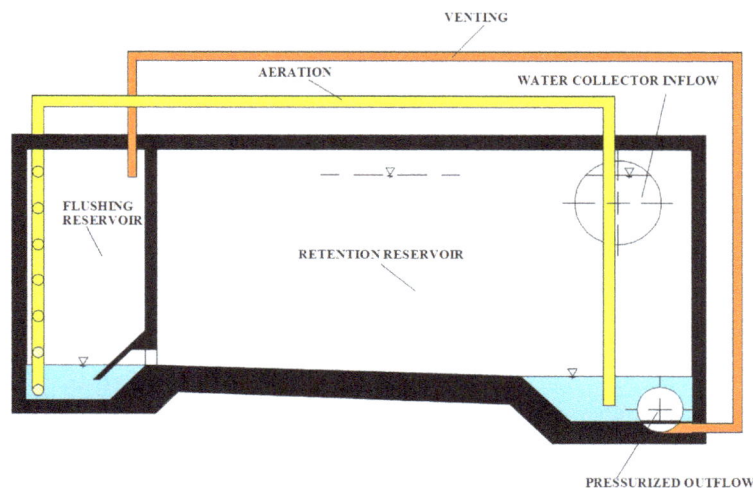

Fig. 5. Functional diagram for the air flow system in conventional retention reservoirs

WORKING PRINCIPLE OF THE FLUSHING RESERVOIR

The retention reservoir, composed of two gravitational chambers, with one of them being flow chamber with low volume compared to the second chamber (retention chamber), is equipped in the flushing container separated from the part of the retention chamber. Filling the flushing chamber occurs using the principle of communicating vessels through the niche on the bottom between the flushing and retention chambers. With this basic version of the container, the vent pipe and aeration pipe are the components of the same venting and aeration system. This solution is less efficient than the proposed one, but represents the idea of operation of such systems [Malmur et al., 2008].

The inlet to the venting and aeration system is located in the roof of the flushing chamber. The

tips of the aeration and vent pipes are installed near the bottom of the retention chamber and in the well (in the standing water) after the outlet hole.

If the inflow of sewage to the flow chamber is greater than its outflow, the chamber is filled. The atmospheric pressure is present in all three chambers of the reservoirs (Fig. 7a). After filling the flow chamber, the sewage overflows the overflow edge infiltrating into the retention chamber through the niche to the flushing chamber. The sewage surface in the submerged vent pipe and the flushing chamber are at the same level, lower than the sewage surface in the retention chamber. The difference (h_x) is equal to the submersion depth of the outlet of the vent pipe (Fig. 7b). The retention chamber is filled until the sewage surfaces are equal in the flow and retention chambers (Fig. 7c). At this point, the flap valve installed in the bottom part of the overflow edge that sepa-

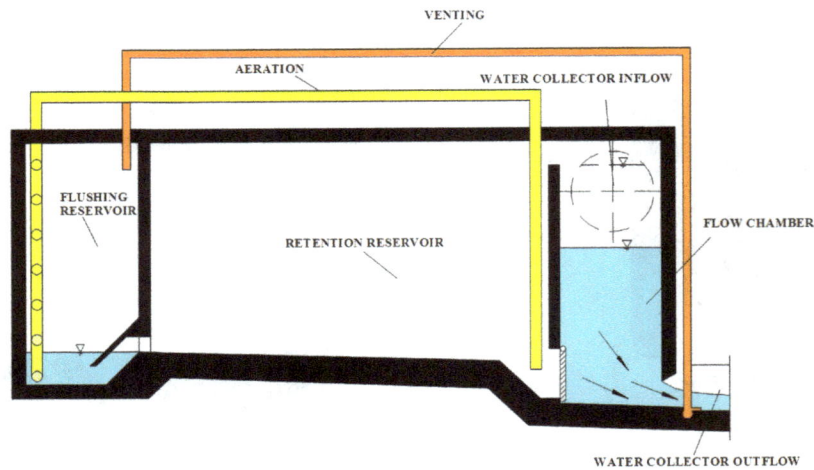

Fig. 6. Functional diagram for the air flow system in the CONTRACT type reservoir with pressure-free outflow

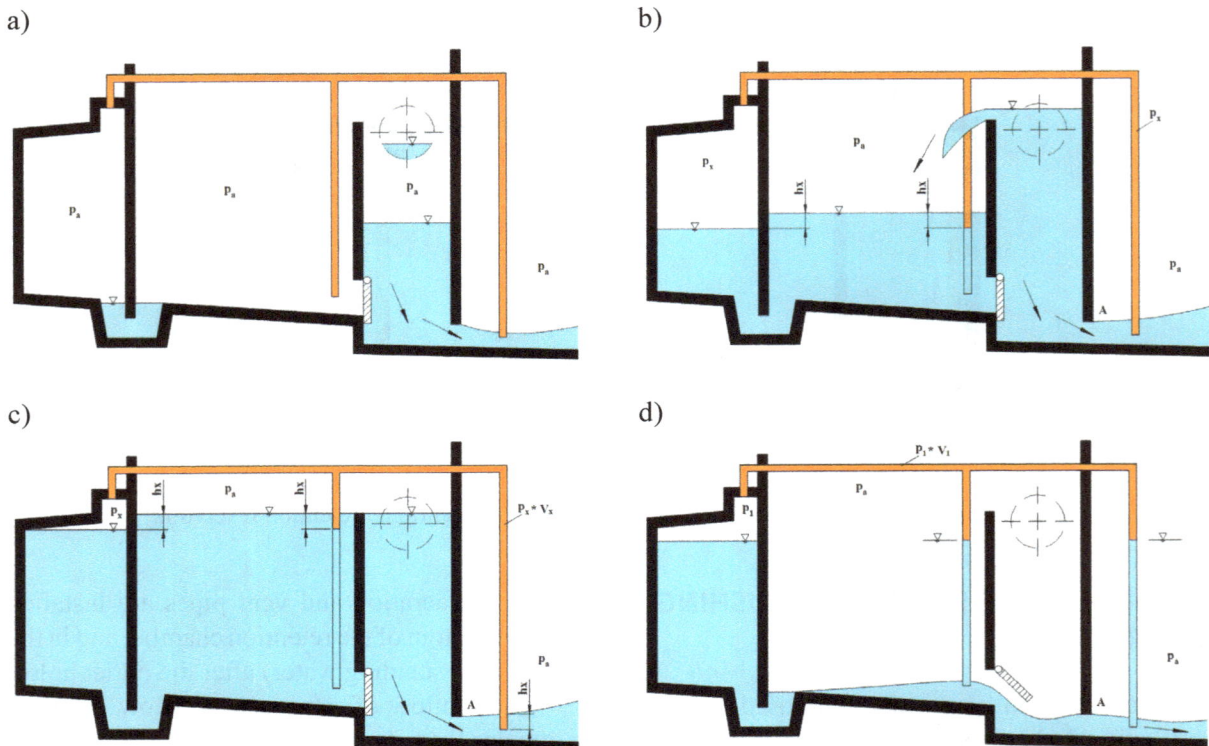

Fig. 7. Phases of operation of the flushing reservoir

rates both chambers of the reservoir opens automatically due to the difference in the hydrostatic pressures. It allows for the flow of sewage only in the direction from the retention chamber to the flow chamber. During the emptying of the reservoir, the level of sewage in the flushing chamber is slightly lowered (Fig. 7d) and the negative pressure sucks the sewage to the vent pipe. After the retention chamber is entirely emptied, the sewage is discharged from the vent pipe and the flushing process starts.

SYSTEMS FOR FLUSHING THE RETENTION CHAMBERS

Previous solutions concerning the method of flushing bottoms of the retention chambers of the reservoirs worked automatically and offered high efficiency and effectiveness of operation [Kisiel et al., 2008; Malmur et al., 2008]. Therefore, they can be recommended for the use in engineering practice. From the hydraulic standpoint, one should emphasize the following aspects:

- in the process of emptying the flushing chamber, the maximal outflow velocity of the flushing stream of liquid is at the initial point when the chamber filling is the biggest. This speed decreases with the reduction in the level of liquid surface in the chamber, substantially reducing the flushing efficiency of the fluid stream.
- due to the decreasing velocity of the fluid outflow from the chamber, the emptying process time is elongated, so that in the final phase the flushing effect is lost.

These observations show that the effective flushing of the retention chamber bottom should be performed in the initial phase of emptying the flushing chamber.

In the solutions presented in the study, the bottom of the flushing chamber is located below the maximal filling of the retention chamber of the reservoir (Fig. 8 and Fig. 9). Consequently, in the final phase of emptying this flushing chamber, the outflow velocity of the fluid is similar to the initial phases in the chamber, the bottom of which is at the level of the flushed bottom of the reservoir retention chamber.

GRAVITY FLUSHING CHAMBER

The retention reservoir (Fig. 8), designed for the stormwater system and the combined sewer system, contains the flow chamber (KP) and gravity retention chamber (KRG). They are separated from each other with the partition wall with overflow edge (PSZ). An outflow hole with a check valve (KL) is located at the bottom zone of the gravity retention chamber (KRG), automatically opened towards the flow chamber (KP). In the upper part, the flow chamber (KP) is equipped in the inflow collector (KD) that supplies sewage to the reservoir, whereas in the lower part, it contains a sewage outflow hole connected with the outflow collector (KO). A flushing chamber (KPŁ) is located inside the retention chamber (KRG), with its bottom is situated at the level over the sewage surface in the entirely filled retention chamber (KRG). In practice, if field conditions are suitable, the flushing chamber (KPŁ) can be located outside the vertical projection of the reservoir. It should also be noted that the capacity of the flushing chamber is included in the total retention capacity of the reservoir.

The flushing chamber (KPŁ) and retention chamber (KRG) are connected with the discharge channel (PZ) adjacent to the wall of the reservoir which is located opposite to the outlet hole (KL). The flushing channel is connected with the discharge channel (PZ) through the overflow pipe (PN) and outlet valve (Z) in the bottom zone. Furthermore, the discharge channel (PZ) is connected through a gap in the bottom zone with the retention chamber (KRG). The flushing chamber (KPŁ) is open to the atmosphere through a vent (W) which allows for venting the air from the flushing chamber during its filling and inflow of air from the atmosphere during its emptying. Filling the flushing chamber occurs through suction and forcing pump (P).

The sewage supplied to the reservoir through the inflow collector (KD) gradually fills the overflow chamber (KP) and, after reaching adequate

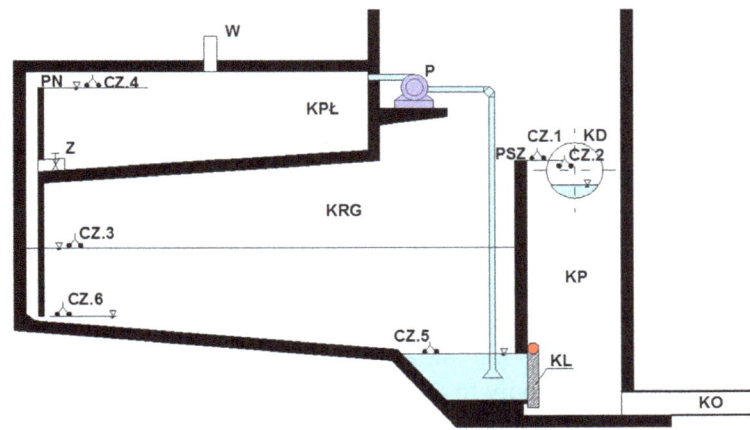

Fig. 8. Diagram of the gravity chamber for flushing the bottom of the retention chamber immediately after emptying from accumulated sewage

filling level, is transported through the overflow edge (PSZ) to the retention chamber (KRG). With gradual reduction in the sewage inflow, the reservoir starts to be emptied. This is signalled by the integrated CZ.1 and CZ.2 sensors located at the level slightly below the top part of the overflow wall (PSZ) in the flow chamber (KP). When sewage reaches the level of the integrated CZ.1 and CZ.2 sensors, it switches off the pump (P) and starts filling the volume of the flushing chamber (KPŁ). Switching on the pump (P) leads to shutting the outlet valve (Z) in the flushing chamber (KPŁ). Shutting off the pump (P) occurs after completion of filling of the flushing chamber (KPŁ), which is signalled with the sensor (CZ.4) situated at the level of the top of the overflow wall (PN). Switching off the pump will also occur at the moment of emptying of the retention chamber, i.e. when the filling reaches the level of the CZ.5 sensor. At this point filling of the flushing chamber (KPŁ) will be incomplete. When the fluid stops flowing into the reservoir and the filling level in the flow chamber (KP) is decreased, the contacts of the integrated CZ.1 and CZ.2 sensors are exposed and the power supply for the pump is cut off.

The process of empting the reservoir occurs with the emptying of its gravity chambers (KP) and (KRG). At the point of nearly total emptying of the retention chamber (KRG), when sewage surface reaches the position of the sensor (CZ.5), the process of flushing the bottom of the retention chamber is started through the opening of the valve (Z) and switching on the pump (P), if it continues operation. The contaminants settled on the bottom of the retention chamber are flushed to the outlet hole (KL) and further to the outlet collector (KO). If, during the flushing, the retention chamber (KRG) is filled to the level of CZ.6 sensor situated at the level of outflow of sewage from the discharge channel (PZ) that signals entire covering of the bottom of the retention chamber (KRG) with sewage, the valve Z is closed and the flushing process is brought to a standstill. After another emptying of the retention chamber (KRG), the signal from the CZ.5 sensor opens the valve (Z) again and flushing is resumed. The process is repeated until the flushing chamber (KPŁ) is entirely emptied. The CZ.1 and CZ.2 sensors are integrated, which means that the power supply to the relay coils depends on the state of contacts of the relays. The only opening of the contacts of both relays for the CZ.1 and CZ.2 sensors can lead to

the conditional opportunity of supplying the power to the pump (P). As mentioned before, the precondition for starting the pump (P) is submersion of the CZ.3 sensor in the sewage of the retention chamber. The CZ3 sensor causes the opening of the contacts of its respective relay, which is maintained by the closed contacts at the relay assigned to the CZ.5 sensor. This means that the pump (P), if it is not switched off by the signal of the CZ.4 sensor, will be switched off using the CZ.5 sensor with opening of the outlet valve (Z), which initiates the process of bottom flushing in the emptied retention chamber (KRG). The pair of integrated CZ.1 and CZ.2 sensors prevents from repeated switching on and off the relay of pump power supply caused by waving of sewage surface in the flow chamber (KP). Furthermore, the integration of the relays of the CZ.3 sensor with the CZ.5 sensor ensures operation of the pump (P), despite surfacing of the CZ. 3 sensor from sewage.

The flushing chamber located at the level of maximal filling the retention chamber ensures high efficiency of the process of flushing the bottom of the retention chamber, since the velocity of the stream of fluid flowing out on the flushing surface is large, even at the final stage of emptying of the flushing chamber. As in all solutions of flushing, the capacity of the flushing chamber is added to the total retention capacity of the reservoir. The location of the CZ.5 and CZ.6 sensors at the bottom of the retention chamber allows for starting the cycle of multiple flushing. Furthermore, the location of the CZ.3 sensor in the retention chamber on the adequate level from its bottom depends, among other things, on the capacity of the flushing chamber. The control over the emptying of the flushing chamber can be changed so that the process of flushing the bottom of the retention chamber occurs at all times, even at a minimal filling.

GRAVITY AND VACUUM CHAMBER FOR FLUSHING

The liquid reservoir (Fig. 9) contains the flow chamber (KP) with the inlet hole (KD) and outlet hole (KO) as well as the gravity retention chamber (KRG), separated from each other with the partition wall with overflow edge (PSZ). An outflow hole with check valve (KL) is located at the bottom zone of the gravity retention chamber, automatically opened towards the flow chamber

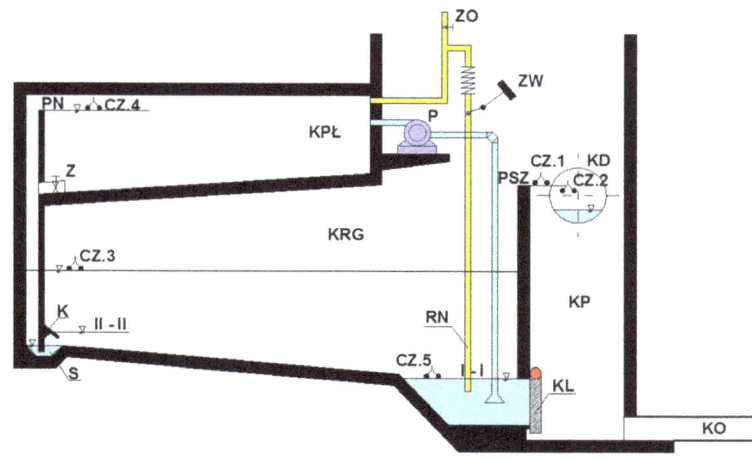

Fig. 9. Design of gravity and vacuum chamber for flushing the bottom of the retention chamber immediately after emptying from accumulated sewage

(KP). A closed flushing chamber is located inside the retention chamber (KRG), with its bottom situated at the level of liquid surface in the entirely filled retention chamber. The flushing chamber (KPŁ) and retention chamber (KRG) are connected with the discharge channel (PZ) adjacent to the wall of the retention chamber which is located opposite the outlet hole (KL). The flushing channel is connected with the discharge channel (PZ) through the overflow pipe (PN) and the outlet valve (Z) in the bottom zone. Furthermore, the discharge channel (PZ) is connected with the drain trap (S) in the bottom zone with the retention chamber (KRG). Its outflow is also equipped in the guide (K) that guides the flushing stream to the bottom of the retention chamber. The flushing chamber (KPŁ) is filled with the suction and forcing pump, with its suction tip located in the retention chamber (KRG) below the level of its bottom in the check valve zone (KL). The under-top zone of the flushing chamber (KPŁ) is connected with the atmosphere through the vent valve (ZO) and the aeration pipe (RN) equipped in the vertical motion support system (ZW). The inlet to the aeration pipe is situated in the retention chamber (KRG) in the check valve zone (KL). The operation of the pump (P) and valves (Z) and (ZO) is controlled by the signals from the liquid level sensors in the container chambers. The system of chambers in the reservoir is consistent with the solution presented in the patent application from the year 2005. However, the solution employed in this research differs substantially in the method of hydraulic control of the flushing of the bottom of the retention chamber.

The sewage flowing to the reservoir through the inflow collector (KD) gradually fills the flow chamber (KP) and retention chamber (KRG) through the top overflow wall (PSZ). Completion of the sewage accumulation process in the retention chamber (KRG) occurs at their gradual decrease in the flow to the reservoir. This is signalled by the CZ.1 and CZ.2 sensors located at the level, and slightly below, the top part of the overflow wall (PSZ) in the flow chamber (KP). Reaching the level of the integrated CZ.1 and CZ.2 sensors during the process of flow chamber (KP) filling, switches on the power supply for the pump resulting from the immersion of the CZ.3 sensor in the accumulated sewage. Switching on the pump (P) and filling the volume of the flushing chamber (KPŁ) occurs in the retention chamber. Switching off the pump occurs after the completion of the flushing chamber (KPŁ) filling through the CZ.4 sensor relatively after reaching the waste level I-I in the retention chamber that is after its emptying, signalled by the CZ.5 sensor. If the liquid stops flowing into the reservoir and the filling level in the flow chamber (KP) is decreased (exposing the integrated CZ.1 and CZ.2 sensors), the power supply for the pump is cut off.

The start of the pump (P) is closely related to the necessary shutting of the outlet valve (Z) and opening the vent valve (ZO). Emptying of the reservoir chambers i.e. flow chamber (KP) and retention chamber (KRG) occurs simultaneously. Switching off the pump (P) leads to the negative pressure in the flushing chamber (KPŁ), since the vent valve (ZO) is shut and the outlet valve (Z) is open at the same time. The level of the negative pressure closed in the flushing chamber (KPŁ) equals the atmospheric pressure minus the current difference in the levels of sewage in the flushing chamber and the retention chamber. Therefore, the maximum value of the negative pressure in the flushing chamber (KPŁ) is reached at the moment of completion of the retention chamber (KRG) emptying. Furthermore, the increasing level of the negative pressure in the flushing chamber (KPŁ) also leads to suction of sewage to the aeration pipe (RN) until the level of sewage surface reaches the level in the flushing chamber. The aeration pipe (RN) is equipped with an elastic joint and the unloading assembly. Before flushing is started, the aeration pipe, containing the sucked sewage, is heavier and its outlet is immersed to the desired depth.

When the waste level in the retention chamber (KRG) drops below I-I, the aeration pipe (RN) is exposed, and after discharge of the sewage from its interior, it is lifted by the support system up to the level II-II as it becomes lighter. The air starts to flow in from the atmosphere to the flushing chamber (KPŁ), causing its intensive emptying by the opened valve (Z). The sewage flowing out of the flushing chamber (KPŁ) through the discharge channel (PZ) to the retention chamber (KRG) flushes the contaminants settled on the bottom, which, are transported to the outflow collector (KO) through the output hole with the flap valve (KL). Filling the retention chamber (KRG) to the level II-II in the process of flushing closes the air supply to the pipe (RN), thus stopping the outflow of sewage from the flushing chamber (KPŁ). The sewage sucked again into the pipe (RN) lower the position of its inlet below the level I-I. Resuming the bottom flushing process occurs, similarly to the previous case, after the level of sewage in the retention chamber (KRG) drops to the level I-I.

Filling of the flushing chamber in the solution presented in this paper occurs during the emptying of the retention chamber, which allows for partial sedimentation of the solid phase suspended in the liquid. The location of the flushing chamber over the flushed bottom leads to the increase in the kinetic energy of the flushing stream, while the stream of the liquid flowing out of the chamber almost does not limit the flow dynamics during emptying of the flushing chamber as it is the case of popular solutions. This solution increases the capacity of the flushing chamber and allows for repeated flushing of the bottom of the retention chamber with the fluid collected in the flushing chamber, effectively preventing from collection of the contaminants on its bottom. Integration of the CZ.1 and CZ.2 sensors consists in that their relays have mutually current-supported electromagnet coils. This means that cutting off the power supply to the pump occurs only if both sensors stop the operation of both relays. In practice, this means that the support current supplied to the coil of the first relay flows through the closed contacts of the second relay. This integration of the sensors CZ.1 and CZ.2 prevents from unsteady operation of pump control operation caused by undulated surface of the sewage. The CZ.3 sensor, located in the retention chamber allows for supplying current to the coils of the respective sensors CZ.1 and CZ.2. The sensor CZ.3 guarantees switching on the pump only if sewage is accumulated in the retention chamber.

CONCLUSIONS

1. Previous solutions concerning the methods of flushing the bottom of the retention chamber, with particular focus on the most recent ones, meet the requirements of effective removal of the sludge accumulated in the process of retention of sewage in sewerage reservoirs. It is remarkable that there is no solution for flushing of the vertical walls of the reservoir that would be recommended for the use in practice. The problems of flushing the sewerage reservoirs also have a hygienic aspect connected with epidemiological safety. Therefore, it is essential to continue the research on flushing the sewerage system reservoirs.

2. The systems for flushing the retention chambers presented in the study are presented only as a general idea, rather than a specific technical solution for the problem. Therefore, the design of the flushing system will be individual for any case, with the principles and aim of its operation remaining the same.

3. The increase in the efficiency of flushing can be obtained by a reliable control of hydraulic parameters of the stream that dynamically affects the contaminants present on the bottom of the reservoir. The design of the retention reservoir should take into account the parameters of the flushing chamber with respect to the retention chamber, in order to achieve the effect of double or repeated flushing. Obtaining the effect of repeated flushing is controlled by the

adoption of the depth of the aeration pipe tip immersion in the retention chamber.

4. The analysis of the hydraulic operation of the presented flushing chamber leads to the conclusion that the most efficient solution is the system where the height of the bottom of the flushing chamber with respect to the bottom of the retention chamber that is cleaned from the sewage sludge is the highest possible. Such relation between the bottoms of the flushing and retention chambers ensures high velocities of the flushing stream, thus enhancing the flushing efficiency and measurably shortening the time of emptying of the flushing chamber.

Acknowledgements

This scientific study was financed within the research project BSPB-401–301/17.

REFERENCES

1. Deska I., Ociepa E., Mrowiec M., Łacisz K. 2016. Investigation of the influence of hydrogel addition on the retention capacity of green roofs. Proceedings of ECOpole, 10(2), 625–633 (in Polish).

2. Dziopak J. 1990. Highly effective methods flushing bottom of the gravitational of retention chamber sewage tank. Gaz, Woda i Technika Sanitarna, 2–3, 68–70 (in Polish).

3. Kisiel A., Kisiel J., Malmur R., Mrowiec M. 2008. Retention tanks as key elements of modern drainage systems. Czasopismo Techniczne, No. 1-Ś, 18(105), 41–63 (in Polish).

4. Malmur R., Kisiel A. 2008. Flushing out sludge settling at the bottom of retention chambers in sewage reservoirs. Inżynieria i Ochrona Środowiska, 11(3), 269–280 (in Polish).

5. Ociepa E., Mrowiec M., Deska I., Okoniewska E. 2015. snow cover as a medium for deposition of pollution. Rocznik Ochrona Środowiska – Annual Set The Environment Protection, 17(1), 560–575 (in Polish).

6. Wolski P. Wolny L. 2011. Effect of disintegration and fermentation of sewage sludge on susceptibility to dehydration. Rocznik Ochrona Środowiska – Annual Set The Environment Protection, 13, 1697–1706.

7. Zawieja I., Wolski P. 2013. Effect of hybrid method of excess sludge disintegration on the increase of their biodegradability. Environment Protection Engineering, 39(2), 153–165.

12

Reagent Technology of Joint Purification of Sewage Water for Paint and Galvanic Production

Anatoli Hurynovich[1*], Uladzimir Maroz[2]

[1] Faculty of Civil and Environmental Engineering, Bialystok University of Technology, ul. Wiejska 45A, 15-351 Bialystok, Poland

[2] Faculty of Environmental Engineering Brest State Technical University, Moskowskaja 265, 224017 Brest, Belarus

* Corresponding author's e-mail: a.gurinowicz@pb.edu.pl

ABSTRACT

There is a need to create a low-cost and effective resource-saving technology for wastewater treatment contaminated with paintwork materials, which would allow reusing water in the production cycle of the enterprise. Studies have been carried out to prove the possibility of using spent technological solutions in the joint purification of sewage from galvanic and paint and varnish industries. The developed technology enables to reduce the costs of the enterprises for its cleaning and negative impact on the environment, reduce the load on urban treatment facilities, and reduce the equipment costs.

Keywords: paintwork materials, waste technological solutions, mixer, reactor, sorption, flocculation, sewage, galvanic production, paint and varnish production.

INTRODUCTION

The enterprises of mechanical engineering, as a rule, besides galvanic manufactures and production of printed circuit boards also include the manufacturing of paints . Various organic substances (acids, alkalis, solvents, surfactans) found in the wastewater of paint industries constitute extremely dangerous pollutants. These substances are characterized by a complex and variable composition, high toxicity, the predominant content of dissolved, rather than suspended substances. Therefore, their isolation is a task of extreme complexity. The famous methods of cleaning of such kind of waste water (ultrafiltration, combustion, ion exchange, etc.) are associated with very high energy costs, high cost of technological equipment, shortage of reagents and the need for significant production areas [Gogina et al., 2012].

The purpose of the work is the creation of an effective resource-saving technology for joint wastewater treatment of paint and varnish and galvanic production [Maroz, 2013].

The research was conducted aimed at creating an effective resource-saving technology for the purification of such wastewater, which would allow not only to clean up the waste water of paint and varnish production to the required standards, but also to reuse a significant part of them [Grinin 2002].

The survey of water supply and sewage systems of paint and varnish production showed that the painting of large products is carried out, as a rule, by the method of pneumatic spraying. At the same time, only 45–75% of the paint is applied to the surface of the products, and the rest is discharged with sewage.

In addition, these production problems create problems in wastewater treatment due to the presence of heavy metals in them, which enter sewage during the preparation for painting the surface of products, as well as when discharging contaminated water from the hydrofilters of the paint and varnish chambers shown in Figure 1.

When painting products, the water in the hydrofilter constantly circulates according to the

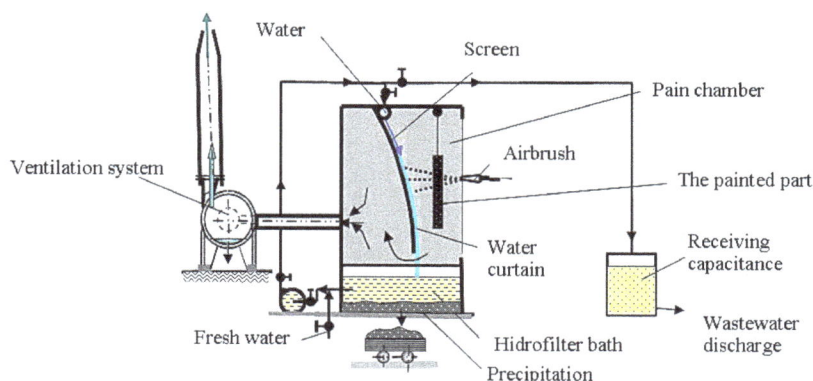

Figure 1. Paint by pneumatic spraying using a hydrofilter

scheme: a water curtain – a hydrofilter bath – a pumping unit – a water curtain. Upon contact with water of paint and varnish material, a partial dissolution of the chemicals entering into its composition occurs, the concentration of which in the water curtain increases with each cycle. After reaching the maximum permissible concentration of pollutants in the water of the hydrofilter bath, the water curtain loses its retention properties.

Therefore, such water from the hydrofilter bath and the formed bottom deposits of paint and varnish are removed; afterwards, the hydrofilter bath is replenished with fresh water. The air cleared of toxic aerosols by means of ventilating systems is taken away into the atmosphere [Subotkin 2017].

METHODS OF RESEARCH

In order to develop a technology for joint cleaning of wastewater paint and galvanic industries:

- production processes of paint and varnish and galvanic industries were investigated;
- the qualitative and quantitative composition of waste technological solutions was studied for the purpose of their use in the process of wastewater treatment;
- the rational formation of sewage streams of galvanic and paint and varnish production at the places of their formation;
- the kinetics of aggregation and sorption of paint and varnish material on oxyhydrate reservoirs in a mixture of wastewater of paint and varnish and galvanic production is investigated; hydrodynamic processes in laboratory and production conditions were studied;
- automated reactor units were developed.

Studies pertaining to the dynamics of flocculation and sorption of paint and varnish material on the oxyhydrate collectors under certain conditions were carried out on a laboratory installation.

In a reactor with the volume of $V = 10$ dm^3 the wastewater of the paint and galvanic production is mixed with a stirrer for 10 minutes and then settled in a graduated cylinder for 30 minutes. The pH value is controlled by an ion meter.

At the first stage, the hydrolysis of the salts of metals Cr (III) occurs, Fe (III) and others, with the formation of micelles and their subsequent aggregation into larger particles of the sol. At the second stage, the construction of chain structures begins and the formation of a huge number of tiny flakes, which are aggregated into larger ones, and, reaching a certain size, they settle down under the action of gravity.

With prolonged mechanical mixing, the flocs formed are destroyed, which as a result of the reduction in their hydraulic size do not settle well. The experimental setup was shown in Figure 2.

In order to assess the reliability of these hydrodynamic processes in an actual apparatus and to reduce the uncertainty that inevitably arises in large-scale transfer of results, studies were carried out in the laboratory setup shown in Figure 3.

The installation consists of a propeller-mixer apparatus, driven with a direct current electric motor (12 volts). The speed of the mixer shaft is recorded by a tachometer. The water in the device is fed by a pump from the tank. The flow rate is regulated by a valve mounted on the pressure line, and is measured by a rotameter. The electrical conductivity of the solution is determined using a conductometer and fixed on the recorder. The used solution was drained into the sewage system. Distilled water was used as a working liquid, whereas potassium chloride (KCl) is an indicator. The liquid was fed into the mixing area.

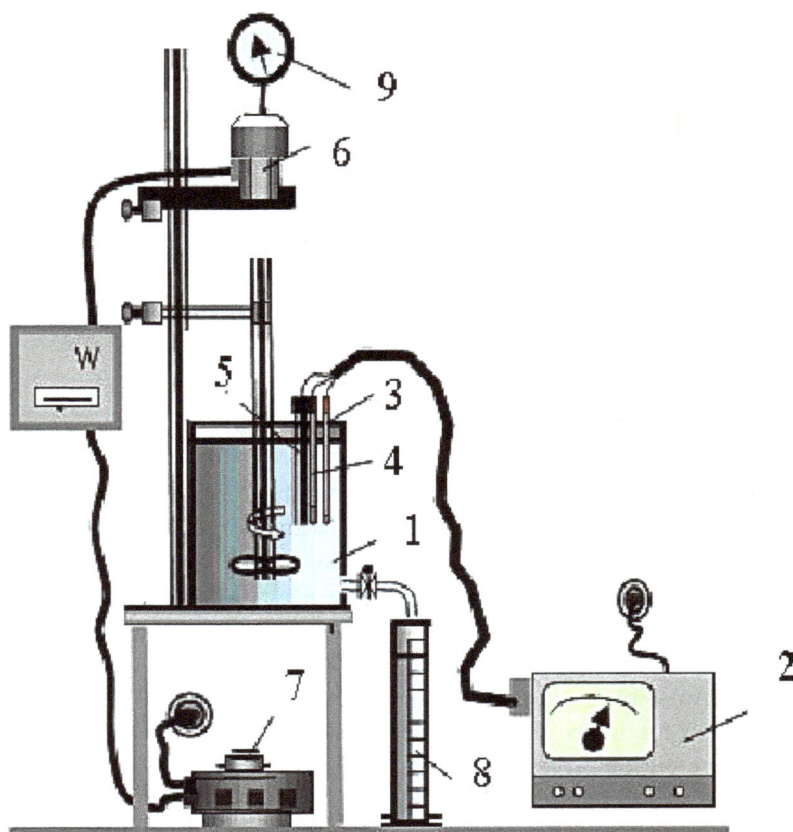

Figure 2. Experimental setup

1 – reactor; 2 – ion meter 3 – measuring electrode; 4 – reference electrode; 5 – thermal compensation electrode; 6 – mixer; 7 – transformer: 8 – measuring cylinder; 9 – tachometer; 10 – wattmeter

The structure of the flows was investigated by analyzing the response of the system to a perturbation [Tishin, 2002; Vinogradov, 2002]. Before the beginning of the experiment, the apparatus was filled with distilled water to the required level, including a stirrer. An indicator was inserted into the apparatus.

At the entrance to the apparatus, an indicator (*KCl* solution) was introduced, the time of entry being taken as the beginning of the experiment, and a corresponding mark was made on the instrument diagram.

During the experiment, a constant flow of water and the speed of the mixer were maintained. The electrical conductivity was measured continuously during the experiment; the experiment was terminated when the value of electrical conductivity became equal to the electrical conductivity of distilled water.

The obtained response curves in relative coordinates are shown in Figure 4.

RESULTS OF RESEARCH

The analysis of the results obtained from the studies showed that the response curves taken at different fluid flows and different mixer speeds (190–300 rpm) corresponding to the revolutions of the mixers of standard chemical devices are identical, i.e., the structure of the flows in the apparatus in this range of mixer speeds practically does not depend on the speed of rotation of the mixer and the flow rate of the fluid. The structure of the flows is close to the ideal mixing model and a similar apparatus can be referred to a reactor-mixer.

For the chosen geometry of the apparatus, in the investigated range of fluid flow rate and rotational speed of the mixer, the hydrodynamic situation in the laboratory apparatus corresponds to the conditions of ideal mixing and is described with sufficient accuracy by the ideal mixing model.

When used instead of the purchased reagents, waste metal solutions containing high *Fe (II)* and

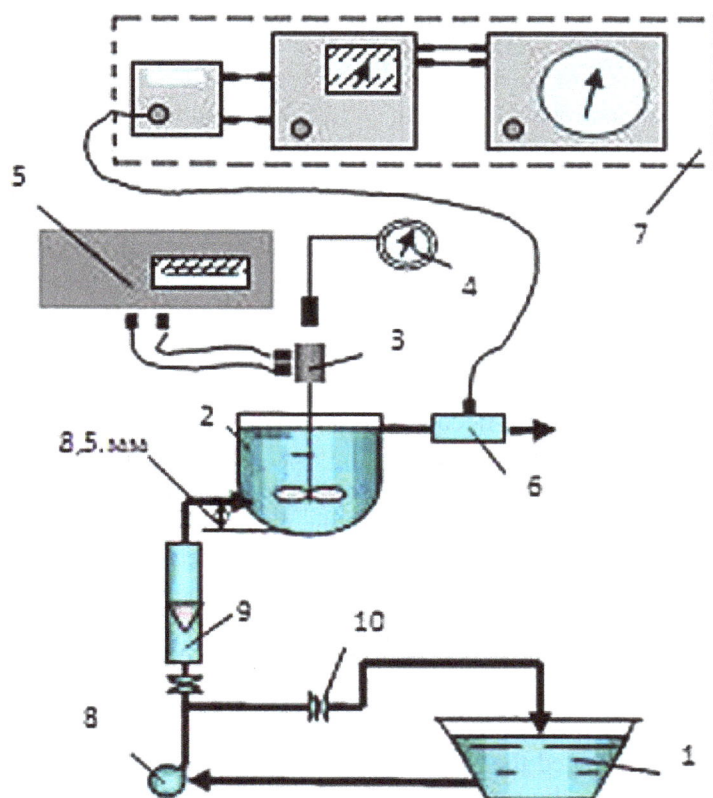

Figure 3. Diagram of an experimental laboratory installation for studying the structure of flows in apparatus equipped with mechanical stirrers

1 – distilled water; 2 – mixer; 3 – electric motor; 4 – tachometer; 5 – speed control system for mixers; 6 – conductometric cell; 7 – conductivity measuring system; 8 – pump; 9 – rotameter; 10 – valves

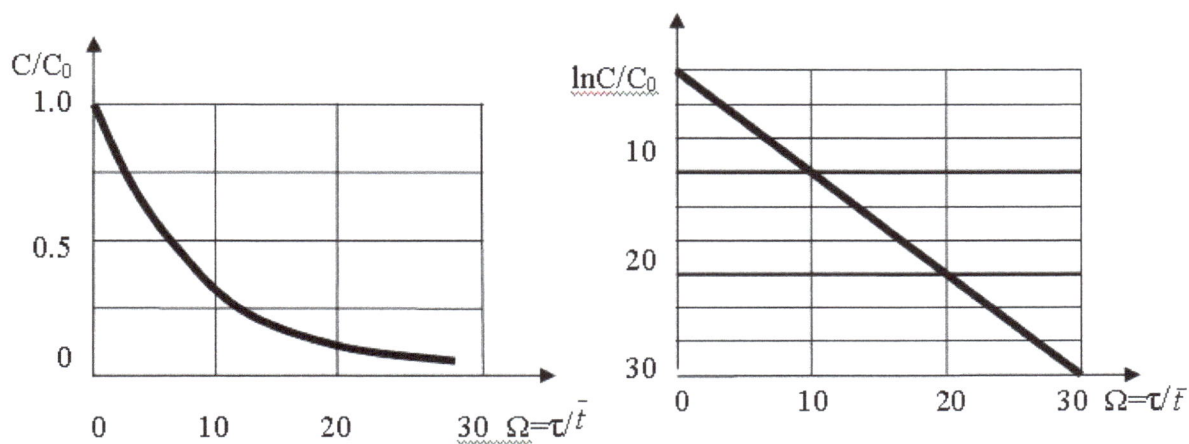

Figure 4. Response curves in relative coordinates

Fe (III) (*Fe (III)* concentrations of more than 20 g / dm^3 for *Cr (VI)* reduction up to *Cr (III)*, the standard system for automatic control of the *Cr (VI)* reduction process proved to be ineffective. This was due to the fact that the sensitive element of the chromium indicator with the *Fe (III)* content of more than 5 mg/dm^3 was ineffective.

The laboratory investigations were carried out to find a technical solution that would exclude the effect of *Fe (III)* ions on the sensitive element of the chromium indicator and allow automatic control of the redox potential. In order to eliminate the "interfering background", the ability of *Fe (III)* ions to form stable complex ions with certain compounds was used. At the same time, the degree of influence of the "interfering background" on the sensitive element of the automatic control system by the reduction process *Cr (VI)* to *Cr (III)* was refined by laboratory studies.

Complexing reagents were introduced into the sample taken from the reactor for processing the sewage mixture containing the paint and chromium, which react readily with *Fe (III)* ions in aqueous solutions.

It can be seen from Figure 5 that the signaling device for the presence of *Cr (VI)*, and consequently the entire automatic control system, becomes operative when a complexing agent is introduced into the reaction medium. When adding a reducing agent to the solution, which is necessary for complete recovery, a sharp potential jump occurs.

Complex formation, causing a change in the concentration of free ions in the oxidized form, causes a change in the redox potential of the system and selectively removes the influence of the "interfering" background in the work of the chromium meter. This feature eliminated the difficulties that arose, but did not solve the problem of the location of the complexing reagent into the technological scheme. If a masking complexing reagent were introduced directly into the medium to be treated, it would require a large number and additional equipment for preparing the working solution.

As a result of the reaction, the *Fe (III)* ions are bound to the complex, i.e., the "interfering background" is eliminated. In the case of the chromates appearance in the detoxified water, an electromotive force appears on the electrodes of the sensing element, which is amplified by the transducer. The amplified signal is fed to the control unit, where it is controlled by a pneumatic valve. As a result, a working reagent that reduces *Cr (IV)* is added to the main reactor from the dispenser. After the complete reduction of *Cr (IV)*, the pneumatic valve closes.

With the use of advanced standard automatic control systems, it became possible to rationally use the created "disturbing background" of spent technological solutions instead of purchased reagents. The studies were conducted on the stage-by-stage treatment of wastewater containing paint and varnish material [Maroz, 2014]. At the first stage, bubbling of wastewater containing LMC was carried out for 20 minutes. At the same time, the value of COD decreased by 20% (Figure 6).

In the second stage, as a result of oxidation of organic contaminants in a mixture of wastewater containing paint and chrome, within 10 minutes, the COD value decreased by another 25% (Figure 7).

At the third stage, after neutralization of all types of wastewater and formation of an oxyhydrate collector sorbing organic and mineral impurities on its surface, followed by clarification in the sedimentation tank, the COD decreased to 25 mg O_2/dm^3 (Figure 8).

The technology is implemented as follows. Wastewater containing paint and varnish contamination is sent to a reservoir in which a purge of sewage with compressed air is carried out for at least 20 minutes with a stirring intensity of 3–5 l / (s-m²). Afterwards, the treated wastewater containing paint-and-lacquer material is dosed to the tank, together with the wastewater containing chromium and acidic wastewater containing the

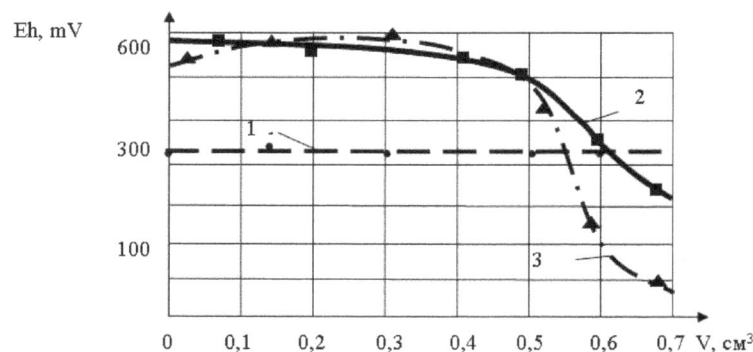

Figure 5. The graph of the change in the oxidation-reduction potential Eh in the process of reduction of *Cr (VI)* by the spent pickling solution containing *Fe (II)* and *Fe (III)*

1 – without complexing agent; 2 – with addition of complexing agent "A"; 3 – with the addition of complexing agent "B"

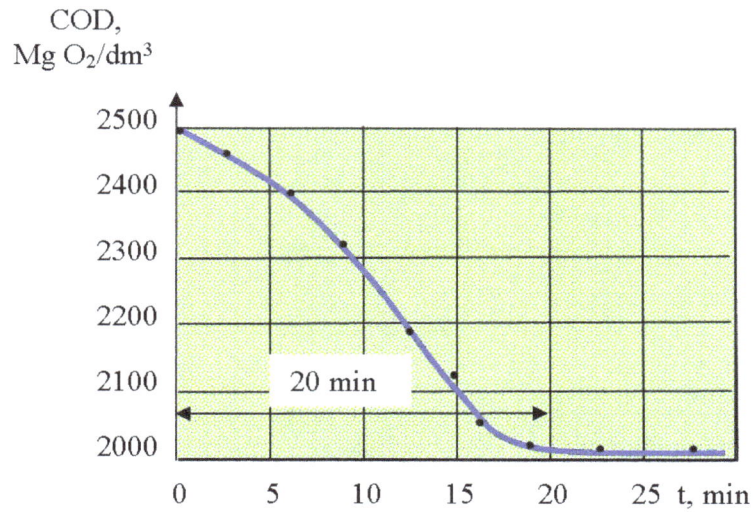

Figure 6. Schedule of COD reduction by bubbling compressed air (1st stage)

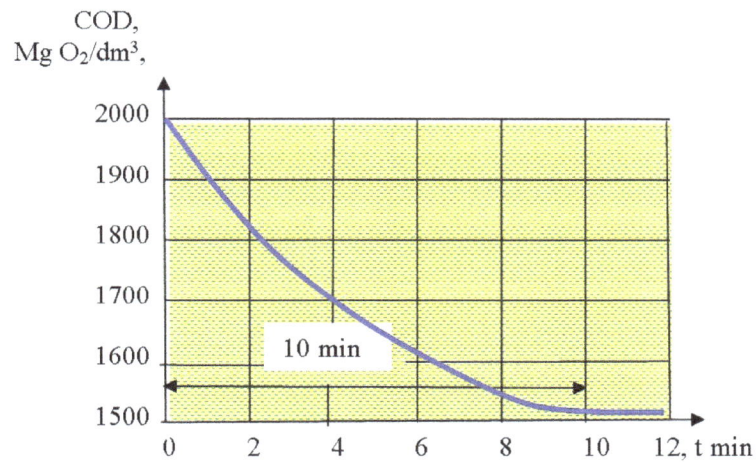

Figure 7. COD reduction graph by chromium-containing stream oxidation (averaged data) (2nd stage)

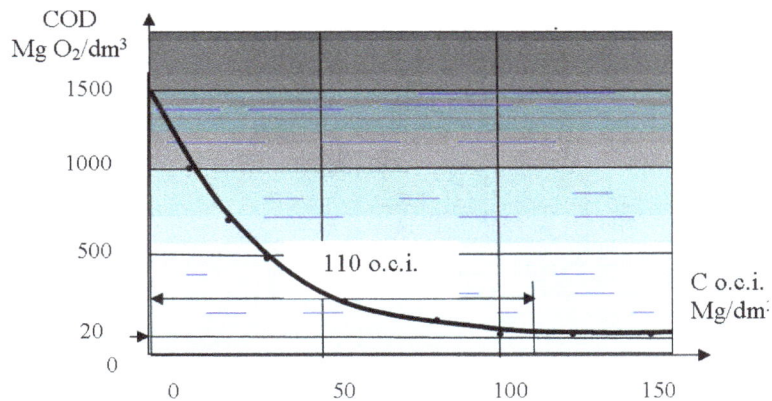

Figure 8. Schedule of COD decrease, sorption by oxyhydrate collector followed by 2-hour sedimentation (averaged data) (stage 3).

* o.c.i – oxyhydrate collector ions

reducing agent *Fe (II)*. The block diagram of the technology of wastewater from paint and varnish materials is shown in Figure 9.

Further, this mixture of wastewater is fed into the reactor by a reducing agent *Cr (VI)*. Maintaining in it pH = 2–3 is carried out by acid waste technological solu-

tions etching of ferrous metals containing up to 90% *Fe (II)*, the rest is *Fe (III)*. In this case, *Cr (VI)* is reduced to *Cr (III)*, as well as the destruction of organic contaminants by dichromates and catalyzing this process with the traditional chromium catalyst. The time of sewage treatment is up to 10 minutes.

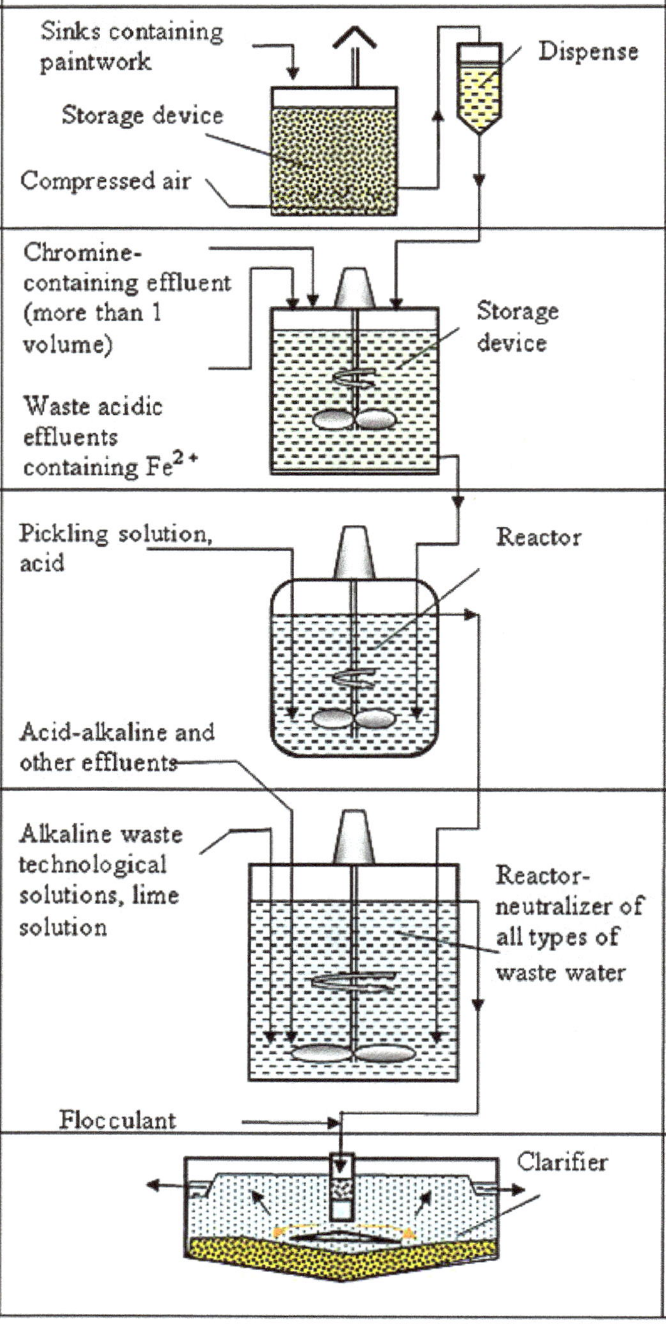

№	Scheme, streams, reagents		Stages of the process	Time of processing, mines
1	Sinks containing paintwork Storage device Compressed air	Dispense	Accumulation drains with paintwork	Purge of sewage sludge air is not less than 20 minutes
2	Chromine-containing effluent (more than 1 volume) Waste acidic effluents containing Fe^{2+}	Storage device	Mixing of drains with LMC in chrome	Averaging, at least 10 minutes
3	Pickling solution, acid Acid-alkaline and other effluents	Reactor	Recovery of Cr^{6+} in a combined flow with paintwork pH = 2.0-3.0	7-10 min
4	Alkaline waste technological solutions, lime solution Flocculant	Reactor-neutralizer of all types of waste water	Neutralization all types of effluents pH = 8-9.0	7-10 min
5		Clarifier	Lightening	The time interval, depending on the type of clarifier

Figure 9. Simplified scheme of "co-ordinated" decontamination of wastewater contaminated with paintwork materials in the framework of sewage treatment plants of galvanic production of reagent type

Table 1. Interval of contamination after cleaning

No.	Ingredient name	unit of measurement	The average. confidence interval of concentrations after the neutralizer reactor	Observed maximum before and after the vertical settler	
				before	after
1	Chromium (VI)	mg / l	-	-	-
2	Chromium Society.	mg / l	10 – 20	not spec.	1.7
3	Zn	mg / l	15 – 28	32.1	0.2
4	Ni	mg / l	2.2 – 4.9	5.20	0.1
5	Fe	mg / l	70 – 195	250	1.5
6	Cu	mg / l	11 – 22	28.0	0.5
7	Acid	mg-eq / l	7.9 – 9.1	10.8	-
8	Alkali	mg-eq / l	3 – 4.2	5.25	-
9	pH		not spec.	not spec.	8.6
10	COD	mg / l	1500 – 2000	2500	21.6
Total content of main ingredients (Fe$_{обш.}$, Cr$_{обш.}$, Cu. Zn. Ni) forming a hydroxyhydrate reservoir					108.2 – 269.9

Following this stage of treatment, the waste liquid is directed to a joint neutralization with other kinds of wastewater to the reactor-neutralizer.

In this reactor, with the aid of neutralizing solutions (alkaline waste technological solutions, and in their absence, the lime solution of $Ca(OH)_2$ is maintained at pH = 8.0–9.0). It carries out the hydrolysis of heavy metals (Fe^{2+}, Fe^{3+}, Zn^{2+}, Ni^{2+}, Cr^{3+}, Cu^{2+}, etc.) present in the neutralized mixture to form the oxyhydrate collector, while the hydroxide of *Fe (II)* and *Fe (III)* are the main components of the oxyhydrate collector on which the paint and other organic contaminants are effectively sorbed.

Further reduction in the concentration of paint and varnish compounds takes place in the clarifier (step 5), resulting from the co-precipitation of suspensions. The efficiency of wastewater treatment is shown in Table 1.

The introduction of "passing" technology of sewage treatment containing paint materials was carried out at the treatment plants of the Brest Electromechanical Plant (BEMZ) on the previously installed equipment for the treatment of waste water from the galvanic production

CONCLUSIONS

On the basis of the experimental research results, a resource-saving technology for joint wastewater purification of paint and varnish and galvanic productions for instrumentation and machine-building at the existing reagent wastewater treatment plants of the BEMZ galvanic production was developed and implemented [Gurinovich et al. 2014].

The technology allows:

1. A 3–5-fold increase in the capacity of existing technological equipment of galvanic treatment plants, to which polluted waste water is discharged, with a corresponding decrease in their material consumption and energy consumption without reducing the efficiency of wastewater treatment.

2. Due to the use of waste technological solutions instead of purchased reagents, to achieve savings no lesser than 80%.

3. Measure the depth of cleaning in the hydrotipers of the spraying chamber waste water discharged from the baths, at the outlet of the treatment plant for COD from 2500 mg/l O_2 to 20–15 mg/l O_2, which is much lower than the value of this indicator (250 mg/l O_2) brought by the controlling companies to the instrument and machine building enterprises in Brest.

4. Utilize the sewage sludge of galvanic production, contaminated with paintwork, in the production of ceramic materials. The environmental safety of the products obtained was confirmed by the Belarusian Scientific Research Institute of Sanitation and Hygiene.

REFERENCES

1. Gogina E.S., Gurinovich A.D., Urecky E.A. 2012. Resource-saving technologies of industrial water supply and water disposal. ASB, Moscow.

2. Grinin A.S. 2002. Industrial and household waste. Fair-Press, Moscow.

3. Maroz U.V., Urecky E.A. 2013. Studies of the kinetics of processes of sorption of organic impuri-

ties on oxyhydrate reservoirs and their aggregation in a mixture of paint-containing and galvanic drains. Bulletin of the Belarusian State Technical University, 2(80), 54–57.

4. Maroz U.V. 2014. Resource-saving technology of wastewater treatment of paint and varnish production in instrumentation and engineering. Bulletin of the Belarusian State Technical University, 2(86), 78–81.

5. Subotkin L.D. 2017. Development and implementation of resource-saving technology for joint sew-

age treatment of galvanic and paint production. Construction and technogenic security. – Simferopol, 7(59).

6. Tishin O.A. 2002. Determination of the conditions providing the distribution of the residence time in apparatuses with mixers. Izvestiya VUZov, 45, 57–73.

7. Vinogradov S.S. 2002. Organization of galvanic production. Equipment, calculation of production, rationing. Globe, Moscow.

13

The Impact of Selected Sewage Treatment Methods on the Change in Parameters of Sewage Sludge Originating from Municipal Sewage Treatment Plants

Hanna Bauman-Kaszubska[1*], Mikołaj Sikorski[1]

[1] Faculty of Civil Engineering, Mechanics and Petrochemistry, Warsaw University of Technology, Łukasiewicza 17, 09-400 Płock, Poland

[*] Corresponding author's e-mail: Hanna.Bauman@pw.edu.pl

ABSTRACT

Sewage sludge is produced in every sewage treatment plant and its properties depend on a number of factors. The type, origin and parameters of the treated sewage and the applied treatment technology are the main factors influencing the sludge properties. The produced sewage sludge can be submitted to treatment processes which condition its final use. For example, the sludge produced in small treatment plants, treating mainly domestic sewage, can be used for agricultural purposes. This paper presents the results of the authors' own research on the susceptibility of sludge originating from several municipal wastewater treatment plants to selected treatment processes and changes in properties as a result of the carried out processes. The obtained results confirm that the ultrasonic field, as a physical method of sewage sludge modification, is a factor intensifying the drainage processes. In most cases, the investigations have shown better results of the dewaterability of sludge treated with ultrasonic field in relation to raw sediments. For the tested sediments, it is justified to use the centrifugation process as a method of mechanical compaction.

Keywords: sewage sludge, dewatering, ultrasonic disintegration, vacuum filtration

INTRODUCTION

Sewage sludge management carried out in municipal wastewater treatment plants is determined by the quantitative (capacity of a plant) and qualitative structure of the generated sludge, legal circumstances and local conditions. Apart from the technological aspects, both the running and investment costs play the most important role here. According to the data provided by the Polish Central Statistical Office (Statistical Yearbooks of the Central Statistical Office, 2017), in 2016 the municipal sewage treatment plants in Poland produced 568.3 thousand Mg of dry weight of sewage sludge, which constitutes a rise of 300 Mg in comparison to 2015 and a rise of 12.3 thousand Mg of dry weight in comparison to 2014. The data contained in the report on the implementation of the National Sewage Treatment Programme in the period from 2014–2015 (National Water Manage-ment Authority, 2016) show that the amount of the produced and deposited dry weight of sewage sludge originating from wastewater treatment plants in 2014 and 2015 amounted to 600 078 Mg and 590 199 Mg, respectively. A sustained growth in the amount of produced sewage sludge is caused by increasing the capacity of municipal wastewater treatment plants and the employment of more and more advanced technologies of increased nutrient removal.

Both the data from the Polish Central Statistical Office (CSO) and National Water Management Authority (NWMA) indicate a very limited degree of sludge use in agriculture, land reclamation and cultivation of the plants intended for compost production. Approximately 20% of sludge in total, produced in municipal wastewater treatment plants, is used for the above mentioned purposes. However, for today's needs, this is still not enough – nearly 15%. One should not expect that the situation in this case will improve quickly, mainly because of high investment costs of installations for incineration of sewage sludge.

Because of their physical and chemical properties and threats it can pose for the environment and human health, the sewage sludge produced in municipal wastewater treatment plants has to undergo proper treatment. It should be disposed of in order to decrease its fouling rate in the stabilization process, destroy pathogenic organisms in the hygienisation process, as well as reduce the volume and weight of sludge in the dehydration and drying processes. Sewage sludge treatment should be carried out in such a way so as to obtain the sludge with the properties enabling its further safe management (Fukas-Płonka et al., 2013). The basis for classification of sludge to treatment methods is chemical composition, sanitary condition and physical properties. Sewage sludge can be a source of valuable nutrients necessary for plant growth (Wiater and Butarewicz, 2014). Its application improves the physical and chemical properties of soil and exerts influence on the increase of organic matter in soil (Grobelak et al., 2016; Obbard, 2001; Świsłowski, 2016). The greatest difficulty connected with using sewage sludge in nature is the high content of heavy metal, highly toxic chemicals and pathogenic organisms (Krzywy et al., 2015). When administering sewage sludge into soil, the excess of heavy metals in sewage sludge can also cause changes in the fertility of soil and decrease the quantity and quality of yield. The initial treatment of sewage sludge can also limit its management. For example, when liming hydrated sewage sludge, one has to take into account high increase in weight and adherence to rules used in agriculture, especially considering that such sludge is of lime and organic nature and should be used on lime fields rather than fertilizing them as it is in the case of regular sludge (Czekała, 2013).

Oxygen stabilization and conditioning are the processes that can be applied to every type of sewage sludge. They aim at changing the structure and properties of sewage sludge, thereby allowing to increase the efficiency of its dewatering (Skarżyński, 2017). The commonly employed method of sediment conditioning is polymer coagulation, which uses the high molecular weight of an organic polymer (polyelectrolyte) (Bień, Bień 2014, Bień 2017). Various methods of sediment conditioning are still being tested to increase the effect of mechanical sludge dewatering (Mohammad et al., 2015). A very effective method of drainage is relatively difficult and despite the use of a given technique, the final product containing less than 65% of dry matter is not achieved (Mahmoud et al., 2011). In addition, the properties of sewage sludge change, for example with the use of ultrasonic disintegration, which affects the intensification of the methane fermentation process (Zawieja, 2010) and has a significant impact on the dispersion of the sediment structure. It is increased by weakening the forces bonding water with the surface of solid phase particles, and in consequence, easier removal of water in the processes of mechanical dewatering (Bień and Wystalska, 2011). However, using aluminium and ferric salts in sludge conditioning enables to obtain the intended technological effect and at the same time increase the content of those elements in sludge. This mainly concerns aluminium, the occurrence of which in soil is especially harmful for plants (Skarżyński, 2017). It confirms that physical and chemical properties determine susceptibility of sludge to a particular type of disintegration; therefore, in the process of pre-project research, it is necessary to adapt both disintegration method and its process parameters (Zielewicz, 2016).

The research on the susceptibility to disintegration, including ultrasonic, (Sridhar et al., 2011) should include also information on participation of substances resolved in sludge since the condition for obtaining significant effects is high participation of a substance that is resolved chemically, thermally or by means of thermo-pressure (designated as COD of substances resolved in sludge liquid) in comparison to total COD (Zielewicz, 2016). At the same time, when membranes are torn to a liquid phase, enzymes and organic substances, which are easily degradable, are released (Trojanowska and Myszograj, 2017). This has a beneficial influence on the acceleration of the hydrolysis process, thereby increasing the effectiveness of the stabilization process (Bień et al., 2013; Tomczak-Wandzel et al., 2009).

Therefore, selecting the right treatment of sewage sludge is not easy. The needs and expectations of a particular sewage treatment plant (Zielewicz, 2016) are of key importance here; however, the obtained results of the carried out processes depend on a number of factors, including mainly the physical and chemical properties of sludge, which, in turn, are dependent on the properties of sewage and the manner in which it is purified.

LEGAL ISSUES

The national formal and legal regulations concerning sewage sludge, and in particular its treatment, result from the legal requirements of the European Union. Among them, the most important are: Council Directive 86/278/EEC of 12 June 1986 on the protection of the environment, and in particular of the soil, when sewage sludge is used in agriculture, Council Directive 91/271/EEC of 21 May 1991 concerning urban waste water treatment, Council Directive 1999/31/EC of 26 April 1999 on the landfill of waste, Directive 2006/12/EC of the European Parliament and of the Council of 5 April 2006 on waste and Council Directive 1999/31/EC of 26 April 1999 on the landfill of waste. In accordance with those acts, discharge of sewage sludge to seas and oceans and deposit of sludge on landfills is forbidden. Legal provisions prohibit the deposit of sludge with heat of combustion higher than 6 MJ/kg of dry weight, which in reality concerns the majority of sludge produced in municipal wastewater treatment plants. The directive on sludge, on the other hand, limits the employment of sewage sludge for agricultural and natural purposes. It determines the conditions that have to be met while using sewage sludge for agricultural and natural purposes. It specifies, inter alia, the maximum values for the concentration of heavy metals in soil and sludge and maximum amounts of heavy metals (Cd, Cu, Ni, Pb, Zn and Hg), which can be applied into soil. What is more, the shortest time periods between the applications of particular sewage sludge into particular types of agricultural land and directions for usage of this land.

Under the above mentioned directives, in Poland the Act of 14 December 2012 on waste (The Act of 14 December 2012) and executive acts to this Act have been passed. The key acts in the subject of sewage sludge include: the Regulation of the Minister of Environment (Regulation of the Minister of the Environment of 6 February 2015; Regulation of the Minister of the Environment of 9 December 2014) and the Regulation of the Minister of Economy (Regulation of the Minister of Economy of 16 July 2015). In accordance with the waste catalog (Regulation of the Minister of the Environment of 9 December 2014), stabilized urban sewage sludge constitutes a group of waste with the following code: 19 08 05, for which the Regulation of the Minister of Economy determines a scope of tests and authorization criteria for disposal in landfills other inert waste. The scope of test and allowable limit values are related to three fundamental criteria:

- total organic carbon (TOC) – 5% of dry weight,
- loss on ignition (LOI) – 8% of dry weight,
- heat of combustion – 6 MJ / kg of dry weight.

Taking into consideration the above mentioned conditions and recommendations of the National Waste Management Plan (Ministry of the Environment, 2016) apart from dumping, other methods of sewage sludge disposal and treatment should be taken into consideration.

Considering the agricultural and natural use of sludge, the provisions of the Act on Fertilizers and Fertilization (The Act of 10 July 2007), which differentiates between and determines different types of fertilizers and means supporting plant cultivation are recommended in particular in the case of sludge originating from small wastewater treatment plants. In accordance with the Act, organic fertilizers are the fertilizers produced from an organic substance or a mixture of organic substances, including compost. The requirements related to any types of fertilizers are regulated above all by the regulation of the Minister of Agriculture and Rural Development on the application of some regulations of the Act on fertilizers and fertilization (Regulation of the Minister of Agriculture and Rural Development of 18 June 2008). The regulation lays down the limit values for pollutants in organic fertilizers and organic-mineral and organic and organic-mineral substances supporting plant cultivation, including above all chrome, cadmium, nickel, lead and mercury. An important requirement is the point which provides that the occurrence of live eggs of intestinal parasites Ascaris sp., Trichuris sp. and Toxocara sp. and Salmonella bacteria is not allowable in fertilizers and substances supporting plant cultivation. According to the requirements of the regulation, solid organic-mineral fertilizers should contain at least 20% of organic substance calculated on the dry matter and organic fertilizers should contain at least 30% of organic substance. In addition, it determines the content of nitrogen, phosphorous and potassium in the above mentioned fertilizers.

MATERIAL AND METHODS

The sludge collected for tests came from rural and urban wastewater treatment plants, namely from a plant of size expressed by population equivalents lower than 2000 – 3 samples and in a range from 2000 to 9999 – 2 samples.

Analytical tests, enabling to characterize sludge, including determining pH, dry residue, organic (expressed as loss on ignition) and mineral substances content (expressed as after ignition description) as well as hydration have been carried out in line with the adopted standardized test methods (Polish Standards). As part of technological test, a dewatering test has been carried out by measuring capillary suction, vacuum filtration test, sludge dewatering by means of centrifugation, concentration and ultrasonic disintegration.

The time of capillary suction consists in analyzing the process in which liquid flows from sludge to capillary of absorbent paper (Whatman -17). Time of capillary suction, expressed in seconds, indicates the transfer of a filtrate layer between designated circles with a diameter of 32 and 45 mm (Fukas-Płonka et al., 2013; Wolny, 2005). The measurement enables to mark changes in sludge filtration abilities, mainly in the conditioning process, and is used more often when selecting and determining doses of chemical substances (Fukas-Płonka et al., 2013; Wolny, 2005).

The tests of vacuum filtration enable to determine the speed of sludge filtration given a particular vacuum value, specific resistance of filtration, compressibility factor and the degree of filtrate contamination. The most important item characterizing vacuum filtration is specific resistance.

Dewatering by means of centrifugation has been carried out with the use of MPW-351 centrifuge for various speeds and times. Gravitational concentration of sludge, on the other hand, has been tested in cylinders with a capacity of 1000 cm^3. After a defined time period (5, 15, 30, 60 and 90 minutes), volume reading at the level of separating layer of concentrated sludge (at the border between sewage sludge and water) has been carried out.

Ultrasonic disintegration is considered as one of sludge conditioning methods prior to dewatering (Podedworna and Umiejewska, 2008). The main aim of disintegration is to destroy the original structure of sludge. Such deep integration in the structure of sludge enables to release the formation water and even the water bound biologically. The role of ultrasonic waves is to improve the process of sludge dewatering in the process of vacuum filtration.

The tests have been carried out with the use of SONICS ultrasonic disintegrator for samples with a volume of 100 m^3, power ranging from 200 to 500 W and frequency of 20 kHz. The times of ultrasonic treatment and vibration amplitude have been determined experimentally for each sample of sludge individually. Particular sewage sludge has been exposed to various concentration substances, mainly Praestol, in order to determine an optimal dosage of flocculant and its type, by means of the method of coagulation.

DISCUSSION OF RESULTS

Designations of sludge from I to IV has been adopted on account of their origin – plants with increasing values of population equivalents:

- Sludge I from a sewage treatment plant with population equivalents = 907;
- Sludge II from a sewage treatment plant with population equivalents = 1090;
- Sludge III from a sewage treatment plant with population equivalents = 1120;
- Sludge IV from a sewage treatment plant with population equivalents = 3491;
- Sludge V from a sewage treatment plant with population equivalents = 5042.

Each sewage treatment plant produces sludge with differing physical and chemical properties, but fundamental physical properties, namely colour, smell and structure of analyzed sludge were similar to each other (see Table 1). The common feature of the analyzed sludge samples was their very high or high hydration, amounting to 96.29 – 99.90 and a high content of organic compounds. The percentage of organic substances varied from 54.30 to 80.49 and is very similar to typical range of organic substances content in a residual sewage sludge, which in literature ranges from 55 to 80%. The pH values were similar to the neutral value and came within the scope of 6–7, which is typical for raw bio film (Bień and Wystalska, 2011). On the basis of time of capillary suction, which is between 25.21 – 187.38 s, good dewatering properties are distinguished in four of five raw bio films.

For the collected sludge, a series of test in various options of employed dewatering, concentration and conditioning processes have been carried out. Due to the fact that the processes carried out did not affect the color, odor, content of organic and mineral substances in the next stage of the research, parameter determinations were

Table 1. Physico-chemical characteristics of the analyzed sewage sludge

Oznaczenie	Unit	Sewage sludge I	Sewage sludge II	Sewage sludge III	Sewage sludge IV	Sewage sludge V
Colour	-	brown gray	dark brown	black	brown	black
Smell	-	specific	putrid	putrid	specific	putrid
Structure (form)	-	fluid	flocculent with a tendency to delaminate	flocculent with a tendency to delaminate	fluid	fluid. uniform
pH	-	6.59	6.39	6.29	7.8	6.33
Hydration	%	96.29	99.13	98.43	99.90	98.55
Dry residue	%	3.71	0.87	1.57	0.10	1.45
Organic matter	% d.m.	66.82	73.25	80.49	54.30	78.30
Mineral matter	% d.m.	33.18	26.75	19.51	45.70	21.70
CST	S	960.24	91.93	78.24	25.21	187.38
Resistivity	m/kg	$2.75 \cdot 10^{11}$	-	-	-	-

made, which are the key in assessing the susceptibility of sediments to the degree of their dehydration. Prior to the commencement of the proper test, the time of ultrasonic treatment and vibration amplitude in ultrasonic disintegration of analyzed sludge was determined and an optimal dosage of flocculant was determined experimentally. The results of test of processes for particular sludge samples have been presented in Table 2. The results presented in the paper are average values obtained from min. three measurements.

In the case of all analyzed sludge (I-V), the impact of the employed treatment processes on the reduction of hydration is visible. The best results were obtained in the case of sludge 1, in particular when vacuum filtration was used, in the course of which the content of dry weight was at the level of 28.45%. Equally good effects were obtained by means of ultrasonic disintegration and a combination of those two processes. What is more, in the case of sludge I, the test of sludge dewatering on a filter confirms a positive effect of disintegration on sludge dewatering. After submitting it to conditioning by means of ultrasonic field energy, the volume of obtained sludge filtrate was after some time significantly greater than the volume of filtrate obtained from raw sludge. Under the influence of the obtained ultrasounds, the time of capillary suction was reduced by approximately 33%. In parallel to those tests, vacuum filtration trials with the use of negative pressure amounting to 60 kPa were carried out. It was shown that the analyzed sludge filtered quicker after a prior conditioning by means of ultrasounds and when pressure was higher. It was also observed that resistance of filtration decreased along with the rise in negative pressure.

In the case of sludge II, the disintegration process, as a result of ultrasonic waves, has significantly influenced the structure and properties of sludge, contributing to an increase in capillary suction time. Its influence on decreasing sludge ability to dewatering in the course of vacuum filtration has also been observed.

In the determined time, the volume of filtrate originating from raw sludge was significantly greater than in the case of the filtrate produced from the sludge which underwent disintegration (Figure 1). Considering the different duration of disintegration, the best effect was obtained for the process carried out in 1 minute. By using disintegration and then centrifuging, one can observe that the most favourable conditions for this combination occurred after 5 minutes of disintegration in a trial in which centrifuging was done with the following parameters: 5000 revolutions per minute and the length of process amounting to 20 minutes. The lower rotational speed and centrifuging time, the softer sludge cake become, which caused a noticeably worse water clarity. When centrifuging at higher speeds, sludge became denser. The concentration substance used for sludge II also caused significant stratification of sludge and improved its concentration ability.

The results of dewatering, carried out in a solid bowl centrifuge for sludge III and V, were obtained successfully and approximate to each other. A linear decline in sludge weight, which was obtained after centrifuging, along with an increasing rotational speed has been observed. Trials were carried out with the parameters considered as optimal were characterized by the best separation of phases. Cake was more dewatered and sludge water clearer. The values of obtained capillary suction times for sludge, after a disintegration lasting

Table 2. The comparison of the results of processes carried out for sewage sludge of rural and municipal sewage treatment plants

The origin and type of sewage sludge		Hydration [%]	Dry residue [%]	CST [s]
Sewage sludge I	Raw	96.29	3.71	960.24
	After dehydration in the centrifuge 5000 rpm, centrifugation time 20 min.	88.57	11.43	-
	After dehydration in a centrifuge with the addition of a Praestol flocculant	88.01	11.99	-
	After dehydration in a centrifuge after ultrasonic disintegration and addition of Praestol flocculant	78.17	21.83	-
	After ultrasonic disintegration in time: 20 s	78.20	21.80	646.82
	After vacuum filtration	71.55	28.45	-
	After vacuum filtration and ultrasonic disintegration (20s)	74.59	25.41	-
Sewage sludge II	Raw	99.13	0.87	91.93
	After ultrasonic disintegration in time: 1 min 2 min 5 min	-	-	2264.45 1711.01 2470.49
	After dehydration in the centrifuge 5000 rpm, centrifugation time: 2 min 5 min 7 min 10 min 20 min	91.65 90.15 89.90 89.51 88.90	8.35 9.85 10.10 10.41 11.10	-
	After dehydration in the centrifuge 2500 rpm, centrifugation time: 2 min 5 min 7 min 10 min 20 min	93.45 92.63 92.51 92.05 91.40	6.55 7.37 7.49 7.95 8.60	-
	After dehydration in the centrifuge 2000 rpm. centrifugation time: 5 min 7 min 10 min 20 min	93.33 93.13 92.79 92.14	6.67 6.87 7.21 7.86	-
	After ultrasonic disintegration and centrifugation (5000 rpm for 20 min), sonication time: 1 min 2 min 5 min	91.35 91.51 89.34	8.65 8.49 10.66	-
	After ultrasonic disintegration and centrifugation (2500 rpm for 20 min), sonication time: 1 min 2 min 5 min	94.31 94.61 91.96	5.69 5.39 8.04	-
	After concentrating the Praestol flocculant	98.05	1.95	369.84
Sewage sludge III	Raw	98.43	1.57	78.24
	After ultrasonic disintegration in time: 10 s	-	-	8521.72
	After concentrating the flocculant	96.73	3.27	-
Sewage sludge IV	Raw	99.90	0.10	25.21
	After draining in a centrifuge for 10 min, rpm: 5000 4000 3000 2000 1000	94.91 95.28 95.70 96.32 96.71	5.09 4.72 4.30 3.68 3.29	-
	After dehydration in the centrifuge 5000 rpm, centrifugation time: 5 min 10 min 25 min 30 min	94.30 94.89 94.10 90.12	5.70 5.11 5.90 9.88	-
	After draining in a centrifuge for 10 min (5000 rpm) with the addition of Flopan flocculant in an amount: 2.5 cm³ 1.25 cm³ 0.25 cm³	96.90 95.82 93.23	3.10 4.18 6.77	-
Sewage sludge V	Raw	98.55	1.45	187.24
	After ultrasonic disintegration in time: 10 s	-	-	1800.29
	After concentrating the flocculant	97.68	2.32	-

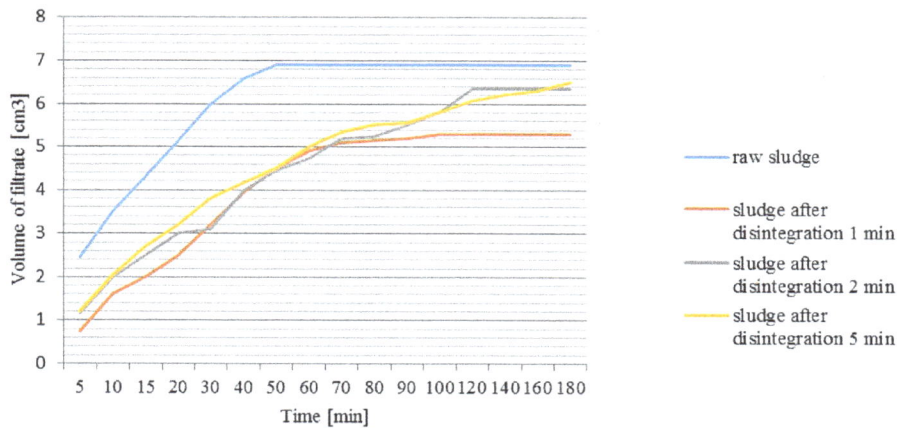

Fig. 1. Changes in the volume increase of sediment filtrate after ultrasonic disintegration

10 s (sludge III and V) are a highly visible sign of positive effect of ultrasonic field on the increase of susceptibility of sludge to dewatering.

In the case of sludge IV the best effects, namely lowering hydration by approximately 9.8%, were obtained by applying centrifuging with the following parameters: 5000 revolutions per minute, centrifuging time: 30 minutes. Shortening the centrifuging time to 10 minute, while keeping the same value of revolutions per minute and adding flocculant at the same time, enabled dewatering sludge by approximately 6.7%.

CONCLUSIONS

Describing the characteristics of sludge help to determine which treatment processes will be used and under what conditions. Each interference in the structure of sewage sludge influences its physical and chemical properties, which entails change of their characteristics that in turn can influence their management options.

Taking a more environmentally sound sewage sludge management – namely using it for agricultural or natural processes – and greater usefulness of sludge originating from small sewage treatment plants for those purposes, it is important to submit it to proper treatment once they are produced. The condition for choosing the right manner of sewage sludge management is its proper preliminary preparation, including dewatering, conditioning etc. The rationale for carrying out the test in various combinations of applied processes is the differing nature of sludge. If sludge is more difficult to dewater, using conditioning prior to concentration and initial dewatering is appropriate.

The research applied a number of different variants of the processes carried out, i.e. with different doses of flocculant, different time of sonication, variable revolutions and time of centrifugation. In the majority of cases, the employed processes of sludge preparation gave the intended results in the form of easier dewatering, concentration, as well as changes in some parameters. To a large extent, the obtained differences in results were due to diversified sludge properties.

1. It is confirmed that ultrasonic field as a physical method for modification of sewage sludge is a factor intensifying the dewatering processes. In the majority of cases, the tests have shown better results of dewatering of sludge submitted to ultrasonic field in comparison to raw sludge.

2. Susceptibility of sludge originating from various wastewater treatment plants to ultrasonic disintegration is very different; however, it must be stated that applying ultrasounds to sludge causes changes in their structure and influences the change in other indicators, including final hydration and capillary suction time. The lowest value of the final sludge hydration and its largest reduction was obtained for sediments subjected to vacuum filtration and ultrasonic disintegration.

3. The use of ultrasonic disintegration and centrifugation gives better results in changing hydration and dry matter than using only one of the methods.

4. Sonication of the studied sediments causes a significant increase in the CSK values. The sludge submitted to initial conditioning by means of the energy of ultrasound field release water quicker and the final amount of filtrate

is greater than in the case of raw sludge. However, it must be remembered that the time of ultrasonic treatment and vibration amplitude has to be determined separately for each type of sludge. Excessive dose of ultrasounds can have a reverse effect than expected, which in consequence can worsen the sludge dewatering.

5. Centrifuging as a manner of mechanical concentration is justified for the analyzed sludge. It makes a significant contribution in decreasing the level of water in sludge, and therefore, increases the level of density. Precise parameters of centrifuging, namely centrifuging time and rotational speed, are of key importance.

REFERENCES

1. Bień B. 2017. The Effect of PIX 123 and Polyelectrolyte Zetag 8160 on the Conditioning and Dewatering of Sewage Sludge. Engineering and Protection of Environment, 20(2), 155–174.

2. Bień B., Bień J.D. 2014. Use of inorganic coagulants and polyelectrolytes to sonicated sewage sludge for improvement of sludge dewatering. Desalination and Water Treatment, Science and Engineering, 52/19–21, 3767–3774.

3. Bień J., Neczaj E., Worwąg M., Grosser A., Nowak D., Milczarek M., Janik M. 2011. Directions of sludge management in Poland after 2013, Engineering and Environmental Protection, 14(4), 375–384 (in Polish).

4. Bień J., Wystalska K. 2011. Sewage sludge. Theory and practice. Publishing House of Czestochowa University of Technology, Częstochowa (in Polish).

5. Czekała J. 2013. Sewage sludge dewatering – selected issues, Water supply – Sewerage, 12, 38–40 (in Polish).

6. Fukas-Płonka Ł., Janik M., Płonka I. 2013. Physical analysis of sewage sludge as a criterion of freeness. In: Manczarski P. Complex management of waste management, Poznań, 711–724 (in Polish).

7. Grobelak A., Stępień W., Kacprzak M. 2016. Sewage sludge as a component of soil fertilizers and substitutes. Ecological Engineering, 48, 52–60 (in Polish).

8. Krzywy E., Wołoszyk C., Możdżer E. 2015. Possibility of Producing Granulated Organic-Mineral Fertilizers from Some Municipal and Industrial Wastes. Chemik, 69(10), 684–697.

9. Mahmoud A., Olivier J., Vaxelaire J., Hoadley F.A. 2011. Electro-dewatering of wastewater sludge: influence of the operating conditions and their interactions effects. Water Research, 45, 2795–2810.

10. Ministry of the Environment. 2016. National Waste Management Plan 2022, Warsaw (in Polish).

11. National Water Management Authority. 2016. Report on the implementation of the National Program for Municipal Sewage Treatment in the years 2014–2015. Warsaw (in Polish).

12. Obbard J. 2001. Ecotoxicological assessment of heavy metals in sewage sludge soils, Applied Geochemistry, 16(11), 1405–1411.

13. Podedworna J., Umiejewska K. 2008. Sewage sludge technology. Publishing House of Warsaw University of Technology. Warsaw (in Polish).

14. Polish Standards. Normalization Publishers, Warsaw.

15. Regulation of the Minister of Economy of 16 July 2015 on the admission of waste for landfill. Journal of Laws, 2015, item 1277 (in Polish).

16. Regulation of the Minister of the Environment of 6 February 2015 on municipal sewage sludge. Journal of Laws, 2015, item 257 (in Polish).

17. Regulation of the Minister of the Environment of 9 December 2014 on the waste catalog. Journal of Laws, 2014, item 1923 (in Polish).

18. Regulation of the Minister of Agriculture and Rural Development of 18 June 2008 on the implementation of certain provisions of the Act on fertilizers and fertilization. Journal of Laws, 2008, No. 119, item 765 (in Polish).

19. Skarżyński S. 2017. Review of devices used in conditioning of sewage sludge with ultrasonic waves. Gas, Water and Sanitary Technique, 06/2017, 262–264 (in Polish).

20. Sridhar P., Puspendu B., Song Y., LeBlanc R.J., Tyagi R.D., Surampalli R.Y. 2011. Ultrasonic pretreatment of sludge: A review. Ultrasonics Sonochemistry, 18, 1–18.

21. Statistical Yearbooks of the Central Statistical Office. 2017. Environmental Protection 2016. Warsaw.

22. Świsłowski M. 2016. Sewage sludge utilization by granulation with lime. Gas, Water and Sanitary Technique, 03/2016, 106–110 (in Polish).

23. The Act of 14 December 2012 on waste. Journal of Laws, 2012, item 21 with later amendments (in Polish).

24. The Act of 10 July 2007 on fertilizers and fertilization. Journal of Laws, 2007, No. 147, item 1033 with later amendments (in Polish).

25. Tomczak-Wandzel R., Mędrzycka K., Cimochowicz M. 2009. The Effect of ultrasonic disintegration on sewage sludge anaerobic digestion. In: Ozonek J., Pawłowska M. (Eds). Polish Environmental Engineering five years after joining the European Union. 331 – 337 (in Polish).

26. Trojanowska K., Myszograj S. 2017. Ultrasonic disintegration of sewage sludge in GSD technolo-

gy – operating experience. Gas, Water and Sanitary Technique, 03/2017, 110–114 (in Polish).

27. Wiater J., Butarewicz A. 2014. Ways of using sludge from Sewage Treatment Plant in Białystok. Engineering and Environmental Protection, 17(2), 281–291 (in Polish).

28. Wolny L. 2005. Ultrasonic support of the process of preparation of sewage sludge for drainage. Publishing House of Czestochowa University of Technology, Częstochowa (in Polish).

29. Zawieja I., Wolny L., Wolski P. 2010. Influence on the modification of food industry excess sludge structure on the effectiveness increase of the anaerobic stabilization process. Polish Journal of Environmental Studies, 2, 261–267.

30. Zielewicz E. 2016. Disintegration of sludge in the context of increased biogas production. Gas, Water i Sanitary Technique, 02/2016, 69–75 (in Polish).

31. Zielewicz E., Tytła M. 2015. Effects of ultrasonic disintegration of excess sludge obtained in disintegrators of different constructions. Environmental Technology, 36(17), 2210–2216.

14

The Influence of Electrical Current Density and Type of the External Source of Carbon on Nitrogen and Phosphorus Efficiency Removal in the Sequencing Batch Biofilm Reactor

Izabella Kłodowska[1], Joanna Rodziewicz[1*], Wojciech Janczukowicz[1]

[1] University of Warmia and Mazury in Olsztyn, Faculty of Environmental Sciences, Department of Environment Engineering, Warszawska 117a, 10-719 Olsztyn, Poland

[*] Corresponding author's e-mail: joanna.rodziewicz@uwm.edu.pl

ABSTRACT

This work presents the results of a study on the effect of electrical current density (53, 105, 158 and 210 mA/m^2), type of the external source of carbon (citric acid, potassium bicarbonate), and C/N_{NO3} ratio (0.5, 1.0 and 1.5) on the effectiveness of nitrogen and phosphorus removal from synthetic wastewater with physicochemical parameters typical of municipal sewage subjected to bio-treatment in the highly efficient system for organic compounds removal ensuring efficient course of the nitrification process. The denitrification efficiency was found to depend on the type and dose of carbon and on the electrical current density. Higher values of this parameter were determined in the reactor with citric acid than in one with potassium bicarbonate used as carbon sources. Total phosphorus was removed in the processes of electrocoagulation and biomass growth. Higher efficiency of dephosphatation was achieved in the reactor with electrical current passage than in the reactor without it. The type of carbon source had little effect on the dephosphatation efficiency. The use of electrical current density of 210 mA/m^2 and citric acid as a carbon source with C/N=1.5 allowed achieving 87.61($\pm$1.6)% efficiency of denitrification and 97.69($\pm$2.1)% efficiency of dephosphatation.

Keywords: bio-electrochemical reactor, external source of carbon, citric acid, potassium bicarbonate, denitrification, dephosphatation

INTRODUCTION

The discharge of wastewater containing biogenic substances to natural waters contributes to an excessive growth of algae, which leads to the eutrophication of water bodies and affects the quality of drinking water resources. Nitrates constitute a major source of contaminants in the underground and surface waters, and their presence in drinking water poses risk to the human health and to other living organisms [Carrera et al., 2004; Lee et al., 2006]. Low effectiveness of conventional methods for water and wastewater treatment has recently led to the development of novel technologies for the removal of biogenic substances. These compounds may be neutralized in biological and physicochemical processes. The biological processes are cost-effective, relatively easy to conduct, highly stable and reliable, but more time-consuming and slower compared to the physicochemical methods [Hiscock et al., 1991; Ghafari et al., 2008]. Therefore, attempts are undertaken to intensify the processes of biological denitrification through ensuring better contact between microorganisms and nitrates. These attempts are mainly focused on the systems with a biofilm which are convenient for denitrification owing to their cohesiveness. Compared to the reactors with suspended sludge and fluidized beds, the reactors with a solid filling are characterized by a greater simplicity of handling, lower volume, lesser sludge production, higher stability and resistance to a shock load [Zhu and Chen, 2002].

Electrical current passage through a sequencing batch reactor with a biofilm immobilized on a carrier in the form of disks (SBBR) enables coupling the processes of biological and physico-chemical treatment. In a reactor of this type, the nitrogen compounds will be removed primarily in the process of autotrophic (hydrogenotrophic) denitrification, whereas the phosphorus compounds – in the process of coagulation [Kłodowska et al., 2013]. Research works mainly provide the results from the experiments conducted in the systems with separate various reactors for denitrification and electrocoagulation processes [Karanasios et al., 2010; Feng et al., 2013; Shalaby et al., 2014; Kuokkanen et al., 2015].

The authors of this manuscript have demonstrated in their earlier works the feasibility of using an SBBR type bio-electrochemical reactor for the treatment of wastewater with high concentrations of nitrates and phosphorus and with low concentrations of organic compounds at selected densities of electrical current and unitary doses of organic and inorganic substrates [Kłodowska et al., 2013; Kłodowska et al., 2014; Kłodowska et al., 2016].

The presented study was aimed at determining the effect of electrical current density, type of carbon source and C/N_{NO3} ratio on the efficiency of nitrogen and phosphorus compounds removal in the bio-electrochemical SBBR from the wastewater with physicochemical parameters typical of municipal sewage subjected to prior bio-treatment in the system, enabling highly efficient removal of organic compounds and highly efficient nitrification.

METHODS

The experiments were conducted simultaneously in vertical sequencing batch biofilm reactors (SBBR) with the volume of 3.0 L each (active volume – 2.0 L), under anaerobic conditions. A set of 12 disks made of stainless steel with the diameter of 0.10 m and total surface of 0.19 m^2 was mounted in each reactor. The distance between disks was 5 mm. They were mounted coaxially on a vertical shaft rotating with the speed of 10 rpm; their submersion rate was 100% (Fig. 1).

The experiment was conducted under the following conditions: without the passage of electrical current (reactors: R_0, R_{CA} and R_{PB}), and with the passage of electrical current (reactors: R_{H2}, R_{CA+H2} and R_{PB+H2}). In the control reactor (R_0), without electrical current flow and without external source of carbon, the synthetic wastewater was subjected to bio-treatment. In another two

reactors (R_{CA} and R_{PB}) without electrical current flow, citric acid and potassium bicarbonate, respectively, were used in the concentrations ensuring the C/N_{NO3} ratios of 0.5, 1.0 and 1.5. In the reactors with the passage of electrical current, no external source of carbon was introduced in the reactor R_{H2} or in R_0, whereas citric acid and potassium bicarbonate were fed as carbon sources to the reactors R_{CA+H2} and R_{PB+H2}, respectively. Wastewater retention time was 24 hours in each reactor. The carbon dose applied ensured the C/N_{NO3} ratio of 0.5, 1.0 and 1.5 in wastewater fed to reactors. When computing the doses of organic and inorganic carbon, consideration was given to the carbon concentration in crude wastewater after dissolution of enriched broth. The broth was added to wastewater to achieve a desired COD value being typical of the municipal sewage treated in the system intended for the aerobic biological removal of organic compounds and for nitrification (e.g. in four-stage biological disk contactor). In the reactors without the external substrate (R_0 and R_{H2}), the wastewater C_{org}/N_{NO3} was at 0.47. In the reactors with electrical current passage, the gas-

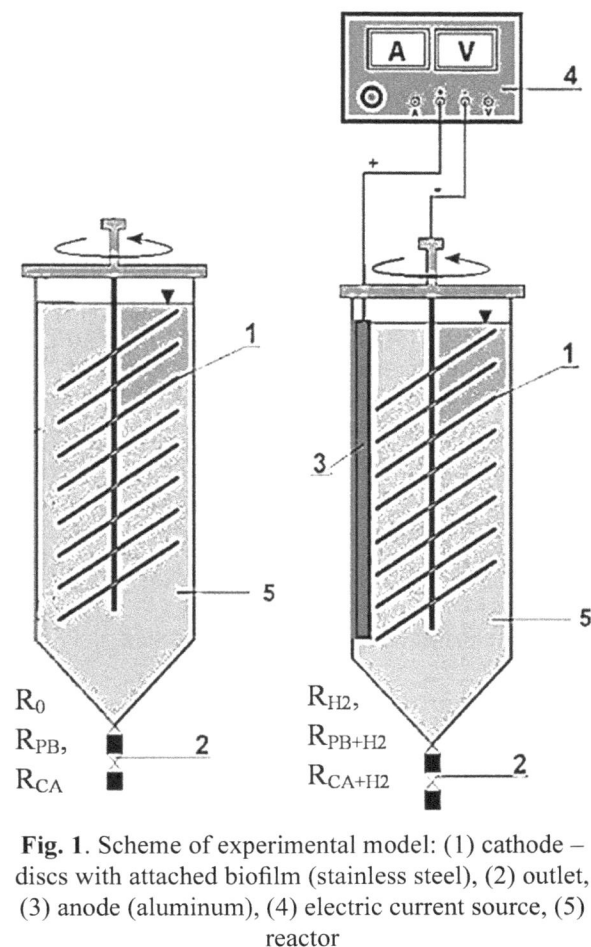

Fig. 1. Scheme of experimental model: (1) cathode – discs with attached biofilm (stainless steel), (2) outlet, (3) anode (aluminum), (4) electric current source, (5) reactor

eous hydrogen emitted from disks was the source of energy in the denitrification process. Hydraulic retention time was 24 h in each reactor. Before the exact experiment, the reactors were adapted for 3 months until appropriate structure of biofilm and stable concentration of nitrogen compounds in the effluent have been achieved.

In the reactors with electrical current passage, provided by laboratory feeders – Programmable DC Power Supply – HANTEK PPS 2116A – (0–5A) (0–32V) and MANSON DC Power Supply – DPD 3030 (0–3A, 0–30V), the disks with immobilized biofilm served as the cathode, whereas an aluminum plate with a total surface area of 0.033 m^2 served as the anode. The cathode and the anode were connected to the laboratory feeder to ensure the desired density of electrical current, i.e. 53, 105, 158 and 210 mA/m^2 (current intensity was 10, 20, 30 and 40 mA, current voltage ranged from 3.0 to 5.0V). The experiment was conducted for 16 weeks under the conditions of controlled pH 7.5–8.0 (by NaOH or HCl application).

WASTEWATER CHARACTERISTICS

The experiments were conducted with synthetic wastewater with physicochemical parameters typical of municipal sewage subjected to bio-treatment in the system enabling highly efficient removal of organic compounds and highly efficient nitrification. Wastewater was characterized by high concentrations of nitrate nitrogen and total phosphorus, and by a low concentration of organic matter expressed by the COD value. In order to prepare the synthetic wastewater, the following compounds were dissolved in 2.0 L of tap water: $NaNO_3$, KH_2PO_4 enriched broth (0.08 g/L) BIOCORP PS 110, KCl (0.021 g/L), $MgSO_4 \cdot 7H_2O$ (0.308 g/L), and $CaCl_2$ (0.021 g/L). Mean values of the physicochemical parameters of the crude wastewater were as follows:

- concentration of nitrate nitrogen – 50.68 ($\pm$1.61) mgN_{NO3}/L,
- concentration of nitrite nitrogen – 0.0 mgN_{NO2}/L,
- concentration of ammonium nitrogen – 0.0 mgN_{NH4}/L,
- concentration of total phosphorus – 5.16 ($\pm$0.20) mgP/L,
- concentration of COD_{Cr} –70.30 ($\pm$10) mgO_2/L,
- electrolytic conductance – 1.74 ($\pm$0) mS/cm,
- temperature – 25.3 ($\pm$0) °C.

Analytical methods

The samples (1.0 L) were collected for analyses once a day. Afterwards, the reactors were emptied (1.0 L) and filled with wastewater (2.0 L).

The physicochemical analyses of crude and treated wastewater included determinations of:
- concentration of nitrate nitrogen with the colorimetric method [ISO 7890–3:1988],
- concentration of nitrite nitrogen with the colorimetric method [ISO 6777:1984],
- concentration of ammonium nitrogen with the colorimetric method [PN-C-04576/04:1994],
- concentration of total phosphorus with the spectrophotometric method based on the procedure developed by Lange LCK company,
- concentration of organic compounds (COD) with the spectrophotometric method based on the procedure developed by Lange LCK company,
- electrolytic conductance using an HI 99301 conductometer by Hanna Instruments,
- pH value using a laboratory pH-meter PL – 700 AL type pH-mV-Cond-TDS-DO stirrer.
- temperature using a digital thermometer by JVTIA.

RESULTS AND DISCUSSION

In this study, we analyzed the effect of electrical current density, substrate type and C/N_{NO3} value on the efficiency of nitrogen and phosphorus compounds removal in the bio-electrochemical SBBR.

The efficiency of denitrification in the reactor without the external source of carbon (R_0) reached 10.31($\pm$2.4)%. Feeding potassium bicarbonate to the reactor (R_{PB}) enabled increasing it to: 43.67($\pm$1.8)%, 13.46($\pm$1.56)% and 15.04($\pm$0.8)% at C/N of 0.5, 1.0 and 1.5, respectively, while feeding citric acid to the reactor led to its higher values, reaching 13.46($\pm$1.82)%, 17.6($\pm$1.48)% and 19.9($\pm$1.67)% at C/N of 0.5, 1.0 and 1.5, respectively (fig. 2).

Under the conditions of passage of electrical current with density of 53 mA/m^2, the feeding of an external source of carbon caused an increase in denitrification efficiency compared to the R_{H2} reactor in which it reached 9.64($\pm$1.33)%.

Higher values of denitrification efficiency were achieved upon the use of citric acid, i.e. 48.89($\pm$1.47)%, 52.75($\pm$2.0)% and 69.43($\pm$2.22)% at C/N of 0.5, 1.0 and 1.5, respectively. In the reactors fed with potassium bicarbonate, the respective values were insignificantly lower and accounted for 44.16(±2.33)%, 47.13(±2.0)% and 53.62(±4.13)% at C/N of 0.5, 1.0 and 1.5 respectively. Increasing the electrical current density to 105 mA/m² caused a further increase in the denitrification efficiency, which was insignificantly higher in the reactors with potassium bicarbonate: 82.25(±4.67)%, 84.47(±2.01)% and 85.54(±2.25)% at C/N of 0.5, 1.0 and 1.5, respectively, than in the reactors with citric acid: 79.4(±1.2)%, 81.93(±1.46)% and 82.5(±1.47)% at C/N of 0.5, 1.0 and 1.5, respectively.

In the R_{H2} reactor, the efficiency of nitrogen compounds removal reached 20.29±(3.15)%. Further increase of electrical current density to 158 mA/m² raised the efficiency of nitrogen compounds removal in the reactors with citric acid, the final values of which reached: 81.54(±1.12)%, 83.03(±3.52)% and 83.34(±5.2)% at C/N of 0.5, 1.0 and 1.5, respectively. In the case of feeding potassium bicarbonate, denitrification efficiency decreased to 74.52(±1.57)%, 78.11(±1.59)% and 80.07(±0.7)% at C/N of 0.5, 1.0 and 1.5, respectively, and higher concentrations of nitrites and ammonium nitrogen were determined in the effluent. Conducting the treatment process only under conditions of electrical current passage, without the external source of carbon being fed to the reactors, allowed obtaining 23.2(±0.8)% efficiency of the process. In turn, the use of electrical current with a density of 210 mA/m² increased the denitrification efficiency in the reactors with citric acid to: 83.05(±1.16)%, 85.87(±2.82)% and 87.61(±1.6)% at C/N of 0.5, 1.0 and 1.5, respectively. In the reactors with potassium bicarbonate

used as the external source of carbon, the removal efficiency of nitrogen compounds was observed to further decrease while the concentrations of nitrites and ammonium nitrogen to increase in the effluent. The efficiency of denitrification obtained in these reactors reached 66.21(±0.39)%, 57.35(±0.75)% and 55.11(±2.84)% at C/N of 0.5, 1.0 and 1.5, respectively.

In the conducted experiment, the nitrogen compounds were removed mainly in the biological processes. As reported by Li et al. (2009), the electrochemical effect – namely the possibility of nitrogen removal in the process of electrochemical reduction on the surface of cathode – may be omitted at electrical current densities of 50–200 mA/m². Zhao et al. (2011) claim that the co-action of autotrophs and heterophytes may be beneficial compared to the autotrophic denitrification alone. The carbon dioxide produced during the heterotrophic denitrification may serve as an additional source of inorganic carbon to autotrophic bacteria (the so-called synergism). Lee et al. (2013) discovered the co-existence of denitrifying autotrophs and heterotrophs at C/N=0.8. In turn, Hao et al. (2013) reported the occurrence of both heterotrophic and autotrophic bacteria responsible for the nitrates removal in the reactor with even higher wastewater C/N, i.e. C/N=1.5. According to Zhao et al. (2011), an increase of C/N>1.0 enhances the growth of heterotrophs. Feeding organic substances to the bio-electrochemical system leads to the intensification of the denitrification process [Feng et al., 2013], which is confirmed by the results obtained in our study. Feeding of an external source of carbon in the form of citric acid increased the denitrification ef-

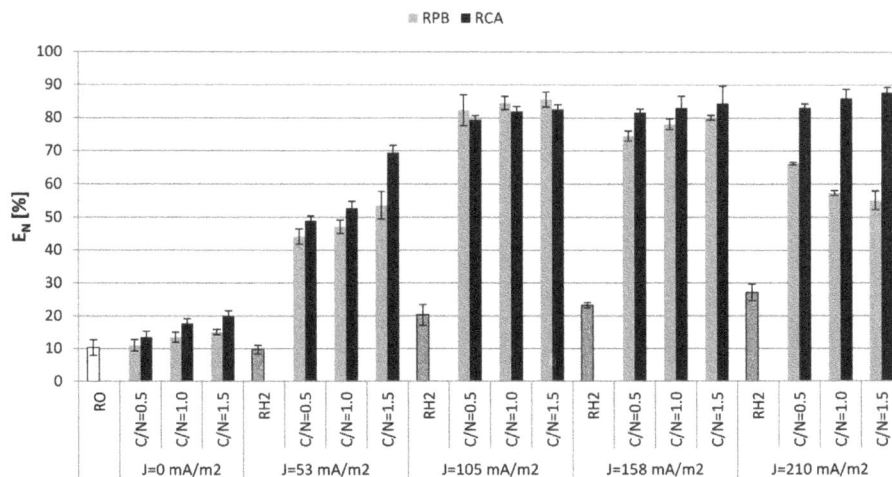

Fig. 2. The efficiency of nitrogen removal depending on electrical current density and C/N$_{NO3}$ ratio

ficiency, which was further intensified by the use of a sufficiently high density of electrical current. The carbon source deficiency may be the major factor promoting accumulation of nitrites [Zhang et al., 2012]; hence, the electrical current densities applied for in-situ hydrogen production must be kept at an appropriate level. Both, too high and too low density of electrical current would cause nitrites accumulation [Islam and Suidan, 1998]. Tong et al. [2013] successively increased the current density to 200 mA/m^2 and obtained respectively lower concentrations of nitrates in wastewater. They managed to achieve the maximum efficiency of nitrates removal equal to 99.9%. The use of electrical current with densities exceeding 200 mA/m^2, led to an increase in nitrates concentration. Raising the current density to 320 mA/m^2 caused a rapid increase in their concentration, which resulted in reduced efficiency of nitrates removal [Tong et al., 2013]. The low density of electrical current does not ensure the sufficient hydrogen production, while its high density may inhibit the activity of microorganisms and even lead to their eradication [Flora et al., 1994].

The use of an inorganic source of carbon in the form of potassium bicarbonate at lower electrical current densities, i.e. 53 and 105 mA/m^2, contributed to an increase in the nitrogen compounds removal efficiency. Ghafari et al. [2009] demonstrated that – compared to carbon dioxide – bicarbonate caused accelerated adjustment of autotrophs and more abrupt growth of denitrifiers. However, the use of higher current densities (158 and 210 mA/m^2) resulted in a decrease of nitrogen removal efficiency and an increase in the ammonium nitrogen concentration in the effluent. The accumulation of ammonium nitrogen in bioelectrochemical reactors may be caused by dissimilative reduction of nitrates to ammonium nitrogen (DNRA) [Zhang et al., 2012]. Zhao et al. [2012] came to the conclusion that ammonium nitrogen concentration decreased along with the decreasing C/N ratio. Similar results were achieved by Zhang et al. [2012] during DNRA in bioelectrochemical systems, who reported that a higher C/N value might lead to enhanced accumulation of ammonium nitrogen.

In this study, the phosphorus compounds were removed mainly in the electrocoagulation process. The dephosphatation efficiency in the reactor without the external source of carbon (R_0) reached 6.94($\pm$1.1)%. Feeding potassium bicarbonate caused no increase in its value

which reached 6.05($\pm$1.5)%, 6.24($\pm$0.85)% and 6.84($\pm$0.47)% at C/N of 0.5, 1.0 and 1.5, respectively. In the case of citric acid fed to the reactors as the source of carbon, the respective values were insignificantly higher and reached 7.12($\pm$0.9)%, 8.03($\pm$0.79)% and 9.04($\pm$0.51)% at C/N of 0.5, 1.0 and 1.5, respectively (Fig. 3).

Wastewater treatment under the conditions of electrical current passage contributed to an increase in the dephosphatation efficiency. Upon the use of electrical current with the density of 53 mA/m^2, the phosphorus removal efficiency in the R_{H2} reactor reached 71.01($\pm$1.4)%. Feeding external source of carbon to the reactors led to a further increase in the process efficiency, i.e. to 82.33($\pm$2.3)%, 84.23($\pm$1.4)% and 85.53($\pm$1.4)% at C/N 0.5, 1.0 and 1.5, respectively, in the case of citric acid, and to insignificantly lower values of 81.96($\pm$3.6)%, 83.13($\pm$2.8)% and 84.05($\pm$2.3)% at C/N 0.5, 1.0 and 1.5, respectively, in the case of potassium bicarbonate. Increasing the electrical current density to 105 mA/m^2 caused a further increase in the efficiency of phosphorus compounds removal. At this density, insignificantly higher efficiency values were also observed in the reactors fed with citric acid as the source of carbon, i.e. 89.23($\pm$2.9)%, 92.31($\pm$1.8)% and 92.88($\pm$1.18)% at C/N 0.5, 1.0 and 1.5, respectively. In the case of potassium bicarbonate, the respective values were as follows: 85.24($\pm$4.4)%, 86.37($\pm$4.3)% and 87.12($\pm$5.2)% at C/N 0.5, 1.0 and 1.5, respectively.

Only slightly lower efficiency was achieved in the R_{H2} reactor – 82.8$\pm$(2.01)%. The use of electrical current with the density of 158 mA/m^2 increased the dephosphatation efficiency in all treatment systems. In the reactors with citric acid fed as the external source of carbon, the process efficiency reached 92.31($\pm$4.35)%, 93.54($\pm$3.2)% and 95.34($\pm$2.8)% at C/N 0.5, 1.0 and 1.5, respectively, whereas in the reactor with potassium bicarbonate it was insignificantly higher and reached 90.58($\pm$4.5)%, 91.39($\pm$4.6)% and 94.09($\pm$2.7)% at C/N 0.5, 1.0 and 1.5, respectively. Conducting the treatment process only under the conditions of electrical current passage, without additional carbon feeding, enabled reaching the process efficiency of 84.2($\pm$1.68)%. The use of electrical current with a density of 210 mA/m^2 affected the successive increase in dephosphatation efficiency in all treatment systems, i.e. to 93.85($\pm$5.3)%, 96.39($\pm$2.42)% and 97.69($\pm$2.1)% at C/N 0.5, 1.0 and 1.5, respectively, in the case of

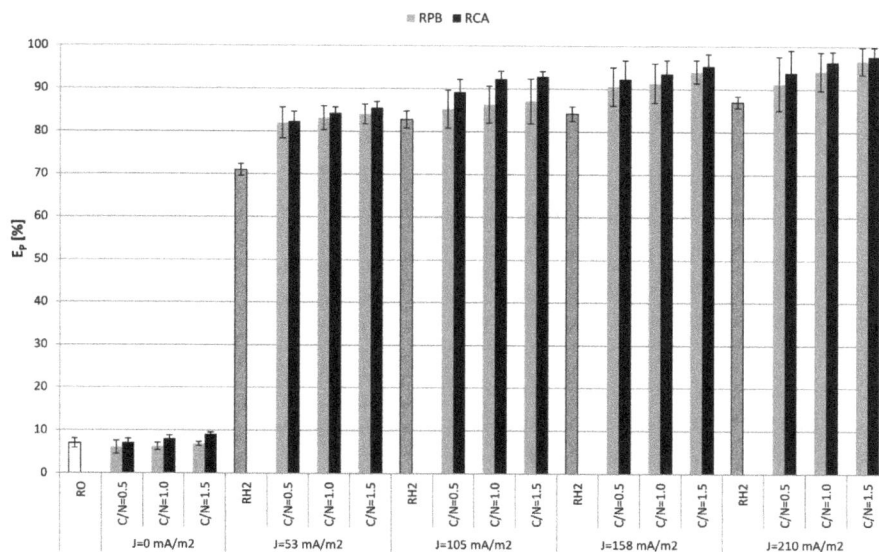

Fig. 3. The efficiency of phosphorus removal depending on the electrical current density and C/N_{NO3} ratio

the reactors with citric acid, and to 91.35(±6.3)%, 94.23(±4.5)% and 96.68(±3.2)% at C/N 0.5, 1.0 and 1.5, respectively, in the case of the reactors with potassium bicarbonate. In turn, the dephosphatation efficiency determined in the RH_2 reactor reached 86.99(±1.37)%.

The use of aluminum anode in the conducted experiment enabled simultaneous removal of phosphorus in the electrocoagulation process and in the process of biofilm growth. The effectiveness of dephosphatation depends on the quantity of aluminum being formed which is, in turn, affected by the reaction duration and electrical current density [Zaroual et al., 2006]. İrdemez et al. [2006] reported the aluminum electrode to be more suitable for dephosphatation conducted via the electrocoagulation method than the iron electrode, considering the rate and efficiency of phosphorus compounds removal. Behbahani et al. [2011] investigated the effect of the type of electrode on the dephosphatation efficiency and observed enhanced phosphorus removal along with increasing electrical current density for both aluminum and iron electrodes; however, obtained higher process efficiency in the case of the aluminum electrode. Besides, the use of the iron electrode led to greater turbidity of the solution being treated, compared to the aluminum electrode. The results of an experiment conducted by Lacasa et al. [2011b] demonstrated the efficiency of electrocoagulation with iron electrodes to be significantly affected by the density of electrical current and to increase considerably at its lower densities. In the case of the aluminum electrode,

the current density had a lesser effect on electrocoagulation; a slight increase in its efficiency was only observed at lower current densities. Shalaby et al. [2014] obtained the phosphorus compounds removal efficiency from 85 to 95% by increasing electrical current density from 1.13 to 4.54 mA/ cm^2 (11.34–45.40 A/m^2). However, they used higher electrical current density and shorter retention time than in our study. The electrical current passage through electrodes results in their dissolution and formation of Al^{3+} ions which bind with the OH^- ions forming aluminum hydroxides which are responsible for the electrocoagulation efficiency. At higher intensities of electrical current, the number of Al^{3+} ions formed upon anode dissolution increases accordingly to the Faraday's law. This results in an increased quantity of aluminum hydroxide which adsorbs a greater part of phosphorus compounds onto its surface. Increasing the electrical current density above the optimal value does not enhance the efficiency of contaminants removal, even despite sufficient amount of metal hydroxide floccules for contaminants precipitation [Shalaby et al., 2014].

CONCLUSIONS

1. The efficiency of nitrogen and phosphorus compounds removal was the lowest in the R_0 reactor without external source of carbon and without electrical current passage.
2. Feeding potassium bicarbonate and citric acid to reactors under the conditions of electrical

current passage contributed to a significant increase in the denitrification efficiency.

3. Citric acid turned out to be a better source of carbon in the process of nitrogen compounds removal.

4. Feeding organic carbon to reactors enabled the removal of nitrogen compounds in a simultaneous process of heterotrophic and autotrophic denitrification with gaseous hydrogen – produced in the electrolysis process – as a source of energy.

5. Compounds of phosphorus were removed in the processes of electrocoagulation and biofilm growth. The highest efficiency of dephosphatation was achieved under the conditions of electrical current passage and citric acid dosage.

6. The use of electrical current with a density of 210 mA/m² and citric acid as a source of carbon with C/N=1.5, allowed achieving 87.61($\pm$1.6)% efficiency of denitrification and 97.69($\pm$2.1)% efficiency of dephosphatation.

Acknowledgments

This study was financed under Project No. 18.610.008–300 of the University of Warmia and Mazury in Olsztyn, Poland. The project was also funded by the National Science Centre, Poland (the decision nr DEC-2013/09/N/ST8/04163).

REFERENCES

1. Behbahani M., Alavi Moghaddam M. R., Arami M. 2011. A comparison between aluminum and iron electrodes on removal of phosphate from aqueous solutions by electrocoagulation process. Int. J. Environ. Res., 5, 403–412.

2. Carrera J., Vicent T., Lafuente J. 2004. Effect of influent COD/N ratio on biological nitrogen removal (BNR) from high-strength ammonium industrial wastewater. Process Biochem., 39, 2035–2041.

3. Feng H., Huang B., Zou Y., Li N., Wang M., Yin J., Cong Y., Shen D. 2013. The effect of carbon sources on nitrogen removal performance in bio-electrochemical systems. Bioresource Technol., 128, 565–570.

4. Flora J.R.V., Suidan M.T., Islam S., Biswas P., Sakakibara Y. 1994. Numerical modeling of a biofilm-electrode reactor used for enhanced denitrification. Water Sci. Technol., 29, 517 – 524.

5. Ghafari S., Hasan M., Aroua M.K. 2008. Bio–electrochemical removal of nitrate from water and wastewater – a review. Bioresour. Technol., 99, 3965–3974.

6. Ghafari S., Hasan M., Aroua M.K. 2009. Effect of carbon dioxide and bicarbonate as inorganic carbon sources on growth and adaptation of autohydrogenotrophic denitrifying bacteria. J. Hazard. Mater., 162, 1507–1513.

7. Hao R.X., Li S.M., Li J.B., Meng C.C. 2013. Denitrification of simulated municipal wastewater treatment plant effluent using a three-dimensional biofilm-electrode reactor: operating performance and bacterial community. Bioresour. Technol., 143, 178–186.

8. Hiscock K.M., Lloyd J.W., Lerner D.N. 1991. Review of natural and artificial denitrification of groundwater. Water Res., 25, 1099–1111.

9. Islam S., Suidan M.T. 1998. Electrolytic denitrification: long term performance and effect of current intensity. Water Res., 32, 528–536

10. İrdemez Ş., Yildiz Y. Ş., Tosunoğlu V. 2006. Optimization of phosphate removal from wastewater by electrocoagulation with aluminum plate electrodes. Sep. Purif. Technol., 52(2), 394–401.

11. Karanasios K.A., Vasiliadou I.A., Pavlou S., Vayenas D.V. 2010. Hydrogenotrophic denitrification of potable water: a review. J. Hazard. Mater., 180(1–3), 20–37.

12. Kłodowska I., Rodziewicz J., Janczukowicz W. 2014. Removal of nitrogen compounds in the process of autotrophic denitrification in a Sequencing Batch Biofilm Reactor (SBBR). Polish J. Nat. Sci., 29 (4), 359–369.

13. Kłodowska I., Rodziewicz J., Janczukowicz W., Cydzik-Kwiatkowska A., Parszuto K. 2016. Effect of citric acid on the efficiency of the removal of nitrogen and phosphorus compounds during simultaneous heterotrophic-autotrophic denitrification (HAD) and electrocoagulation. Ecol. Eng., 95, 30–35.

14. Kłodowska I., Rodziewicz J., Janczukowicz W., Filipkowska U. 2013. Effect of electrochemical process on the outflow from the reactor with immobilized biofilm (in Polish). Annual Set The Environment Protection., 15, 1952–1964

15. Kuokkanen V., Kuokkanen T., Rämö J., Lassi U., Roininen J. 2015. Removal of phosphate from wastewaters for further utilization using electrocoagulation with hybrid electrodes – Techno-economic studies. J. Water Process Eng., 8, 50–57.

16. Lacasa E., Cañizares P., Sáez C., Fernandez F.J. Rodrigo M.A. 2011. Electrochemical phosphates removal using iron and aluminium electrodes. Chem. Eng. J., 172, 137–143.

17. Lee S., Maken S., Jang J.-H., Park K., Park J.-W. 2006. Development of physicochemical nitrogen removal process for high strength industrial wastewater. Water Res. 40, 975–980.

18. Lee D.J., Pan X., Wang A., Ho K.L. 2013. Facul-

tative autotrophic in denitrifying sulfide removal granules. Bioresour. Technol., 132, 356–360.

19. Li M., Feng C.P., Zhang Z.N., Lei X.H., Chen R.Z., Yang Y.N., Sugiura N. 2009. Simultaneous reduction of nitrate and oxidation by-products using electrochemical method. J. Hazard. Mater., 171, 724–730.

20. Shalaby A., Nassef E., Mubark A., Hussein M. 2014. Phosphate removal from wastewater by electrocoagulation using aluminium electrodes. Am. J. Environ. Eng. Sci., 1(5), 90 – 98.

21. Tong S., Zhang B., Feng C., Zhao Y., Chen N., Hao C., Pu J., Zhao L. 2013. Characteristics of heterotrophic/biofilm-electrode autotrophic denitrification for nitrate removal from groundwater. Bioresour. Technol., 148, 121–127.

22. Zaroual Z., Azzi M., Saib N., Chainet E. 2006. Contribution to the study of electrocoagulation mechanism in basic textile effluent. J. Hazard. Mater., 131, 73–78.

23. Zhang J.M., Feng C.P., Hong S.Q., Hao H.L., Yang Y.N. 2012. Behavior of solid carbon sources for biological denitrification in groundwater remediation. Water Sci. Technol., 65(9), 1696–1704.

24. Zhao Y., Feng C., Wang Q., Yang Y., Zhang Z., Sugiura N. 2011. Nitrate removal from groundwater by cooperating heterotrophic with autotrophic denitrification in a biofilm-electrode reactor. J. Hazard. Mater., 192(3), 1033–1039.

25. Zhu S., Chen S. 2002. The impact of temperature on nitrification rate in fixed film biofilters. Aquacult. Eng., 26, 221–237.

26. Zhao Y.X., Zhang B.G., Feng C.P., Huang F.Y., Zhang P., Zhang Z.Y., Yang Y.N., Sugiura N. 2012. Behavior of autotrophic denitrification and heterotrophic denitrification in an intensified biofilm-electrode reactor for nitrate contaminated drinking water treatment. Bioresour. Technol., 107, 159–165.

15

Design of Biofilter Odor Removal System for Conventional Wastewater Treatment Plant

Ali Hadi Ghawi[1]

[1] Department of Civil Engineering, Collage of Engineering, University of Al-Qadisiyah, Iraq
* Corresponding author's e-mail: ali.ghawi@qu.edu.iq

ABSTRACT
Control of odor removal and air pollution in wastewater treatment plants has become critical because of the negative impacts of invasive pollutants that are no longer limited to the working environment of sewage treatment plants, but extend to nearby residential areas when appropriate weather conditions are present. Residents of the city of Al-Nasiriyah in Iraq suffer from the the foul odors from the Jazeera Wastewater Treatment Plant in the city of Al-Nasiriyah, located in the province of Dhi Qar in southern Iraq. Therefore, efforts must be intensified to reduce the risks they pose to the ecosystem and the serious damage to human health. In this study, a biofilter system was designed to remove the odors emitted (hydrogen sulfide and ammonia gas) from the conventional activated sludge Al-Nasiriyah- Jazeera wastewater treatment plant. The biofilter odor removal system is designed for inlet screw pumping station, screening station, aerated grit chamber and parshall flume inlet. The result of the design of the biofilters, which meets the environmental requirements of Iraq according to the law of environmental protection No. 27 of 2009 and its instructions No. 3 of 2011 was the inlet screw pumping station (volume of buildings to be treated – 400 m³, number of changes per hour >5 n/h, design flow rate to be captured and treated – 2000 m³/h, H_2S removal efficiency >98%, volume of substrate media – 24 m³, height of surface media – 1.4 m, and overall dimension (length – 8.9 m, height – 2.27 m, width – 2.13 m)), as well as the preliminary treatments screening station, aerated grit chamber and parshall flume inlet (volume of buildings to be treated (each biofilter) – 400 m³, number of changes per hour – 12 n/h, design flow rate to be captured and treated (each biofilter) – 5000 m³/h, H_2S removal efficiency >98%, volume of substrate media – 60 m³, height of surface media – 1.4 m, and overall dimension (length 10.6 m, height 2.27 m, width m 2.13 m)).

Keywords: Biofilter, Wastewater, Hydrogen Sulfide, Odor

INTRODUCTION

The biological treatment of municipal wastewater is typically based on the so-called activated sludge process, i.e. on the use of suspended biomass capable to oxidize organic compound (Pedro Cisterna, 2017). Experts on wastewater treatment have found the need to address odor as a primary concern in the design and operation of collection and treatment facilities. Any place or process in which wastewater is collected, transported or treated has the potential to generate and release nuisance odors to the surrounding area. However, most odor problems occur in the collection system, in primary treatment facilities and in solids handling facilities. In most instances,

the odors associated with collection systems and primary treatment facilities are generated as a result of an anaerobic or "septic" condition. This condition occurs when the oxygen transfer to the wastewater is limited, such as in a force main. In the anaerobic state, the microbes present in the wastewater have no dissolved oxygen available for respiration. This allows the microbes known as "sulfate-reducing bacteria" to thrive. These bacteria utilize the sulfate ion (SO_4^-) that is naturally abundant in most waters as an oxygen source for respiration. The byproduct of this activity is hydrogen sulfide (H_2S). This byproduct has low solubility in the wastewater and a strong, offensive, rotten-egg odor. In addition to its odor, H_2S can cause severe corrosion problems as well.

Due to its low solubility in the wastewater, it is released to the atmosphere in such areas as wet wells, head works, grit chambers and primary clarifiers. There are typically other "organic" odorous compounds, such as mercaptans and amines, present in these areas, but H_2S is the most prevalent compound (Ben Jaber et al., 2016).

Solids handling facilities are another significant problem area connected with odor. In biosolids dewatering and treatment processes, the biosolids commonly undergo extreme turbulence, pH adjustment and/or thermal treatment. Depending on the nature of the biosolids stream and the treatment used, the odor compounds released can consist of any combination of the following compounds in a wide range of concentrations: ammonia (NH_3), amines, hydrogen sulfide, organic sulfides and mercaptans. Additionally, the anaerobic digestion of sludge creates the anaerobic conditions under which sulfate-reducing bacteria thrive, causing the formation of hydrogen sulfide that is vented with the digester "biogas" formed from the digestion of sludge. There are many different technologies that can be applied to control the odors from wastewater collection and treatment systems. These technologies can be split into two main groups: vapor-phase technologies, used to control odorous compounds in the air or gas; and liquid-phase technologies, used to control odorous compounds in the liquid wastewater itself. The vapor-phase technologies are typically used in point-source applications such as wastewater treatment plants and pump stations or for the treatment of biogas. On the other hand, the liquid-phase technologies are typically used in the collection systems where control of both odors and corrosion are concerns and/or where multiple point odor control is objective (Fletcher et al., 2014).

The objective of this study is to design the Odor Control System for the Jazeera wastewater treatment plant in Nasiriyah, Dhi Qar Governorate, in southern Iraq, according to the environmental requirements under the Environment Protection Law No. 27 of 2009 and the Instructions No. 3 of 2011. The Dhi Qar sewerage department took the necessary measures to protect the population and the environment from the negative effects of sewage projects. The most important negative effects of wastewater treatment plants include the emission of gases such as hydrogen and ammonium gases that constitute aromatic and undesirable odors in the environment and health.

In this study, the SERECO S.r.l. technology was used in the design of the odor removal system.

The removal of these odors is based on the elimination of emitted gases using specialized technological systems that absorb these gases and prevent them from spreading in the surrounding environment of the wastewater treatment plants, especially taking into account the fact that the sites of these plants are sometimes within the residential neighborhoods imposed by the design determinants of sewage networks. Where the smells emitted from the sewage treatment plant are concentrated in the inlet of a screw pumping station, screening station, aerated grit chamber and parshall flume inlet, the preferred technical and environmental solution is to cover these systems technically using a GRP cover with the air withdrawn through effective biological filters (Estrada et al., 2011). The aim of this study is to describe the selection and design of the air treatment units for odor removal from selected areas of the Al-Nasiriyah wastewater treatment plant (NWWTP).

MATERIAL AND METHODS

Description Al-Nasiriyah Wastewater Treatment Plant (NWWTP)

The sewage treatment plant in the city of Al-Nasiriyah in the province of Dhi Qar is one of the strategic projects of the giant infrastructure, serving 300,000 people. The district of Nasiriyah is located about 375 km to the southeast of the city of Baghdad. It processes wastewater to minimize its negative impact and dumps it into the Euphrates River. The area of wastewater treatment unit is (140×428) m. The location of the plant is within the limits of the city's basic design and is close to densely populated areas. The Wastewater Treatment Plants (WWTPs) have the capacity of 60000 m^3/day. They are of conventional activated sludge process type, as shown in Figures 1 and 2. In this solution, the raw wastewater passes three stages of treatment; primary treatment where suspended solids settled or retained, secondary treatment where the biological process is taking place to transform dissolved and suspended organics into simple compounds, then the third stage where sludge stabilization takes place in digestion systems. The Iraqi National Standards set by the Regulation 25 of 1967 for the design and specifications of wastewater raw and treated sew-

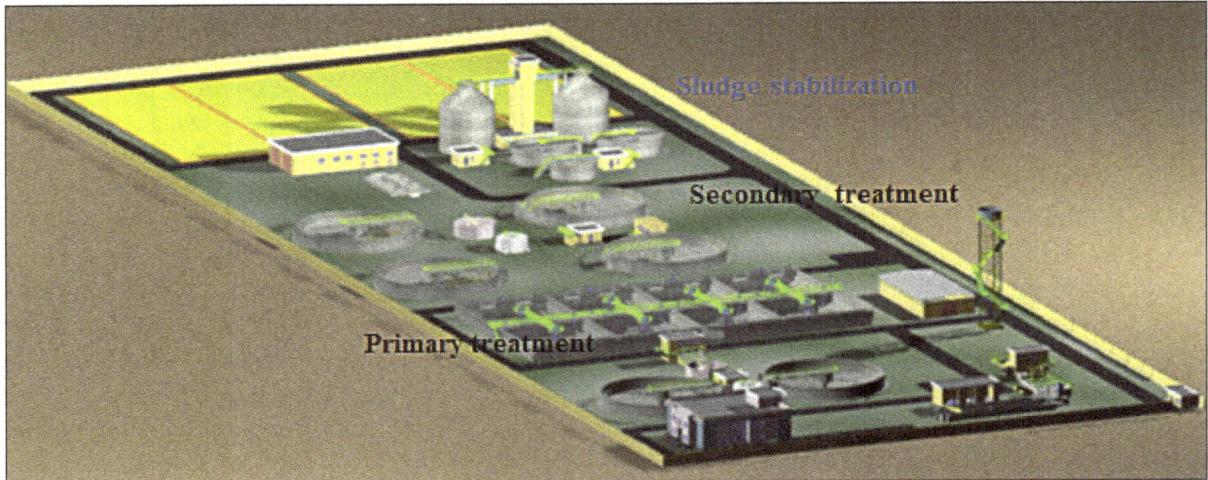

Figure 1. General layout of Al-Nasiriyah WWTP (Water line: Screening- Aerated grit chamber and Parshall flume inlet – Primary settling – Biological Treatment – Secondary Settling –Disinfection. Sludge line: Thickening – Blending – Anaerobic digestion – Drying beds)

Figure 2. The Study Area

age from a domestic wastewater treatment unit also the capacity and technological parameters of this study are as follows (Table 1):

PROCESS SELECTION

In general, the odor control facilities at the WWTP consist of selected plant sections, aspiration of exhausted air and treatment for odor-ants degradation or removal. Typically, the designed air flow rate must guarantee 4–6 volume exchanges per hour in covered sections were the access for operators is restricted and 10–12 volume exchanges where the presence of operators is required for routine operations (WEF., 2004). Continuous air aspiration must guarantee the required ventilation rates and maintain a negative differential pressure across openings in the covered area. Other odor control strategies are

Table 1. Influent and effluent criteria of Al-Nasiriyah WWTP

Parameter	Values
Equivalent inhabitants	316083 inha.
Design Flow rate:	
Average	2897 m³/h
Max	4708 m³/h
Min	1738 m³/h
Characteristic of influent sewage:	
TSS	400 mg/L
COD	460 mg/L
BOD5	350 mg/L
TKN	40 mg/L
NH3-N	20 mg/L
NO3-N	< 1 mg/L
TP	8 mg/L
pH	7
BOD5 loading	24340 kg/d
Treated effluent limits:	
TSS	30 mg/L
COD	100 mg/L
BOD5	20 mg/L
TKN	7 mg/L
TP	2 mg/L
pH	7
Wastewater Temperature:	
Design	20°C
Min	15°C
Max	35°

related to the addition of oxidizing chemicals to the influent wastewater or spraying on scum, screened materials, etc. Moreover, masking or neutralization of volatile odor compound can be performed by spraying suitable liquid mixtures of fragrances and surfactants. However, the efficiency of these techniques is somewhat uncertain and difficult to control.

A range of technologies is available to treat the odorous air aspired from confined sections at wastewater treatment plants and sludge handling facilities. These can be classified according to the main process involved in the odor removal:
- Biochemical: biofilters, bioscrubbers, activated sludge;
- Chemical: scrubbers, thermal oxidation, catalytic oxidation, ozonation;
- Physical: condensation, adsorption (activated carbon), absorption (clean water scrubbers).

The selection of a particular technology or combination of technologies is dependent on factors such as: site characteristics including operation and maintenance capabilities, treatment objectives, foul air flow rates, contaminant loading

patterns, as well as the characteristics and strength of odorous air.

Waste gases from industry have traditionally been treated using the physicochemical processes (scrubbing, adsorption, condensation, and oxidation). However, for municipal wastewater treatment plants, the biological methods of odor treatment are widely applied in Europe and are increasing the USA, owing to their efficiency, cost-effectiveness, and environmental acceptability (Burgess et al., 2001).

The advantage of the biological treatments over physicochemical techniques relies on reduction of the operating costs. For example, yearly biofiltration operating costs are reported to be around one-tenth of those of an absorption process and one-fourth of those of a wet chemical scrubber (Gabriel and Deshusses, 2004). Moreover, adoption of biological treatments eliminates the need of purchasing dangerous chemicals and handling them in concentrated solutions.

Among other biochemical technologies, the biofilters offer simpler operations and maintenance, with very limited use of chemicals and with the main operational cost related to the replacement of the biofilter media (once every 2–5 years). Moreover, biofilters are particularly suited for the odor removal in wastewater treatment plants characterized by constant contaminant loads. For these reasons, the biofiltration technology has been preferred for foul air treatment at the Al-Nasiriyah – Al-Jazeera Wastewater Treatment Plant (NWWTP).

BIOFILTER DESCRIPTION

A biofilter is simply a bed of organic material (media) with a suitable porosity, allowing for air passage, humidity trapping and biomass growth. As the foul air passes upward through the media, the odor causing compounds are adsorbed onto a film of water and microbes developed on the organic material. These microbes then convert the adsorbed compounds by oxidizing them to carbon dioxide, water and inorganic salts. The biofilter is a self-regulating ecosystem, and is therefore likely to function for long time, without excessive control. It is very important, however, to ensure that the moisture levels are controlled in the biofilter for it to function properly. The pH levels are also self-regulating within the ecosystem, and are assisted by proper choice of media. The effectiveness of the biofilter is primarily a function of the amount of time the odorous air spends in the biofilter (contact time) and the moisture content of the filtering material. The biofiltration technology has been developed for over 100 years and its many applications generally differ for the filtering media (soil, compost, wood chips and various mixture) and for the biofilter construction (in-ground or in vessel above the ground, covered or open air).

A typical biofilter has the following basic constructed components: 1) the containment 2) the biofilter media, 3) the humidifier and wetting systems, 4) biofilter air diffusion system and 5) the air ducts and fans (Figure 3).

The technical solution adopted for the Al-Nasiriyah–Al-Jazeera Wastewater Treatment Plant (NWWTP) is a modular self-contained biofilter. This system requires a smaller footprint than the open-bed designs and it is easy to install on a simple concrete basement.

The biofilter is completely pre-assembled in a tank containing the full system, with the dimensions compatible for shipment inside standard containers. The filtering bed substrate is composed of calibrate mixing of high quality wood chips, characterized by high grade of porosity, high retention of humidity. The biofilter coverage ensure the optimal life conditions of the bacteria, avoiding the direct exposition to the sun light that would be responsible for the drying of the bed. Moreover, the biofilter is equipped with an automatic humidification system composed of spray nozzles, solenoid valves, temperature and relative humidity sensor, as well as a control panel. The biofilter is completed by a centrifugal blower for the aspiration of the foul air from the confined sections of the plant and a system of steel piping to connect the blower to the tank. During the start-up, the biofilter can be inoculated with dedicated bacteria or with activated sludge. After the acclimation period, the biofilter reach high efficiency in removing H2S (>98% at inlet maximum concentration of 100 mg/l), NH3 (95% at inlet maximum concentration of 30 mg/l), and reducing smells. The acidification trend of the bed, due to the air pollutant concentration, is soon eliminated by the action of the biofilter media, then the check of the bed pH has to be rarely executed, with simple laboratory examinations or with field instrumentations. The temperature and humidity sensor verify the right working of the system, by regulating the opening time of the solenoid valve depending on the humidification of the bed. Under normal operating conditions, the expected duration of the bed is more than 2 years.

RESULTS AND DISCUSSIONS

Odor generation and treatment and Design Criteria

Generally, the odorous compounds released from a wastewater treatment plant are not directly related to sanitary problems for plant operators

Figure 3. Typical biofilter

and people living/working in the neighbor area. However, the odor release from wastewater treatment plant can cause significant public concern and control of odor has become a standard consideration in design and operation of wastewater and sludge treatment and disposal facilities. Odorous compounds can be generated during the wastewater collection and transport into the sewer as well as during the various operations in the treatment plant. Within the wastewater treatment plant, the following sections are generally agreed as being the main sources of odors release:

• Preliminary treatments (including pumping station, influent channels and distributors, screens, sand and oil/grit/grease removal units, equalization and pre-aeration systems).

The odors released from preliminary treatments are mainly generated into the sewer and are typically dominated by hydrogen sulfide (H_2S), with other organic sulfur compounds (mercaptans, dimethyl sulfide, etc.) also present at lower concentrations. Odors with similar origin and composition can be released from the large emitting surface of the primary settling tanks. Minor on-site odor generation can be associated with the collection/storage system for screened materials, grease and scum (especially in the summer periods, if this highly putrescible material is not

rapidly disposed of). Ammonia and amines can be released from the latter process and especially in the case of alkaline sludge stabilization. The typical H_2S concentrations measured in the exhausted air are reported in the Table 2 (WERF 2003):

Process design

The following stations have been identified as main potential sources of odor at the NWWTP:

• Preliminary treatments (inlet screw pumping station, screening station, aerated grit chamber and inlet parshall flume) (Figure 4).

On the basis of the geometry and the typical ordinary operation for the selected stations, the presence of operators in the confined volume has been excluded. This results in the minimization of

Table 2. Typical H_2S concentration measured in the exhausted air

Section	Typical Range (H_2S ppm)	Typical Average (H_2S ppm)
Inlet screw pumping station	5÷150	32
Screening station	3÷20	9
Aerated grit tanks and parshall flume	0.01÷4	-

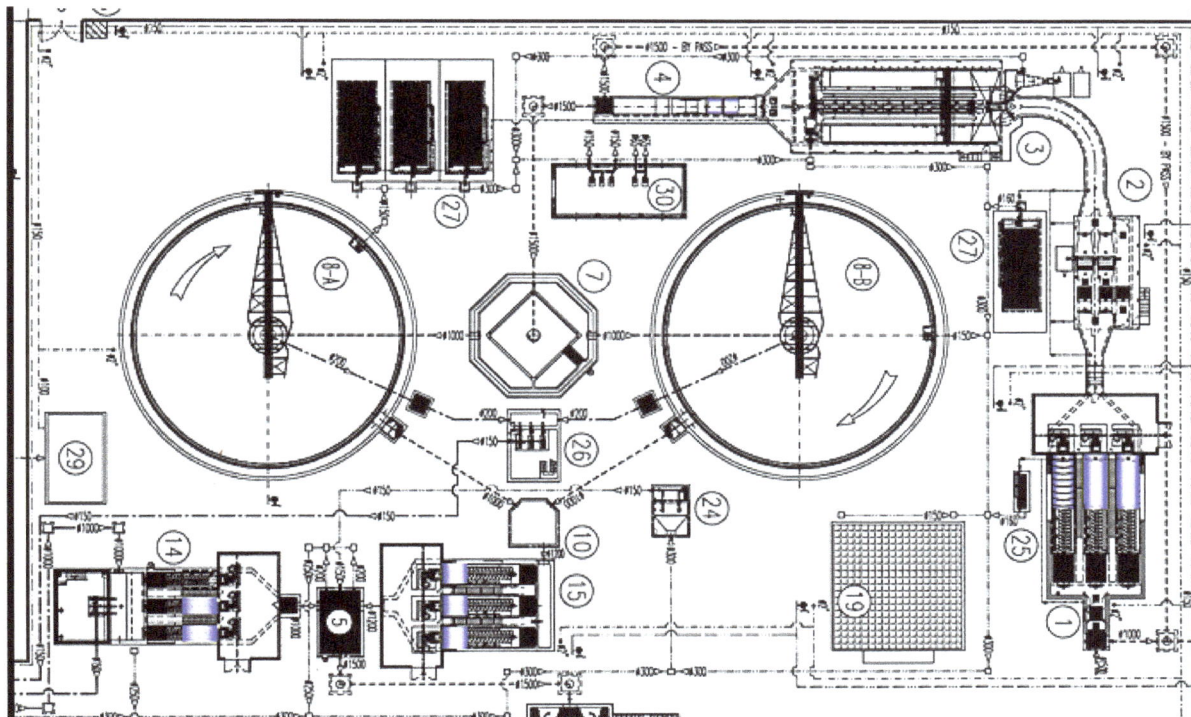

Figure 4. Biofilter for inlet screw pumping station (25), screening station biofilter (27), aerated grit chamber and parshall flume inlet biofilter (27).

the confined volumes and in the selection of the lower range of ventilation ratios, with consequent reduction of the capital and operational costs of the odor treatment units.

Specific air flow rates for each covered section have been designed assuming minimum 4 volume changes plus any forced air inlet (i.e. the air flow rate required for the operation of the sand and grit removal unit), according to the equation (1):

$$Min\ air\ flow\ rate =$$
$$Total\ confined\ volume*Volume\ changes + \quad (1)$$
$$+ Inlet\ air\ flow\ rate$$

The results of the calculations connected with confined volumes and the required air flow rate are reported in the Table 3:

The number and location of the biofilters to be installed at the NWWTP have been designed to minimize the distance from the confined sections, with subsequent minimization of the air pumping head losses, and maximize the operation

flexibility. On the basis of the above-mentioned considerations, the installed odor treatment capacity is presented in the Table 4:

The description and design of the adopted odor treatment units is included in Table 5 (WEF 2004, and Metcalf and Eddy 2003):

Typical elimination capacity is in the range of 20–130 g/m^3h for H_2S and other odorants. Considering that the maximum H_2S concentration of 100 ppm corresponds to about 136 mg/m^3 and considering the removal efficiency of at least 98%, the minimum required elimination capacity can be estimated as follows:

- Required H_2S elimination capacity (Biofilter Dimensions-1(8.9×2.27×2.13) m) = 2000*136*98/24*1000*100 = 11.3 g/m^3 h
- Required H_2S elimination capacity (Biofilter Dimensions-2 (10.6×2.27×2.13)m) = 5000×136×98/60×1000×100= 11.1 g/m^3 h

As results from the above-mentioned calculation, the required elimination capacity for

Table 3. Results of the calculations connected with confined volumes and the required air flow rate

Section	Total confined volume (m³)	Design Volume changes	Minimum required air flow rate (Nm³/h)
Inlet screw pumping station	400	>5	2000
Screening station	400	12	4800
Aerated grit – oil-grease removal tank	800	12	9600
Parshall flume	400	12	4800

Table 4. Biofilter odor treatment capacity

Section	Total required air flow rate (Nm³/h)	Installed capacity (Nm³/h)	Actual volume Exchange per hour	No. of biofilters
Inlet screw pumping station	2000	2000	5	1
Screening station	4800	5000	12.5	1
Aerated grit – oil-grease removal tank and Parshall flume	14400	15000	12.5	3

Table 5. Biofilter odor treatment capacity

Parameter	Unit	Suggested range	Biofilter Dimensions-1 (8.9×2.27×2.13) m	Biofilter Dimensions-2 (10.6×2.27×2.13) m
Filtering bed height	m	1–1.8	1.4	1.4
Average bed porosity	%	35–50	40	40
Filtering bed surface	m²	-	17.4	42
Filtering bed volume	m³	-	24	60
Air flow rate	Nm³/h	-	2000	5000
Surface loading rate	m³ / m² h	10–500	115	128.5
Volume loading rate	m³ / m³ h	10–100	83	90
Empty bed residence time	s	30–60	42	39.2
Design H₂S inlet oncentration	ppm	-	50	50–100

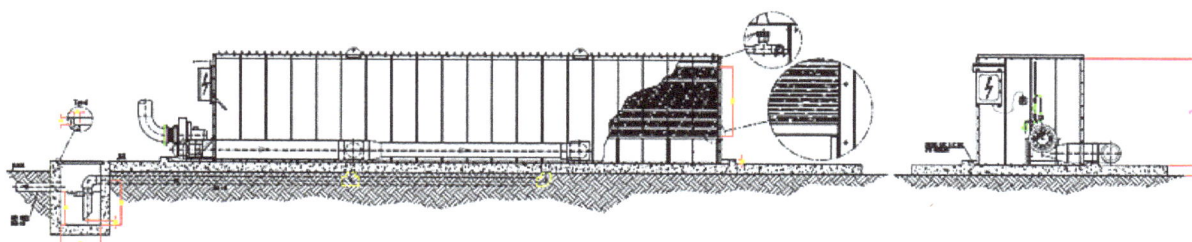

Figure 5. Front view of biofilter

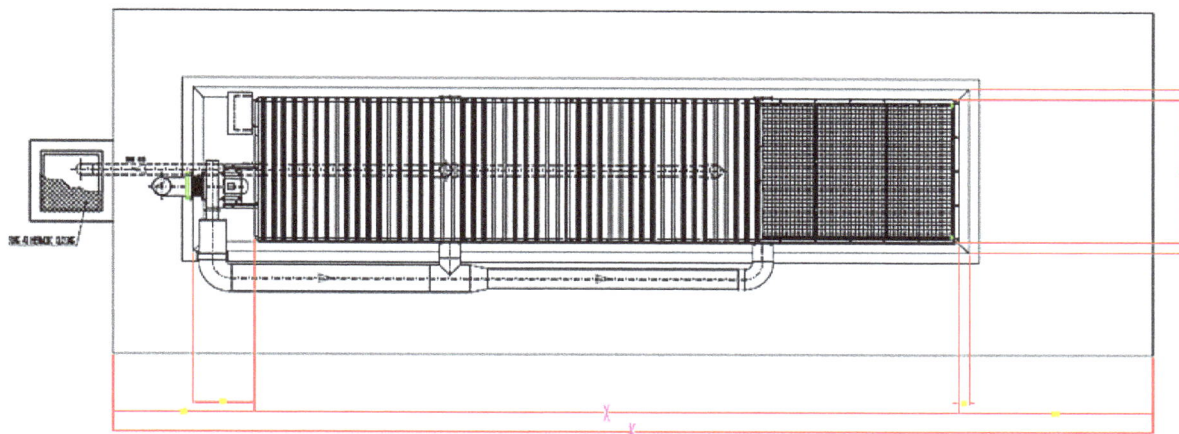

Figure 6. Plan view of biofilter

H_2S is far below the typical elimination capacity reported for this type of odor treatment units (20–130 g/m³h); therefore the treatment efficiency will be reached only within the initial part of the filtering bed (Figures 5 and 6).

CONCLUSIONS

The unwanted and harmful emissions and odors from wastewater treatment plants such as hydrogen sulfide gas and ammonia gas, which cause damage to the population living near these stations, result in the establishment of closed, zero-emission plants. The biofilter filter is designed to remove odors from the wastewater treatment plant in Al-Nasiriyah city, according to the Iraqi environmental regulations. As results from the calculation, the required elimination capacity for H_2S is far below the typical elimination capacity reported for this type of odor treatment units (20–130 g/m³h); therefore the treatment efficiency will be reached only within the initial part of the filtering bed.

REFERENCES

1. Ben Jaber, M., Couvert, A., Amrane, A., Rouxel, F., Le Cloirec, P., Dumont, E. 2016. Biofiltration of high concentration of H2S in waste air under extreme acidic conditions. New Biotechnology 33, 136–143. doi:10.1016/j.nbt.2015.09.008,

2. Burgess J.E., Parsons S.A., Stuetz R.M., 2001. Developments in odour control and waste gas treatment biotechnology: a review. Biotechnology Advances 19, 35–63.

3. Estrada J.M., Bart Kraakman N.J.R., Muñoz R., and Lebrero R., 2011. A Comparative Analysis of Odour Treatment Technologies in Wastewater Treatment Plants. Environ. Sci. Technol., 2011, 45 (3), pp 1100–1106.

4. Fletcher, L.A., Jones, N., Warren, L., Stentiford, E.I. 2014. Understanding biofilter performance and determining emission concentrations under operational conditions, Final Report – Project Number ER36.

5. Gabriel D. and Deshusses M.A., 2004. Technical and economic analysis of the conversion of a full-scale scrubber to a biotrickling filter for odor control. Water Science and Technology 50, 4, 309–318.

6. Metcalf and Eddy 2003. Wastewater Engineering: Treatment and Reuse. Fourth Edition, McGraw-Hill, New York.

7. Pedro Cisterna, 2017. Biological Treatment by Active Sludge with High Biomass Concentration at Laboratory Scale for Mixed Inflow of Sunflower Oil and Saccharose. Environments 2017, 4, 69; doi:10.3390/environments4040069

8. WEF Manual of Practice no. 25, 2004., Control of Odors and Emissions from Wastewater Treatment Plants, Water Environment Federation. 601 Wythe Street – Alexandria, VA 22314–1994 (USA).

9. WERF 2003. Identifying and Controlling Odor in the Municipal Wastewater Environment. Phase 1: Literature Search and Review. Water Environment Federation, 601 Wythe Street – Alexandria, VA 22314–1994 (USA).

16

Cultivation of Oleaginous Microalgae *Scenedesmus obliquus* on Secondary Treated Municipal Wastewater as Growth Medium for Biodiesel Production

Mohamed F. Eida[1*], Osama M. Darwesh[1], Ibrahim A. Matter[1]

[1] Agricultural Microbiology Department, National Research Centre, Giza, Egypt
[*] Corresponding author's e-mail: mf.eida@nrc.sci.eg

ABSTRACT

Local single cell microalgae isolated from a wastewater swamp and identified as *Scenedesmus obliquus* was used to determine its applicability for utilization of domestic wastewater for biomass and lipid production. Secondary treated domestic wastewater with or without mixing of growth medium was used to cultivate *S. obliquus* for the biomass and lipid production as a renewable feedstock for biodiesel. *S. obliquus* showed the highest OD when grown in 100% Bold's basal medium (BBM). *S. obliquus* utilized 95.2% and 78.5% of P and N contents, respectively, when grown in 25% WW+75% BBM mixture and the utilization efficiency of both elements decreased with the increasing wastewater portion in the mixture. Although the BBM displayed the highest dry biomass and lipid production (25.15% of the cell dry biomass). The lowest values were recorded for the uninoculated wastewater, followed by 100% wastewater enriched with *S. obliquus*. The obtained data revealed that the lipid classes of *S. obliquus* differs according to the cultivation medium and conditions. The highest percentage of C16-C18 fatty acids (54.76% from total lipids) were recorded in case of algae cultivated in 100% wastewater, followed by 46.96% in case of 100% BBM medium. These results suggest the utilization of mixtures containing a higher portion of secondary treated wastewater, such as 75% WW+25% BBM or 50% WW+50% BBM, could increase the economical production of the lipid-rich microalgae *S. obliquus* for biodiesel through saving water and nutrients.

Keywords: biodiesel; microalgae; *Scenedesmus*; municipal wastewater; biomass; oil contents; fatty acid compositions

INTRODUCTION

An acceptable alternative fuel should be readily available, competitive with fossil fuels, and environmentally accepted (Meher et al., 2006). Biodiesel has been the leading unconventional liquid biofuel for the past two decades (Meng et al., 2009). Additionally, most of the current commercially available biofuels are crop-based and classified as "first generation biofuels" which compete for arable land and freshwater (Maity et al., 2014). Microbial lipids are highly accumulated, with a ratio of more than 20% of their cell mass, by fast grown oleaginous microorganisms, such as microalgae and fungi, which are known as single cell oils (SCOs). SCOs are considered to be one of the most promising sources for bio-diesel production (Meng et al., 2009). Microalgae represent an auspicious renewable alternative third-generation sustainable bioenergy source. It can efficiently produce huge amounts of biomass (7–20 times greater than soy or corn per land unit) with high lipid content (Maity et al., 2014; Bharathiraja et al., 2015; Lyon et al., 2015).

Implementing various microalgae as a sustainable feedstock for biodiesel production was introduced by Ahmed et al. (2011). They stated that several algal groups, including members of *Diatoms*, *Green algae*, and *Red algae* from both marine and freshwater sources, were reported to be suitable for utilization in the biodiesel production, according to their high biomass productivity and their lipid content. Among the microalgae groups, *Scenedesmus* sp. were found to be appropriate candidates for biodiesel production based on their high lipid productivity (24.66 mg/L/day), high biomass yield (0.9 g/L), as well as appropriate fatty acid profile (Jena et al., 2012). The accumulated lipids in *S. obliquus* were evaluated by Mandal and Mallick (2009) under several growth conditions.

On the other hand, microalgae can grow under various aquatic environments, such as fresh, brackish, marine, and wastewaters (municipal, industrial, agricultural, and domestic wastewaters), when adequate amounts of required nutrients are present (Mobin & Alam, 2014). Growing microalgae consumes a huge amount of water and inorganic nutritional elements that increase the production cost (Xin et al., 2010). Furthermore, the addition of organic carbon for many microalgal strains prompts their growth (Xu et al., 2006), but increases the biomass cost as well (Li et al., 2007). Many obstacles are faced by the mass production of microalgae, including the harvesting and nutrients cost. The harvesting challenges could be overcome using efficient low-cost methods, such as flocculation (Matter et al., 2016), while the nutrient cost could be reduced by recycling nutrients from wastewaters (Lv et al., 2017). Thus, sustainable and economically acceptable culturing medium for algae-based biodiesel production is still needed (Mandal & Mallick, 2011). Accordingly, several previous studies regarding the use of swine wastes, dairy manure, and other animal residues for the cultivation of microalgae have been carried out (Wilkie & Mulbry, 2002). Cultivation of microalgae in a nutrient-rich effluent as an inexpensive, readily available and cheap medium is expected to overcome the economic dilemma of biodiesel production as well as decrease the environmental problems arising from discharging nutrients into bodies of water. Utilization of wastewater for the cultivation of microalgae has several advantages, including providing an alternative source for the huge amounts water required for growing algae, which presents a supplemental source of nutrients, as well as decreases the load of contaminants, according to Ansari et al. (2017). On the other hand, numerous challenges are faced by the cultivation of microalgae in wastewater, such as the unbalanced N/P ratio, the extraordinary biological pollutants and competitors, low biomass and lipid content production, and the low level of effective nutrient elimination (Xin et al., 2010; Min et al., 2011; Zhang et al., 2012). Additionally, the wastewater

treatment methods differ from one country to another, which make the quality of effluents differ as well. Thus, the aim of the current study was to evaluate the suitability of secondary treated municipal wastewater as a cheap growth medium for algal cultivation, lipid content and the composition of the lipid-rich microalgae *Scenedesmus* sp. for biodiesel production under the conditions found in Egypt.

MATERIALS AND METHODS

Wastewater collection

The wastewater used in this study was obtained from the abundant effluent of the Zenen municipal wastewater treatment plant in Giza, Egypt. The municipal wastewater in this plant was subjected to sedimentation and aeration as a secondary treatment step before being discharged into the irrigation canal. The secondary wastewater treatment sample was collected in a clean plastic container and transferred immediately to the lab. The Zenen municipal wastewater plant effluent has the following characteristics: pH, 7.9; COD, 95.8 mgL^{-1}; BOD, 71.5 mgL^{-1}; TSS, 81 mgL^{-1}; TOC, 1.8 mgL^{-1}; TP,1.28 mgL^{-1} and TN, 21.96 mgL^{-1}.

Microalgae isolation, purification and identification

The single-cell green microalgae were isolated from a swamp contaminated with wastewater located in the Gharbia governorate in Egypt. BBM was used for the isolation and maintenance of the microalgal isolate (Pizarro et al., 2006). It contained (per liter) 175 mg KH_2PO_4, 25 mg $CaCl_2$. $2H_2O$, 75 mg $MgSO_4.7H_2O$, 250 mg $NaNO_3$, 75 mg K_2HPO_4, 25 mg NaCl, 11.42 mg H_3BO_3, 1 mL from Microelement stock solution (which consist of: 8.82 g $ZnSO_4.7H_2O$, 1.44 g $MnCl_2.4H_2O$, 0.71 g MoO_3, 1.57 g, $CuSO_4.5H_2O$ and 0.49 g $Co(NO_3)_2.6H_2O$, per liter), 1 mL from Solution 1 (which consist of: 50 g Na_2EDTA and 3.1 g KOH, per liter) and 1 mL from Solution 2 (which consist of: 4.98 g $FeSO_4$ and 1 mL concentrated H_2SO_4, per liter), and final pH of 6.8.

The algae isolation and purification were performed by serial dilution and culturing on liquid BBM followed by plating on solid BBM. The incubation during the isolation process was conducted at ambient temperature under continuous illumination with white fluorescent light (at intensity of 2000 Lux). The purity of the culture was ensured by regular observation under light microscope (Zeiss, Oberkochen, Germany). After purification of the microalgal isolate, microscopic examination was performed to identify the genus according to Bellinger and Sigee (2015). The identification of the species level was confirmed using molecular-based techniques.

DNA extraction, PCR amplification, sequencing, and phylogenetic analysis

The total genomic DNA was extracted using a modified enzymatic lyses method (Bellinger & Sigee, 2015). The microalgae cells were harvested and dried using liquid nitrogen before incubation in CTAB buffer at 60°C for 1 h; afterwards, the lysozyme (20 mg/mL) and Proteinase K (1 mg/mL) were applied. The total genomic DNA was purified using isopropanol precipitation (Darwesh et al., 2014).

The extracted DNA was used as a template for PCR amplification using a Bio-Rad T100 Thermal Cycler (Bio-Rad Laboratories Inc., Hercules, CA, USA). PCR was performed with a pair of universal primers targeting the 18S rRNA gene; EUK1 (5′-AGCGGAGGAAAAGAAACTA-′3) and EUK2 (5′-ACTAGAAGGTTCGATTAGTC-′3) as the forward and reverse primers, respectively, according to Rainer et al. (2007).

The final 50 μL reaction mixture contained $1\times$ PCR buffer (NEB, England), 1 nmol of dNTPs, 1 pmol of 2 mM $MgSO_4$, 0.25 pmol of forward and reverse primers, 1 unit of Taq DNA polymerase (NEB, England) and 5 μL template DNA.

The PCR amplification included an initial denaturation of the DNA at 95°C for 5 min, followed by 35 cycles of 95°C for 30 sec, 55°C for 30 sec, and 72°C for 45 sec. Then, the mixture was kept for 10 min at 72°C for complete extension. The PCR product was purified by means of the QIA quick purification Kit (Qiagen, USA) and ran on an agarose gel to evaluate the purified 18S rRNA fragments for sequencing. Sequencing was performed using the BigDye Terminator v3.1 Cycle Sequencing Kit (Applied Biosystems) on a 3730xl DNA analyzer (Applied Biosystems). The obtained 18S rRNA sequence was deposited in the GenBank database (http://blast.ncbi.nlm.nih.gov/Blast.cgi) and compared to those of other known species using the BLASTN program. The sequences were aligned using Jukes Cantor Model. The phylogenetic reconstruction was built using the neighbor-joining (NJ) algorithm with bootstrap values. The obtained sequences were deposited in GenBank under the accession no. KY621475.

Experimental design

The *Scenedesmus* strain was grown in a 250-mL conical flask containing 150 mL of media and incubated on an orbital shaker under continuous illumination to be utilized as the inoculum. Five mL of the inoculum was used for inoculation of the experimental flasks containing 150 mL of different treatments. In order to evaluate the capacity of the isolated microalgae and utilize wastewater as the growth medium for biomass and lipid production, the microalgal isolate was cultivated in different mixtures of wastewater and growth medium (BBM). A hundred percent of the BBM growth medium served as a control, as well as 100% wastewater inoculated with microalgae, which were implemented in different combinations. The percentages of wastewater/BBM mixtures were 25:75, 50:50 and 75:25, respectively. A wastewater treatment sample without *Scenedesmus* inoculation was consecutively carried out with other treatments to evaluate the native algal load of the utilized wastewater. All treatments were carried out in triplicate on an orbital shaker incubator (Gallenkamp, Germany) with a shaking rate of 120 rpm at ambient temperature under continuous illumination using white fluorescent light (1000 Lux). The samples were initially withdrawn for analysis at 2-day intervals.

Analytical methods

The growth rate of the microalgae was routinely assessed through measuring the optical density of the samples at 680 nm using a spectrophotometer (SHIMADZU UV-2401PC, Japan) as reported by Lee (2008). The chlorophyll *a* content was determined according to Richmond (2008) and Hosikian et al. (2010) with some modifications: 1 mL of algal culture was centrifuged at 10000 rpm for 10 min, the supernatant was discarded and the cells were re-suspended in 1 mL of 90% methanol and sonicated in an ultrasonic water bath at maximum power for 15 min. The mixture was vortexed for 5 min, centrifuged and the supernatant (contains chlorophyll) was transferred into a new tube. Additional extraction steps were performed with 1 mL of 90% methanol to extract almost all of the chlorophyll from the cells. After the second extraction, the mixture was centrifuged again and the supernatant was transferred to the first portion of the extract. Finally, the volume was increased to 2 mL with 90% methanol, and the absorbance at 650 and 665 nm was recorded spectrophotometrically to calculate the chlorophyll *a* content according to the following equation:

$$\text{Chlorophyll } a \text{ (mgl}^{-1}) = (16.5 \times A_{665}) - (8.3 \times A_{650}) \tag{1}$$

The nitrate-nitrogen was estimated using the pyrogallol method with a modification of the method described by De-Nardo (1929). In brief, 1 mL of cells-free culture after centrifugation was placed in a test tube with 1 mL of pyrogallol solution (prepared by dissolving 3 g of Pyrogallol in 100 mL DW). Then, 2 mL of conc. sulfuric acid was added to the mixture and left for 30 min to allow for color formation. After adding 5 mL of DW, the developed color was measured on the spectrophotometer at 606 nm against the stander concentrations of sodium nitrate. Furthermore, ammonium nitrogen was measured colorimetrically using Nessler's method (Basova et al., 2011) and the generated color was measured at 410 nm using a spectrophotometer (SHIMADZU UV-2401PC, Japan). Standard solutions of ammonium sulfate were measured and used to design the standard curve. Additionally, P content in all the samples was determined following the ascorbic acid method, in accordance with Akpor et al. (2007). The absorbance values of the preformed blue color were read at a wavelength of 880 nm using the spectrophotometer. The Potassium dihydrogen phosphate standards were read together with the samples.

Total lipid was determined according to the method described by Bligh and Dyer (1959) with the following modifications: 100 mL of microalgae culture were harvested by centrifugation at 6000 rpm for 20 min and re-suspended in 1 mL distilled water. The sample was then mixed with 2.5 mL chloroform. Afterwards, 5 mL methanol (1: 2 v/v) was subjected to sonication for 30 min at the maximum power. After sonication, the tubes were shacked with extraction solvents overnight. The next day, an additional portion of chloroform (2.5 mL) was added to each tube and the mixture was sonicated again for 30 min and vortexed. Then, 2.5 mL of distilled water was added to separate the chloroform and aqueous methanol layers by centrifugation at 4000 rpm for 10 min. The chloroform layer was gently re-

moved from the bottom and a second extraction step was performed by adding 5 mL of chloroform and then the suspension was vortexed. The chloroform portions were collected and washed with 5 mL of 5% NaCl solution, after which the chloroform was evaporated in the oven at 50°C. The total lipids were measured gravimetrically.

GC characterization of lipid profile for microalgae samples

The fatty acid methyl esters were prepared for the GC analysis according to the method used by Ichihara and Fukubayashi (2010). In brief, the lipid samples were dissolved in toluene (0.20 mL), placed in screw-capped glass test tubes and mixed with 1.8 mL of 8.0% methanolic HCl (w/v) followed by heating at 100°C for 1 h. After cooling to room temperature, 1 mL of hexane was vortexed with the mixture for the extraction of FAMEs followed by the addition of 1 mL of water to enable the layer separation. The hexane layer was pipetted out and purified through a membrane filter for the GC analysis.

The methyl esters of the fatty acids were analyzed by a Gas Chromatography system (Hewlett Packard, HP 6890 series, United States) equipped with an Alltech BPX70 Capillary Column (60 m X 320 μm X 0.25 μm) coated with 70% poly silphenyl-siloxane supplied with flame ionizing detector. Fatty acid content was expressed as a percentage of the total fatty acids identified in the oil.

Statistical analysis

All experimental trials were performed in triplicate and the results are presented as the means ± standard deviation. Statistical analyses were performed using SPSS software 20.0. The comparisons of the mean values were conducted by means of one-way analysis of variance (ANOVA) followed by a Duncan's new multiple-range test for statistical significance. The differences were considered significant at $P < 0.05$.

RESULTS AND DISCUSSION

Isolation and identification of microalgae

The isolation source was a sample of wastewater-contaminated swamp which showed algal growth that could be observed with the naked eye. A preliminary microscopic examination of the water sample showed the domination of the genus *Scenedesmus*. Using the microscopic characterization, a microalgae isolate was collected and primarily identified as belonging to the genus *Scenedesmus,* which was confirmed using molecular biology tools. The data derived from the BLAST analyses and phylogenetic tree (Fig. 1) were constructed based on 18S rRNA gene sequences of isolated microalgae, which indicated that the obtained isolate was close to *S. obliquus* with 99% similarity. The sequences of this strain were uploaded to GenBank and recorded under the accession number of KY621475. Many reports recorded the potential of using *Scenedesmus* microalgae as a feedstock for biodiesel production and the ability of this genus to grow on wastewater as a growth medium (Álvarez-Díaz et al., 2015; Mata et al., 2013). Hodaifa et al. (2008), Ji et al. (2015) and Álvarez-Díaz et al. (2015) reported the capacity of *S. obliquus* to grow under mixotrophic conditions of different wastewaters under illumination.

Growth rates of S. *obliquus* in different wastewater-medium mixtures

The growth of *S. obliquus* on Bold Basal Medium and wastewater, as well as their mixtures, were assessed using cultural optical densities (OD) and the chlorophyll *a* content. This experiment was performed to examine the suitability of such wastewater mixtures to give the maximum growth for the studied microalgae. The obtained results showed different growth parameters according to the mixtures of medium and wastewater (Fig. 2 and 3). The data obtained during the current study revealed that the optimum growth rate, dry biomass production, lipid content, as well as the percentage of C16-C18 fatty acids of the microalgae all belong to different types of medium. These results could be due to the different behavior of each parameter in response to the degree of stress, which is the percentage of WW in the growth medium. Previous studies showed that divers and the individual responses of the studied parameters affected the factors. Madkour et al. (2012) found the highest dry material, chlorophyll

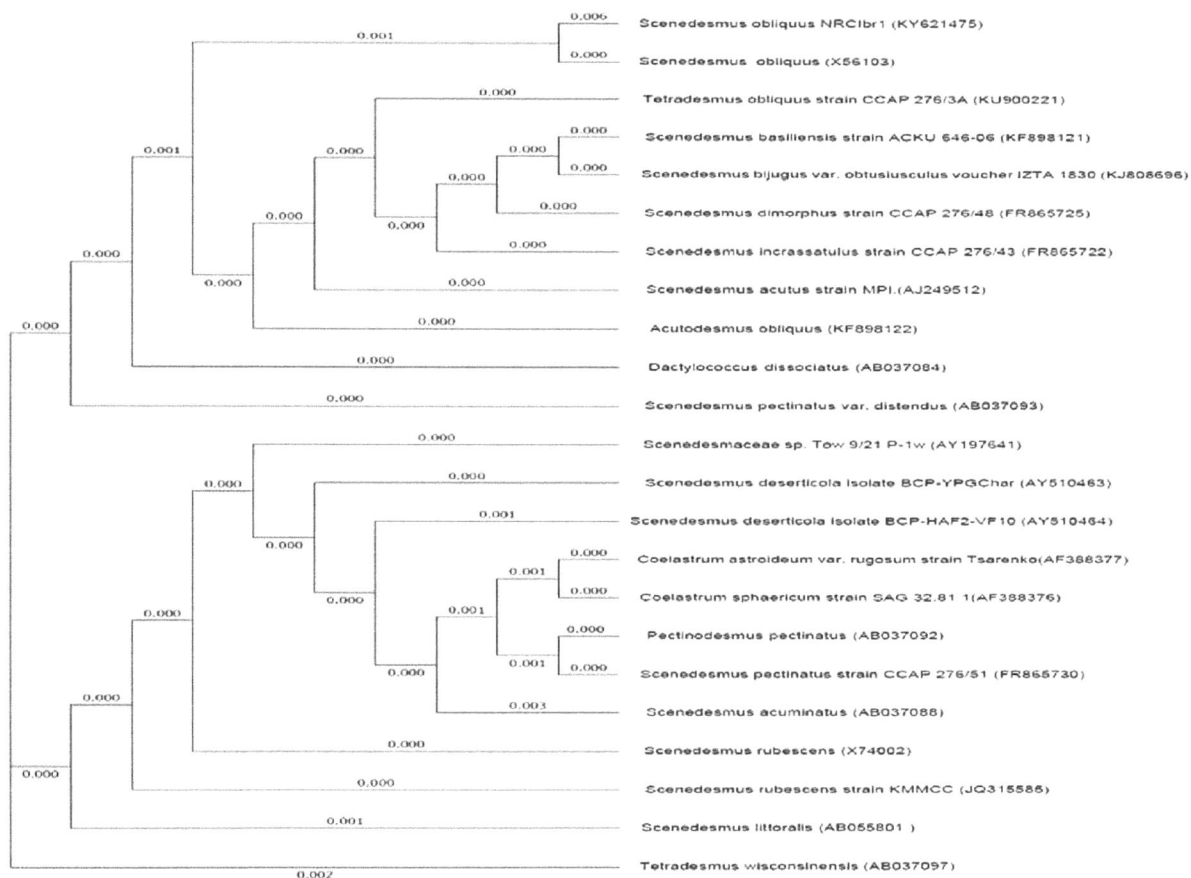

Fig. 1. Phylogenetic relationship between the obtained 18S rRNA sequence of *S. obliquus* and their related strains in GenBank. The scale bar represents the number of changes per nucleotide position (substitution/site)

content and protein yields were achieved via SM medium while utilizing urea and NH_4NO_3 as the N sources, providing the maximum lipid and carbohydrate content of *Spirulina* sp, respectively.

Optical density

During the 14 days of the experiment, the growth rate of *S. obliquus*, which was represented by the OD at 680 nm, showed a steady increase in BBM as the mono cultivation medium (Fig. 2a). By using wastewater as a component in the cultivation medium at 25% and 50%, the cultures showed a slightly higher density during the first 6 days than BBM (control). Afterwards, the growth curve became lower than BBM, 8 days after the experiment started (Fig. 2b and c). In the case of cultivation in 75% wastewater amended with 25%

BBM, the growth trend was similar to the previously mentioned mixtures, but with lower values at the final 4 days of incubation (Fig. 2d). On the other hand, cultivation of *S. obliquus* on 100% wastewater showed a normal trend compared to BBM during the first 6 days; then, the growth decreased and its OD became approximately 28.7% of the BBM (Fig. 2e). However, the wastewater itself contained some native microalgae species that showed negligible growth when subjected to long term aeration and illumination (Fig. 2f). At the end of the cultivation period (day 12 and 14), the OD values of the medium treatment reached 100%, which was significantly higher than other treatments. A similar result was found by Toyub et al. (2008) when growing *S. obliquus* in BBM compared with sweetmeat factory waste media.

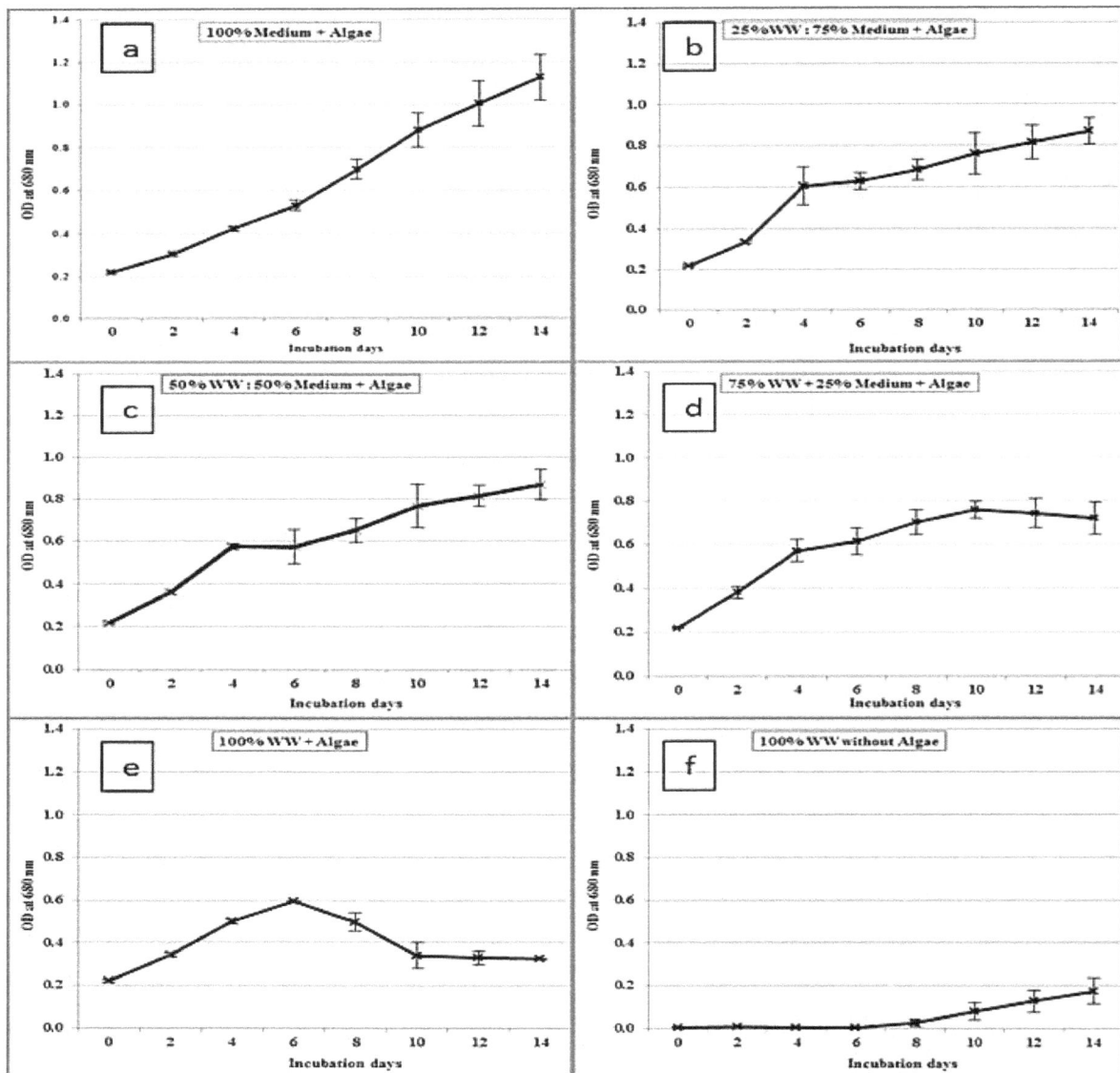

Fig. 2. Optical densities (OD_{680nm}) of *S. obliquus* cultivated on BBM, wastewater or their mixtures

The aptness of BBM for vegetative growth of *S. obliquus* was reported by Martinez-Jeronimo and Espinosa-Chavez (1994).

Chlorophyll a content

Chlorophyll *a (Chl a)*, a photosynthetic pigment, is the principal photochemically active compound, which functions as a receiver of light for photosynthesis. Therefore, the content of this pigment in microalgae reflects the photosynthetic activity and could indicate the growth rate (MacIntyre et al., 2002). In the current study, the chlorophyll *a* content has been implemented as a parameter to evaluate the growth of *S. obliquus* on different mixtures of pretreated municipal wastewater (WW) and BBM (Fig. 3a-f). Following the inoculation of *S. obliquus* into WW (100%), its chlorophyll content increased for 8 days and then declined until the end of the incubation period (Fig. 3a). The maximum chlorophyll *a* content (1.95 mgl⁻¹) was recorded when the WW was mixed with BBM growth medium in the ratio of 1:1 (V/V) (Fig. 3c). The chlorophyll *a* content reached its peak (1.44 mgl⁻¹) when changing the mixing ratio to 1:3 (WW: BBM), then started dropping up to the 10th day before starting to increase. Although the chlorophyll *a* content of *S. obliquus* grown on 100% BBM exceeded its value in 100% WW treatment media, it unexpectedly was not the leading treatment (Fig. 3a and e). On the other hand, the incubation of WW without inoculation showed almost zero chlorophyll *a*, which reflects the very low native algal content of the wastewater used in this experiment (Fig. 3f). Although the *Chl a* content in the 25% WW+75% BBM treatment were significantly higher compared to the experimental setup, after 10 days there was no significant difference between the BBM treatment and their wastewater mixtures.

The chlorophyll *a* content could be applied as a measurement for the algal weight and volume, which reflects the empirical link between the nutrient concentration and other biological phenomena in aquatic ecosystems (Berkman & Canov, 2007). The *Chl a* content is widely accepted as a parameter for algal biomass world-wide (Moore

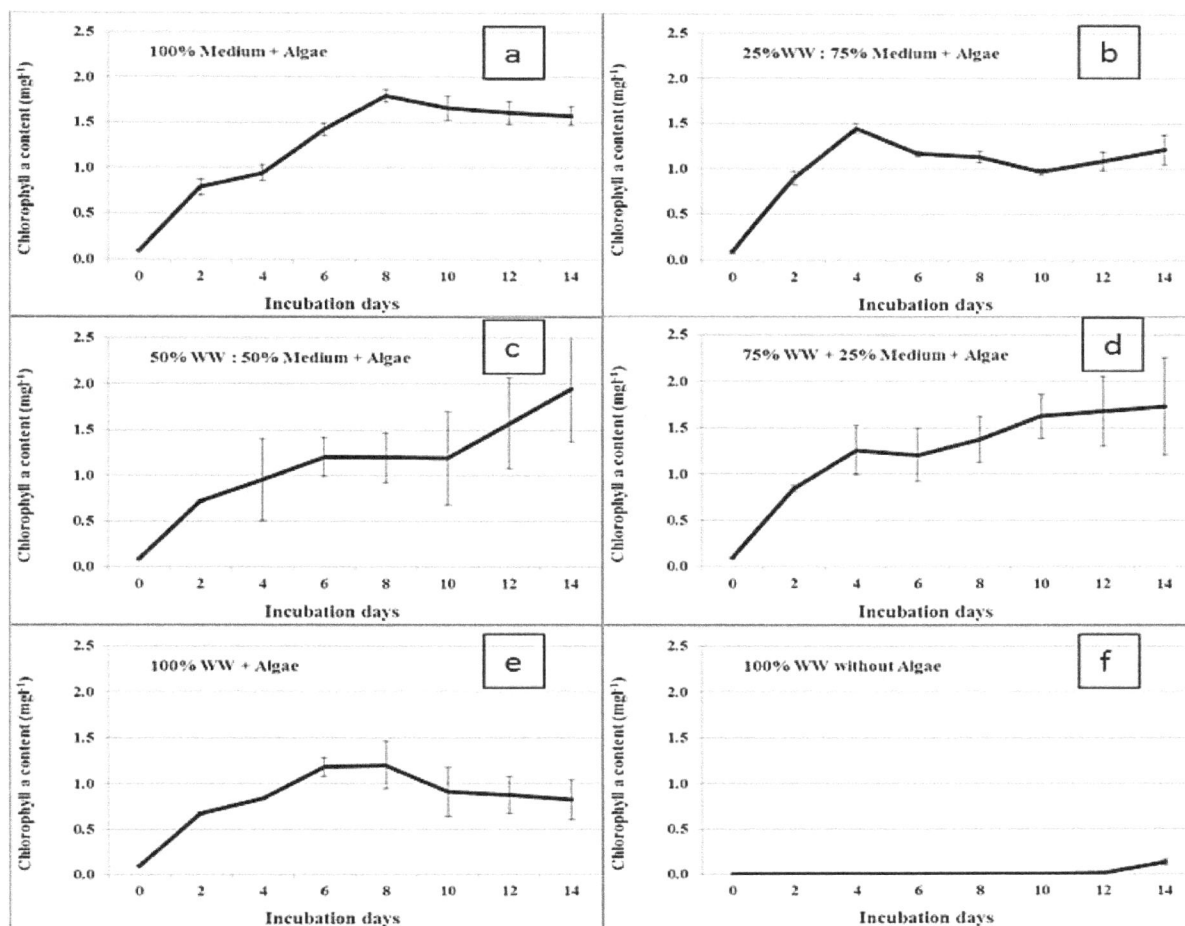

Fig. 3. Chlorophyll *a* content of *S. obliquus* cultivated on BBM, wastewater or their mixtures

& Schindler, 2008; Boyce et al., 2013). In the current study, the differences in the *Chl a* content among the treatments (Fig. 3) may refer to the utilization of nutrients during the growth period. The noted decline in the chlorophyll *a* content was associated with the reduction in the nutrient content (P and N) in the growth medium (Fig. 4 and 5). These results are in accordance with Juneja et al., (2013) who reported the influence of nitrogen and phosphorus deficiency on reducing the chlorophyll *a* and protein content of several microalgae genera, including *Scenedesmus*.

The presented data are the mean ± sd and the letters show significance at P≤ 0.05.

Nutrients utilization of wastewater-media cultivation mixtures

The nutrient content of the growth medium is one of the main factors affecting the microalgal growth, biomass and lipid production. As phosphorus and nitrogen are the major nutrients that can influence the growth and productivity of microalgae, their effect and utilization were studied in the present study.

Phosphorus content

The phosphorus utilization by *obliquus* for different wastewater-medium mixtures was presented in Figure 4. The data obviously declared that the concentration of P in the wastewater used in this experiment was very low, which may be owing to the primary treatment of municipal wastewater. The initial P concentration in some treatments varied due to the mixing rate of WW with the growth medium, which contained P in its composition. The P concentration in the WW was 1.27 ppm, which decreased to 0.3–0.4 ppm in 100% WW after 14 days of incubation with or without the algal inoculation.

The highest P content was in the 100% growth medium and decreased according to the mixing ratio. Generally, the P content in the mixed growth media sharply declined in all the treatments during the period of 2–4 days, except the mixing ratio of 25% WW+75% BBM, which declined quickly starting after the 4th day of incubation (Fig. 4). These results can explain the drop in the OD after 4–6 days of incubation in the WW-BBM mixtures (Fig. 3). Almost the entire P content was utilized when mixing the WW and BBM by the ratio of 75%: 25% (Fig. 4d), providing the most economic initial P concentration

during the growth period. This result suggested that the P concentration was slightly higher than 15 ppm P (between 15 – 25 ppm), which could be enough for growing *S. obliquus* for 14 days. In his work on *Chlorella* microalgae that grow on municipal wastewater, Wang et al. (2013) stated that microalgae can grow well even in low P concentrations, suggesting that P may not be the limiting factor for its growth. Accordingly, the total nitrogen and phosphorus content are considered to be essential factors and directly influence the growth of the algal community composition and richness (Xu et al., 2010). Nitrogen and phosphorus have been reported to be major nutritional elements that govern the potentiality of a water source for algal growth (Abou-Shanab et al., 2013).

Nitrogen content

The Nitrate-N and ammonium-N are the major forms of nitrogen in wastewater that could influence the microalgal growth. These two forms of nitrogen were examined in this study to investigate the nitrogen content of the culture during growth. The ammonium nitrogen was not detected either in the initial wastewater or in the mixtures during the experiments. This result may be owing to the pretreatment of the wastewater utilized in this study. The removal of ammonium throughout the pretreatments of the municipal wastewater has been reported by Klieve and Semmens (1980).

The initial nitrate content in different wastewater-media mixtures varies because of the percentage of wastewater in the mixtures. *S. obliquus* utilized N after cultivation led to a gradual decrease in its content during the incubation period (Fig. 5).

The lowest N content was observed at the end of the incubation time while inoculating the microalgae strain in 100% wastewater. In this case, *S. obliquus* consumed almost all the N (92.17%) in the wastewater. On the other hand, the native community in the WW without inoculation utilized only 59.8% of the N content during the growing period. This result may be due to the low P content in the wastewater (Fig. 4-e and f), which causes the lower growth rate as shown in the OD and chlorophyll *a* content data (Fig. 2-e and Fig. 3-f). On the other hand, growing microalgae in the BBM for 14 days resulted in the reduction of its N content by 70%. The reduction in the N content when mixing WW with BBM was higher than its value when cultivating

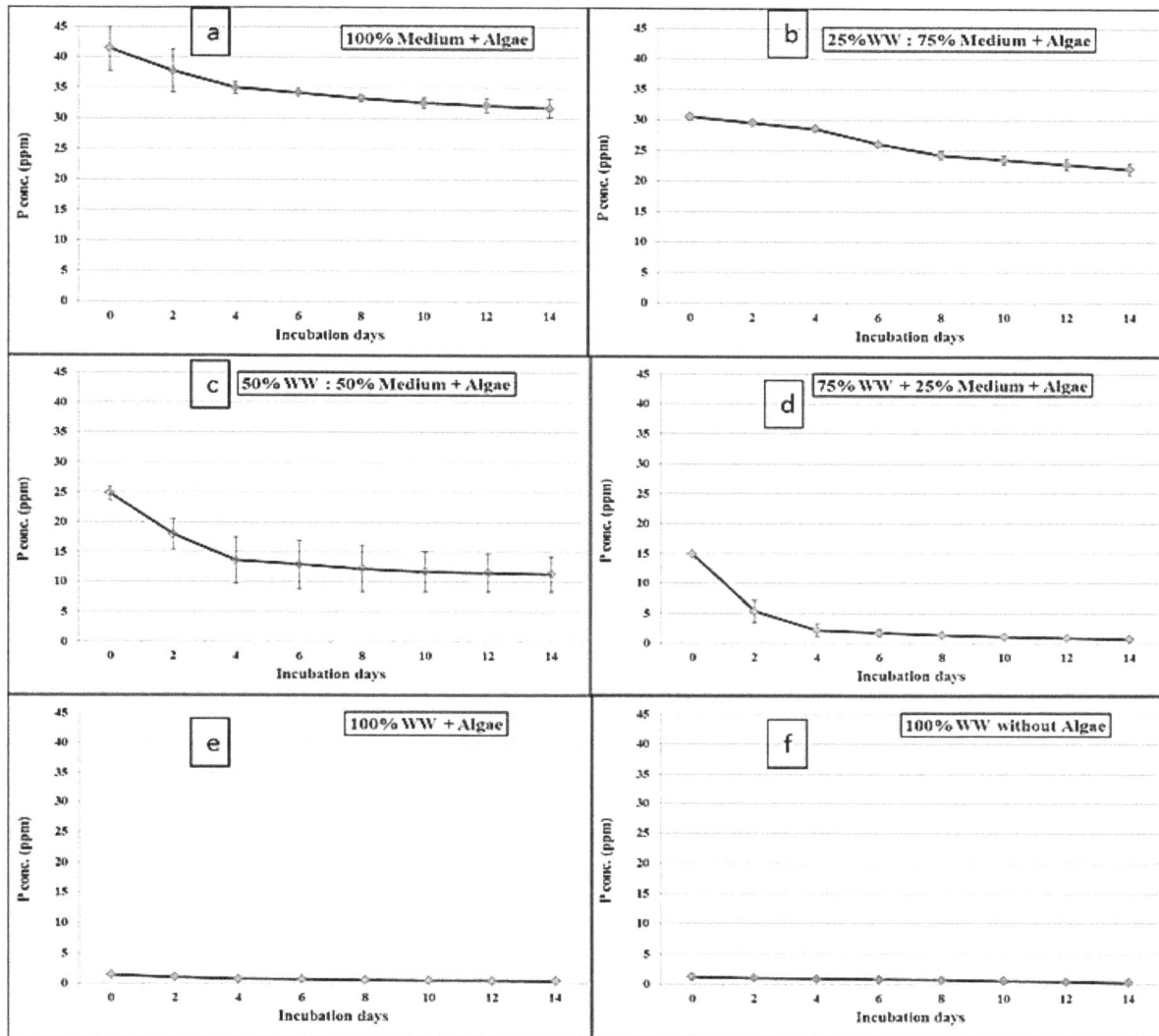

Fig. 4. Phosphorus content of different wastewater-media combinations during the cultivation of *S. obliquus*

microalgae in BBM. This finding is in accordance with the results of growth parameters (OD and chlorophyll *a*) in the first stage of the growth curve (6–8 days), suggesting that the abundance of N and P in the WW-BBM mixture improved the growth of *S. obliquus*. In this way, Vasileva et al. (2016) reported the importance of the nitrogen source and concentration in the growth media for the production rate and the biochemical composition of *Scenedesmus* sp.

Dry biomass production

The growth rate and biomass productivity of microalgae differed according to the growth condition (Fig. 6). The highest dry biomass was produced while growing *S. obliquus* on a medium composed of 75% WW and 25% BBM and 100% BBM, which reached 0.529 and 0.524 gl⁻¹,

respectively. As expected, the lowest biomass production was recorded with 100% WW without algae inoculation treatment. The obtained data revealed that there were no significant differences (P ≤0.05) in dry biomass among BBM and all the BBM-WW mixtures. These results indicate the suitability of wastewater for partial replacement of growth media for the production of adequate algal biomass.

The adequate existence of nutrients in both the 75% waste water and 25% algal growth medium, as well as 100% algal growth medium treatment, could explain the superior dry biomass production from the microalgae. While the algal growth medium is normally designed to provide the growing algae with sufficient and balanced nutrients, the wastewater could introduce several nutrients into the cultivated microalgae in satisfactory amounts (Vasileva et al., 2015; 2016).

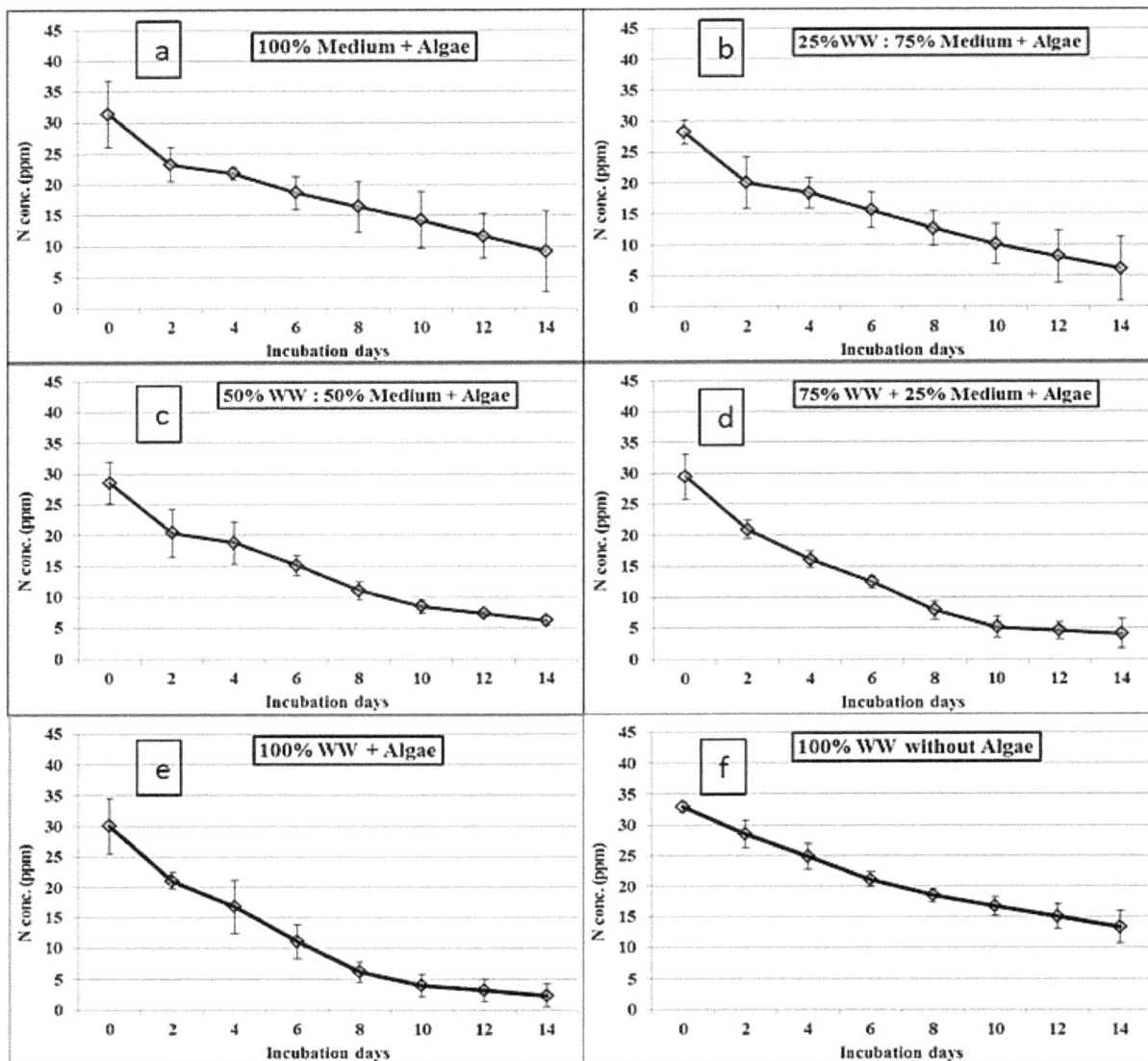

Fig. 5. Nitrogen (Nitrate) content in different wastewater-media combinations used for cultivation of microalgae *S. obliquus*

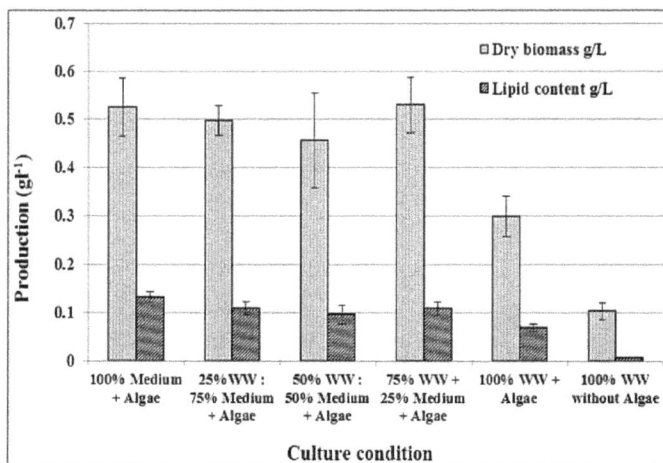

Fig. 6. Dry biomass and lipid content of *S. obliquus* cultivation on different wastewater-media combinations after 14 days of incubation

Lipid content

The algal biomass produced from the different treatments was utilized for lipid extraction. The amount of lipids produced by *S. obliquus* cultivated on WW, BBM and their mixtures are presented in Figure 6 and the percentage of the lipids contents are presented in Table 1. Table 2 compares the lipid content of some *Scenedesmus* spp. from the current and previous studies (Griffiths & Harrison, 2009; Rodolfi et al., 2009; Mata et al., 2010; Tang et al., 2011; Hakalin et al., 2014). The highest lipid accumulation was recorded when growing *S. obliquus* in 100% BBM growth medium, which may due to the highest accumulated biomass from this treatment at the end of growth period. The lipid content in this treatment reached 25.2% of the cell dry biomass. The second lipid accumulation percentage (22.7%) was recorded when the microalgae *S. obliquus* were grown on 100% wastewater. Out of the wastewater and growth medium mixtures, mixing 25% WW+75% BBM had the highest lipid content. Among the recorded variances in the lipid content of *S. obliquus*, there were no significant differences in its values among all the treatments, except for the 100% WW without inoculation, which exhibited the lowest lipid content. A higher lipid accumulation rate in the WW treatment corresponds to the nutrient-deprived conditions and resulted in nitrogen stress on the algal cells leading to increases in the lipid content. Such effects have been stated by Guldhe et al. (2014) and Ra-manna et al. (2014), who reported high levels of lipids under nutrient stress.

Fatty acid composition of S. *obliquus* cultivated in different wastewater mixtures

The lipids extracted from the microalgae *S. obliquus* that were cultivated in different wastewater-media mixtures, as well as in BBM, were subjected to esterification and analyzed using GC. Table 3 explores the fatty acid profiles of algal lipids from different treatments.

The obtained results referred to the lipid profile of *S. obliquus*, which differs depending on the cultivation medium and conditions. The highest percentage of C16-C18 fatty acids (54.76% from total lipids) was recorded when the algae were cultivated in 100% wastewater, followed by 46.96% in the case of 100% BBM. Culturing *S. obliquus* on 25:75 WW-BBM produced the lowest C16–18 fatty acids among all the treatments. The increase in the appropriate lipid fraction (C16-C18) in the wastewater may refer to the starvation condition, while the increase in its content in BBM could be attributed to the balance in the composition of this medium.

Amin et al. (2013) stated that C16-C18 fatty acids are the ideal fraction for biodiesel production. El-Baz et al. (2016) studied the performance of the biodiesel obtained from *S. obliquus* in the diesel engine and recommended it as an environmentally-friendly biofuel.

Table 1. Lipid content (%) of *Scenedesmus obliquus* cultivation on different wastewater-media combinations

Culture condition	Lipid content (%)
100% Medium + Algae	25.2[a] ± 1.8
25% WW: 75% Medium + Algae	21.9[a] ± 1.4
50% WW: 50% Medium + Algae	21.1[a]± 3.6
75% WW + 25% Medium + Algae	20.4[a] ± 3.9
100% WW + Algae	22.7[a] ± 1.9
100% WW without Algae	6.8[b] ± 1.3

Table 2. Lipid content of some *Scenedesmus* spp.

Scenedesmus species	Lipid content [% of dry weight]	Reference
Scenedesmus obliquus	20.4–25.2	Current study
	21 – 42	Griffiths & Harrison (2009)
	11 – 55	Mata et al. (2010)
	15.2- 24.4	Tang et al. (2011)
Scenedesmus dimorphus	26	Griffiths & Harrison (2009)
Scenedesmus quadricauda	18.4	Rodolfi et al. (2009)
Scenedesmus rubescens	18.5 – 23.2	Hakalin et al. (2014)

Table 3. Fatty acid composition of *S. obliquus* cultivation on different wastewater-media combinations

Fatty acid	Algal cultivation medium				
	100% BBM	25% WW: 75% BBM	50% WW: 50% BBM	75% WW: 25% BBM	100% WW
Palmitic acid C16 (0)	16.45	10.90	12.33	18.40	16.03
Palmitoleic acid C16 (1)	2.54	nd*	nd	nd	nd
Stearic acid C18 (0)	6.96	3.30	7.70	7.56	11.79
Oleic acid C18 (1)	10.18	6.60	11.85	12.53	19.46
Linoleic acid C18 (2)	7.65	3.20	3.78	4.80	5.60
Linolenic acid C18 (3)	3.18	1.60	0.90	3.00	1.88
Eicosanoic acid C20 (1)	2.90	1.50	2.20	2.90	3.03
Erucic acid C22 (1)	6.75	3.90	9.58	10.54	16.04
Lignoceric acid C24 (0)	4.76	26.97	16.02	nd	nd
Nervonic acid C24 (1)	8.61	nd	nd	11.40	nd
Hexacosanoic acid C26	13.57	23.90	16.60	11.55	10.73
Heptacosanoic acid C27	11.27	11.36	15.18	13.98	13.58
Undefined	5.18	6.77	3.86	3.34	1.86
Total	**100**	**100**	**100**	**100**	**100**

* nd – Not detected.

CONCLUSION

Secondary treated municipal wastewater represents a huge portion of wastewater that is discharged to water bodies. Thus, finding a practical way to utilize it to produce an eco-friendly product(s) is an important step for optimizing the benefits from one of the most abundant lost resources of water. Due to its relatively low organic content, light permeability and the existing mineral nutrients, secondary treated municipal wastewater could be utilized by mixotrophic microalgae as a low-cost growth medium to produce biomass-based biofuels. The results of the current study indicated no significant difference in the biomass production and lipid content of *S. obliquus* grown in either BBM or BBM-WW mixtures. These results suggest the utilization of mixtures containing a higher proportion of secondary treated wastewater, such as 75% WW+25% BBM or 50% WW+50% BBM, which could increase the economical production of microalgae for biodiesel. It saves water and nutrients as well. The growth of *S. obliquus* in wastewater could be improved by enhancing its nutrient content either by mixing wastewater and algal medium in certain ratios or by adding nutrients to wastewater. Further studies are needed to enhance the growth and lipid productivity of microalgae grown on wastewater as abundant cheap cultivation medium either in the lab scale and large scale.

REFERENCES

1. Abou-Shanab R.A., Ji M.K., Kim H.C., Paeng K.J., Jeon B.H. 2013. Microalgal species growing on piggery wastewater as a valuable candidate for nutrient removal and biodiesel production. J Environ Manage; 115, 257–264.

2. Ahmad AL, Yasin NM, Derek CJ, Lim JK. 2011. Microalgae as a sustainable energy source for biodiesel production, a review. Renew Sust Energ Rev; 15(1), 584–593.

3. Akpor OB, Momba MN, Okonkwo J. 2007. Phosphorus and nitrate removal by selected wastewater protozoa isolates. Pak J Biol Sci; 10(22), 4008–14.

4. Álvarez-Díaz PD, Ruiz J, Arbib Z, Barragán J, Garrido-Pérez MC, Perales JA. 2015. Wastewater treatment and biodiesel production by Scenedesmus obliquus in a two-stage cultivation process. Bioresour Technol; 181, 90–96.

5. Amin NF, Khalafallah MA, Ali MA, Abou-Sdera SA, Matter IA. 2013. Effect of some nitrogen sources on growth and lipid of microalgae Chlorella sp. for biodiesel production. J Appl Scis Res; 9(8), 4845–4855.

6. Ansari FA, Singh P, Guldhe A, Bux F. 2017. Microalgal cultivation using aquaculture wastewater, Integrated biomass generation and nutrient remediation. Algal Res; 21, 169–177.

7. Basova EM, Bulanova MA, Ivanov VM. 2011. Photometric detection of urea in natural waters. Mosc Univ Chem Bull; 66(6), 345–50.

8. Bellinger EG, Sigee DC. 2015. Freshwater algae, identification and use as bioindicators. John Wiley and Sons.

9. Berkman JAH, Canov MG. 2007. Algal biomass indicators. Version 1.0 8. U.S. Geological Survey TWRI Book.

10. Bharathiraja B, Jayamuthunagai J, Chakravarthy M, Kumar RR, Yogendran D, Praveenkumar R. 2015. Algae, Promising Future Feedstock for Biofuels. In: Singh B, Bauddh K, Bux F, eds. Algae and Environmental Sustainability. Springer India; 1–8.

11. Bisht TS, Pandey M, Pande V. 2016. Impact of different nitrogen sources on biomass growth and lipid productivity of Scenedesmus sp. for biodiesel production. J Algal Biomass Utln; 7(4), 8–36.

12. Bligh EG, Dyer WJ. A rapid method of total lipid extraction and purification. Can J Biochem Physiol 1959;37(8), 911–7.

13. Boyce DG, Lewis MR, Worm B. 2010. Global phytoplankton decline over the past century. Nature;466(7306), 591–596.

14. Darwesh OM, Moawad H, Wafaa MA, Olfat SB, Sedik MZ. 2014. Bioremediation of textile Reactive Blue (RB) azo dye residues in wastewater using experimental prototype bioreactor. J Environ Health Scis Eng; 5(4), 1203–1219.

15. De-Nardo L.U. 1929. The pyrogallol method for the determination of nitrates in soil and waters. Giorn Chim Ind Appl; 11, 107–9.

16. El-Baz FK, Gad MS, Abdo SM, Abed KA, Matter IA. 2016. Performance and exhaust emissions of a diesel engine burning algal biodiesel blends. Int J Mech Mech Eng; 16(3), 150–157.

17. Griffiths MJ, Harrison ST. 2009. Lipid productivity as a key characteristic for choosing algal species for biodiesel production. J App Phyc; 21(5), 493–507.

18. Guldhe A, Singh B, Rawat I, Bux F. 2014. Synthesis of biodiesel from Scenedesmus sp. by microwave and ultrasound assisted in situ transesterification using tungstated zirconia as a solid acid catalyst. Chem Eng Res Des; 92(8), 1503–1511.

19. Hakalin NL, Paz AP, Aranda DA, Moraes LMP. 2014. Enhancement of cell growth and lipid content of a freshwater microalga Scenedesmus sp. by optimizing nitrogen, phosphorus and vitamin concentrations for biodiesel production. Natu Scie; 6, 1044–1054.

20. Hodaifa G, Martínez ME, Sánchez S. 2008. Use of industrial wastewater from olive-oil extraction for biomass production of Scenedesmus obliquus. Bioresour Technol; 99(5), 1111–1117

21. Hosikian A, Lim S, Halim R, Danquah MK. 2010. Chlorophyll extraction from microalgae: a review on the process engineering aspects. Int J Chem Eng doi:10.1155/2010/391632.

22. Ichihara KI, Fukubayashi Y. 2010. Preparation of fatty acid methyl esters for gas-liquid chromatography. J Lipid Res;51(3), 635–40.

23. Jena J, Nayak M, Panda HS, Pradhan N, Sarika C, Panda PK, Rao BV, Prasad RB, Sukla LB. 2012. Microalgae of Odisha coast as a potential source for biodiesel production. World Environ; 2(1), 11–16.

24. Ji MK, Yun HS, Park YT, Kabra AN, Oh IH, Choi J. 2015. Mixotrophic cultivation of a microalga Scenedesmus obliquus in municipal wastewater supplemented with food wastewater and flue gas CO_2 for biomass production. J Environ Manage; 159, 115–120.

25. Juneja A, Ceballos RM, Murthy GS. 2013. Effects of environmental factors and nutrient availability on the biochemical composition of algae for biofuels production: a review. Energies; 6(9), 4607–4638.

26. Klieve JR, Semmens MJ. 1980. An evaluation of pretreated natural zeolites for ammonium removal. Water Res; 14(2), 161–168.

27. Lee RE. Phycology. Cambridge University Press; 2008.

28. Li X, Xu H, Wu Q. 2007. Large- scale biodiesel production from microalga Chlorella protothecoides through heterotrophic cultivation in bioreactors. Biotechnol Bioeng; 98(4), 764–771.

29. Lv J, Feng J, Liu Q, Xie S. 2017. Microalgal cultivation in secondary effluent: Recent developments and future work. Int J Mol Sci; 18(1), 79. doi:10.3390/ijms18010079.

30. Lyon SR, Ahmadzadeh H, Murry MA. 2015. Algae-based wastewater treatment for biofuel production: processes, species, and extraction methods. In: Moheimani N R, McHenry M P, de Boer K, Bahri P A, eds. Biomass and Biofuels from Microalgae. Springer International Publishing, 95–115.

31. MacIntyre HL, Kana TM, Anning T, Geider RJ. 2002. Photoacclimation of photosynthesis irradiance response curves and photosynthetic pigments in microalgae and cyanobacteria. J Phycol; 38(1), 17–38.

32. Madkour FF, Kamil A, Nasra HS. 2012. Production and nutritive value of Spirulina platensis in reduced cost media. Egypt J Aqua Res; 38, 51–57.

33. Maity JP, Bundschuh J, Chen CY, Bhattacharya P. 2014. Microalgae for third generation biofuel production, mitigation of greenhouse gas emissions and wastewater treatment: Present and future perspectives–A mini review. Energy; 78, 104–113.

34. Mandal S, Mallick N. 2009. Microalga Scenedesmus obliquus as a potential source for biodiesel production. Appl Microbiol Biot; 84(2), 281–291.

35. Mandal S, Mallick N. 2011. Waste utilization and biodiesel production by the green microalga

Scenedesmus obliquus. Appl Environ Microb; 77(1), 374–377.

36. Martinez-Jeronimo, F, Espinosa-Chavez, F. 1994. A laboratory-scale system for mass culture of fresh water microalgae in polyethylene bags. J Appl Phycol; 6, 423–425.

37. Mata MT, Melo AC, Meireles S, Mendes AM, Martins AA, Caetano NS. 2013. Potential of microalgae Scenedesmus obliquus grown in brewery wastewater for biodiesel production. Chem Eng Trans; 32, 901–906.

38. Mata TM, Martins AA, Caetano, NS. 2010. Microalgae for biodiesel production and other applications: a review. Renew Sustainable Energy Rev; 14(1), 217–232.

39. Matter I.A, Darwesh O.M, El-baz F.K. 2016. Using the natural polymer chitosan in harvesting Scenedesmus species under different concentrations and cultural pH values. Int J Pharm Bio Sci, 7(4), 254–260.

40. Meher LC, Sagar DV, Naik SN. 2006. Technical aspects of biodiesel production by transesterification–a review. Renew Sust Energ Rev; 10(3), 248–268.

41. Meng X, Yang J, Xu X, Zhang L, Nie Q, Xian M. 2009. Biodiesel production from oleaginous microorganisms. Renew Energ; 34(1), 1–5.

42. Min M, Wang L, Li Y, Mohr MJ, Hu B, Zhou W, Chen P, Ruan R. 2011. Cultivating Chlorella sp. in a pilot-scale photobioreactor using centrate wastewater for microalgae biomass production and wastewater nutrient removal. Appl Biochem Biotechnol; 165, 123–137.

43. Mobin S, Alam F. 2014. Biofuel production from algae utilizing wastewater. in: 19th Australasian Fluid Mechanics Conference Melbourne, 2014, Australia. Australia: RMIT University, 8–11.

44. Moore JW, Schindler DE. 2008. Biotic disturbance and benthic community dynamics in salmon-bearing streams. J Anim Ecol; 77(2), 275–284.

45. Pizarro C, Mulbry W, Blersch D, Kangas P. 2006. An economic assessment of algal turf scrubber technology for treatment of dairy manure effluent. Ecol Eng; 26(4), 321–327.

46. Rainer S, Arne N, Diethard T. 2007. An evaluation of LSU rDNA D1-D2 sequences for their use in species identification. Front Zool; 4(1), 6. doi:10.1186/1742–9994–4-6.

47. Ramanna L, Guldhe A, Rawat I, Bux F. 2014. The optimization of biomass and lipid yields of Chlorella sorokiniana when using wastewater supplemented with different nitrogen sources. Bioresour Technol; 168, 127–135.

48. Richmond A. 2008. Handbook of microalgal culture: biotechnology and applied phycology. John Wiley and Sons.

49. Rodolfi L, Chini Zittelli G, Bassi N, Padovani G, Biondi N, Bonini G, Tredici MR. 2009. Microalgae for oil: Strain selection, induction of lipid synthesis and outdoor mass cultivation in a low-cost photobioreactor. Biotechnol bioeng; 102(1), 100–112.

50. Tang D, Han W, Li P, Miao X, Zhong J. 2011. CO2 biofixation and fatty acid composition of Scenedesmus obliquus and Chlorella pyrenoidosa in response to different CO2 levels. Bioresour Technol; 102(3), 3071–3076.

51. Toyub MA, Miah MI, Habib MAB, Rahman MM. 2008. Growth performance and nutritional value of Scenedesmus obliquus cultured in different concentrations of sweetmeat factory waste media. Bang J Anim Sci; 37(1), 86–93.

52. Vasileva I, Marinova GV, Gigova LG. 2015. Effect of nitrogen source on the growth and biochemical composition of a new Bulgarian isolate of Scenedesmus sp. J. Biosci Biotechnol;125–129.

53. Wang C, Yu X, Lv H, Yang J. 2013. Nitrogen and phosphorus removal from municipal wastewater by the green alga Chlorella sp. J Environ Biol; 34(2), 421–425.

54. Wilkie AC, Mulbry WW. 2002. Recovery of dairy manure nutrients by benthic freshwater algae. Bioresour Technol; 84(1), 81–91.

55. Xin L, Hong-ying H, Ke G, Ying-xue S. 2010. Effects of different nitrogen and phosphorus concentrations on the growth, nutrient uptake, and lipid accumulation of a freshwater microalga Scenedesmus sp. Bioresour Technol; 101(14), 5494–5500.

56. Xu H, Miao X, Wu Q. 2006. High quality biodiesel production from a microalga Chlorella prothecoides by heterotrophic growth in fermenters. J Biotechnol; 126(4), 499–507.

57. Zhang Y, Su H, Zhong Y, Zhang C, Shen Z, Sang W, Yan G, Zhou X. 2012. The effect of bacterial contamination on the heterotrophic cultivation of Chlorella pyrenoidosa in wastewater from the production of soybean products. Water Res; 46, 5509–5516.

17

Changes of the Content of Heavy Metals and PAH's in Sewage Sludge Treated with Reed Bed Lagoons

Dariusz Boruszko[1]

[1] Faculty of Civil and Environmental Engineering, Bialystok University of Technology, 45A Wiejska St., 15-351 Bialystok, Poland, e-mail: d.boruszko@pb.edu.pl

ABSTARCT

In Poland, low-input methods such as composting, vermicomposting, reed beds, willow energy and solar driers are increasingly often being used in the processing of sewage sludge. The northeastern part of Poland has operated successfully for years using these methods. This paper presents the results of using low-cost methods of sludge treatment in the wastewater treatment plant located in Zambrow, Podlaskie Province. The results of sewage sludge studies on the PAHs and heavy metals content after treatment in a reed bed system are presented. Among the 16 examined PAHs, the lowest concentration was obtained for the dibenz (a,h)anthracene. Not a single sample exceeded a concentration of 100 µg/kg d.m. The highest concentration was exhibited by fluoranthene, benzo(g,h,i)perylene and indeno(1,2,3-c,d)pyrene. The concentration of these compounds exceeded 1000 µg/kg d.m. The obtained results for the PAHs in sewage sludge from the reed lagoon at the treatment plant in Zambrow showed that the average content of PAHs studied was approximately 8300 µg/kg d.m. The lowest concentration, below 1.3 mg/kg d.m. of the seven heavy metals examined was obtained for mercury (Hg). On the other hand, the highest concentration, exceeding 1300 mg/kg d.m. was found in the case of zinc (Zn). The obtained results for the heavy metals in sewage sludge from the reed bed lagoons in Zambrow show that the average content of heavy metals studied is approximately 1620 mg/kg d.m. The results of the study demonstrate a high efficiency of the low-cost methods used in the Zambrów WWTP in terms of the quality of processed sludge. The sewage sludge from the lowest layer of the reed lagoon (13–14 years of dewatering and transformation) are characterized by the lowest PAHs and heavy metals content. The higher a sediment layer lies, i.e. the shorter the time of processing, the higher are the PAHs and content of heavy metals content. This indicates a great role of reeds in the accumulation of these compounds.

Keywords: polycyclic aromatic hydrocarbons, sludge treatment reed bed, speciation of heavy metals

INTRODUCTION

The low-input methods of processing sewage sludge are characteristic first of all by their simple design and technology as well as by their easy operation. These methods allow using only a limited number of process equipment resulting on low consumption of electrical energy. Utilisation of natural processes occurring in the environment allows maintaining low operation costs.

Reed lagoons are successfully used for the long-term stabilization and dehydration of sludge from waste water treatment. They are easy to construct and operate and consume small amounts of energy [Zwara & Obarska-Pempkowiak 2000; Nielsen et al. 2012]. What is perhaps the most im-portant in STRB is that any kind of sewage sludge of organic character can be stabilized [Uggetti et al. 2010]. This technology has been applied with success for more than thirty years, among others in Denmark. The initial content of organic matter in the sediment between 50% and 65% and low load of the lagoon surface allow for drainage of sewage sludge to a degree no less as when using filter presses up to the value of 20–30% DM [Nielsen 2003, 2011; Nielsen et al. 2012].

Reed lagoon is compulsorily equipped with a drainage system that allows ventilation of the processed sewage sludge and effluent discharge. The efficiency of dehydrating is also provided by gravity drainage and evaporation. In addition, the long process of sludge stabilization (7–15 years)

remarkably reduces the organic matter content and causes the processes of sludge humification (Peruzzi et al. 2009). Many authors acknowledge that the sludge stabilized in STRB is similar in its chemical composition to the compost (high content of organic matter and nitrogen) [Kołecka & Obarska-Pempkowiak 2008].

From the ecological point of view, the PAHs collection and transformation within individual ecosystems is more important. Since nearly 90% of the PAHs contained there come from soil, specific actions and means should be implemented to protect these ecosystems. Moreover, the studies conducted so far demonstrate differences in the degree and the speed of the PAH transformation in the soil, which proves their long life and the possibility of bioaccumulation in this environment. The PAH (and other organic compounds) also occur in plants, thus being a great concern, as the mechanism of absorbing the PAHs and other organic compounds is not fully understood. It is assumed that accumulation of harmful substances on the surface of leafs is possible during atmospheric precipitations or due to contamination with soil particles. Irrespective of that in many countries studies are conducted in order to determine the limit values for organic substance content in sewage sludge, composts and in potable water. In Europe, when sewage sludge is used in agriculture, the limits on the content of PAHs in naturally used municipal sewage sludge have been proposed [WD, 2000; WD, 2010]. Some of the EU countries have introduced internal limitation of PAHs contents in sewage sludge intended for use in agriculture [Aparicio et al. 2009; Gomez et al. 2002; Kapanen 2013; Langenkamp et al. 2001; Vacha et al. 2006]. In Germany, there are no binding regulations concerning the PAHs content in sludge, however, a designation of 9 PAH is recommended; the sum of which should not exceed 6000 µg/kg d.m. In the proposed amendment to the Sediment Directive, the sum of 11 hydrocarbons should not exceed 6000 µg/kg d.m.

The Polish legislation does not provide any regulations on the maximum PAHs content in sewage sludge designed for use in the agriculture. There is no uniform procedure of analytic analysis of PAHs in sludge either. Determination of micro-contaminations in the sewage sludge is one of the most complicated analytical tasks connected with the heterogeneity of these materials and with differentiated PAH concentrations [Smolik 2005].

Although in Poland the PAHs content in the sediments used in agriculture is not subject to norms, these substances are taken into account in the draft of alterations/amendments to the UE Directive 75/442/EWG according to which a permissible summary concentration of 11 aromatic compounds (acenaphtene, phenatrene, fluorene, fluorantene, pyrene, benzo(k)fluorantene, benzo(a)pyrene, benzo(ghi)perylene, indeno(1,2,3-cd)pyrene) should be 6 mg·kg-1 (converted to dry mass).

The content of heavy metals in the stabilized sewage sludge is often crucial for the possibility of their use in agriculture. Their excessive content in the sewage sludge may be toxic and dangerous to the environment. The knowledge on their overall content, but also their fractional distribution is highly important. The spread of BCR method enabling to determine the heavy metals present in sewage sludge in 4 fractions was of great importance for science: ion exchange and carbonate fraction (I – the most mobile), hydroxide fraction (bound with oxide and hydroxide iron and magnesium) (II), organic fraction (III), residua fraction (IV – the most stable) [Boruszko 2013a; Peruzzi et al. 2011; Kolecka & Obarska-Pempkowiak 2008].

This paper presents selected examples of experience and the study of one of the first wastewater treatment plants in the Podlaskie Province in Zambrów run by the Zambrowskie Ciepłownictwo i Wodociągi sp. z o.o. (District Heating and Waterworks Zambrów Ltd), which has successfully treated sewage sludge by low-cost methods for a dozen or so years.

During the research, the following values were determined: dry matter and organic matter concentration as well as concentration of nitrogen, phosphorus, 16 polycyclic aromatic hydrocarbons PAHs, total and selected heavy metals: Cd, Cr, Cu, Ni, Pb, Zn.

METHODS

Analyzed systems

In the municipal wastewater treatment plant of Zambrów, approximately 1 ton DM of raw sewage sludge is generated per day. The sludge management is based on sludge lagoons, reed beds and earthworm fields. Outside of the winter season, 70% of the sludge is carried away directly to a reed bed with an area 3 times 3500 m². In winter, the sludge is directed onto sludge lagoons provided with a vertical drainage, ensuring a high efficiency of the sludge dewatering while filling the lagoon. In the wastewater treatment plant there are 2 sludge lagoons with a capacity of 2500 m³ each. Approximately 80–90% of the sludge collected on sludge lagoons is pumped in summer time onto earthworm fields with an area of about 1 ha. The remaining sludge is utilised in the agriculture, for reclamation in autumn (after harvest). This solution allows shortening the time of the sludge removal from the wastewater treatment plant down to two weeks, i.e. during emptying the sludge lagoon and after removing the sludge from earthworm fields during a period when this product is needed.

Sedimentary lagoons – reeds bed stayed partly in the excavation and partly in the embankment. The height from the top of the embankment is taking 3.2 m flat out into the bottom of the lagoon. The primer base of lagoons is built of clenched clay. Sequences were arranged in the bottom drainage (ϕ 100) in the filter ballast taken to the pumping station located in the lagoon. The channel is being depumped into the production line to sewers reaching the sewage treatment plant, i.e. a local pumping station. The supplying of sludge was solved by laying a crown escarpments of pipelines from PVC pipes of built in three-way adapters. In the upper layer of the filter ballast in the first year of the use they effected planting the reed in the quantity of 8 cuttings to the square metre. In the winter period, the lagoons are not being fed with sludge. The ground part of the reed was not removed for the entire operating period. In early spring, wither rushes of the reed constitute the additional organic substances. Long-term storing of reed "transfers" dead ground parts of plants every year, creating similar conditions to of the ones in a compost heap.

Sampling strategy

The samples were collected at 4 points, evenly spaced along the symmetry axes for each basin. The samples of the sludge from the same layers were mixed and their quality was determined.

The equipment

From 2012–2014, the sludge samples were taken from the middle part of the oldest reed bed (still in service) in the 3-meter profile of the sediment fill. Sampling was performed using a special probe to enable the extraction of sewage sludge from the full depth of the lagoon. The samples were averaged and sludge samples from each 0.5 m section were tested. In 2013, the service of the oldest lagoon was finished after 14 years.

The studies reflect the following STBR cycles:

- end of 2012 – after last 14th year of STBR supply with sewage sludge at full load,
- end of 2013 – after transitional year at STBR supply with sewage sludge at minimum load,
- end of 2014 – year after STBR closing.

Analitycal procedure

The analysis of the obtained material included, among others, carrying out the following examinations: heavy metal content, fertilizing values, liquefication, dry mass, and polycyclic aromatic hydrocarbons (PAHs). The examinations were carried out in the Department of Technology in Engineering and Environmental Protection laboratory, according to the valid norms.

Dry matter and organic matter were determined by using standard methods [APHA 2005].

Kjeldahl nitrogen (N), the sum of organic and ammonia nitrogen, was determined in the analyzed sludge. The sludge sample was dried and homogenized. It was then alkalized using a 35% solution of NaOH and mineralized in the presence of the catalyst $CuSO_4 + K_2SO_4$ using ammonium destilation [APHA 2005]. Determination of ammonia nitrogen was carried out using the distillation method.

For determining the phosphorus concentration, the sample was dried, homogenized and then mineralized using a mixture of the concentrated acids $HCLO_4$ and HNO_3. In the obtained solution, the PO_4^- ions were determined calorimetrically in the reaction with ammonia molybdate in the presence of glycerin with dissolved $SnCl_2$ [APHA 2005].

The study of polycyclic aromatic hydrocarbons (PAHs) contents in sewage sludge was carried out using gas chromatography coupled with mass spectrometry detection. In purified and concentrated extracts from sewage sludge, 16 PAHs were identified (according to US EPA): naphthalene (NAPH), acenaphthylene (ACY), acenaphthene (ACE), fluorene (FLU), phenanthrene (PHE), anthracene (ANT), fluoranthene

(FLA), pyrene (PYR), benzo[a]anthracene (B[a]ANT), chrysene (CHR), benzo[b]fluoranthene (B[b]FLA), benzo[k]fluoranthene (B[k]FLA), benzo[a]pyrene (B[a]PYR), dibenzo[a,h]anthracene (D[a,h]ANT), indeno[1,2,3-c,d]pyrene (Ind[1,2,3-c,d]P), benzo[g,h,i]perylene (B[g,h,i]PER), as well as their sum [APHA 2005].

Extraction procedure

From 10 g to 20 g of sludge sample with a field moisture content was weighed and placed in a conical flask with a capacity of 200 ml. Then, 50 ml of acetone was added; the flask was sealed and subjected to extraction on a shaker for 1 hour. Petroleum ether wit the volume of 50 ml was added and agitated again for 1 hour. The supernatant was decanted and shaken with another 50 ml petroleum ether portion. The supernatant liquid was subsequently decanted. The extracts were combined and acetone and other polar compounds were removed by shaking twice with 400 ml of water. The water was discarded. The organic layer was dried over anhydrous sodium sulfate (VI) and then filtered through a filter. The dried extract was transferred to a concentrating device; 100 ml of isooctane was added as a stabilizer and concentrated to the volume of about 2 ml. The resulting solution was analyzed by means of GC-MS.

Calibration

Five solutions containing the calibration standard mixture and the test sample matrix were analyzed. The calibration function was calculated by linear regression of the corrected peak areas. The current sensitivity of the method was estimated from the calculated regression function. The calibration was performed on the measurement day. Additionally, determination of a certified reference material was carried out in order to investigate the recoveries of PAHs. The values of PAHs recoveries ranged between 80% and 120%, which is permissible for chromatographic methods [Oleszczuk & Baran 2004].

The PAHS extraction was conducted with the use of acetone and petroleum ether. Two extraction methods are described in International Norm: two-phased and one-phased extraction. Two-phased extraction was used in the research.

The sludge samples were treated with mineralization in HACH mineralizator with the use of sylphuric acid and hydrogen peroxide in a mixture of nitric and hydrochloric acid in the ratio of 1:3. For further analysis, the mineralizators were filtered through MN 616 G paper filter.

Determination of the cadmium, nickel and total chromium content was done in the samples of mineralizators with the use of Perkin-Elmer 4100 ZL atomic absorption spectrometr with transversely heated graphite cuvette and Zeeman-effect background correction.

Determination of the mercury content was performed in the samples of mineralizators by means of cold steam technique with the use of Perkin-Elmer 4100 ZL atomic absorption spectrometr equipped with FIAS-200 add-on device.

Determination of the zinc, lead and copper content was carried out in the samples of mineralizators using Varian SpectrAA 20 Plus atomic absorption spectrometr by means of flame atomization.

The determination of individual fractions of heavy matals was completed using the sequential extraction method according to The Community Bureau of Reference (now Standarts measurement and testing Programme) – BCR [Scancar et al. 2000; Kazi et al. 2005]. Four solutions were used in the test: 0.11 molar CH_3COOH (Fraction I), 0.5 molar $NH_2OH \cdot HCL$; pH 1,5 (Fraction II), 1 molar NH_3COONH_3 ; pH 2 (Fraction III) and mixture H_2O_2 and HNO_3 (Fraction IV), respectively. Using these extractions, the following four fractions of metals were separated: ion exchange and carbonate fraction (I – the most mobile), hydroxide fraction (Bound with oxide and hydroxide iron and magnesium) (II), organic fraction (III), residual fraction (IV – the most stable).

Data analysis

Determinations of the test samples were performed in triplicate. The mathematical-statistical calculations presented in the paper were made using common computer software and spreadsheets. In order to study the relationship between the quantitative parameters and to describe the strength of the correlation in the case of a small number of observations, the Spearman's rank correlation coefficient was applied [Sobczyk, 2002]. In order to verify whether there was a statistically significant change, ANOVA rank Kruskala-Wallisa and Friedman were used for each pair of observations [Siegel 1956].

RESULTS AND DISCUSION

A change in the basic parameters of stabilized sewage sludge in reed lagoons is shown in Figure 1. The sludge is stabilized and dehydrated. The degree of sludge dehydration is low and ranges from 2.9% in 2012 to 3.1% in 2014. The lowest dry matter content was recorded in the top layer of a vertical profile – 12.6%. The highest content of dry matter characterized sludge that was deposited on the reed lagoon for the longest period of time (the lowest layer in vertical profile) reached 14.5%. In the middle section of the vertical profile of sludge at a depth of 1.5–2.0 m, a remarkable reduction in the dry matter content was observed: to 10.4% in 2012 and 13.2% in 2014. The dry matter content in STRB at different depths of vertical profile is a measure of dehydration. The degree of dehydration is on the one hand dependent on the degree of water evaporation from the lagoon surface and on the other hand on the gravitational concentration and uptake by reed roots [Kolecka & Obarska-Pempkowiak 2008]. Definitely higher dry matter contents above 29% in the sludge stabilized in STRB were achieved in the studies by Matamoros et al. [2012].

The content of organic matter oscillated within a wider range. The lowest organic matter content of 49.0% DM was recorded in the deepest layer of sludge in 2014, i.e. one year after closing the reed lagoon. The highest organic matter values were observed in the highest sludge layer (sludge that was drained and stabilized for the shortest time); this value ranged from 58.1% DM to 62.2% DM. The organic matter accumulated in the sludge over the years becomes biodegraded and stabilized, leading to a reduction in dry matter content. Similar results of the organic matter

reduction by 10–12% were obtained by Nielsen [2005] and in the studies conducted in Poland in Darżlubie [Zwara & Obarska Pempkowiak 2000].

The average nitrogen content in the sludge stabilized over three years ranged from 3.5% DM to 3.8% DM. The highest nitrogen concentration characterized the top layers of the sludge, ranging from 3.7% DM to 4.1% DM, while the lowest nitrogen contents were found in the sludge from the deepest layers, ranging from 3.3% DM to 3.4% DM. There was a constant, yet slight, nitrogen loss during its stabilization in the reed lagoon. It was described in earlier studies [Boruszko 2015; Peruzzi et al. 2009]. A decrease in the nitrogen content in STRB-stabilized sludge is a result of the element uptake by plants and microbial conversions occurring around the root system (nitrification and denitrification).

Mean phosphorus content in the stabilized sludge was similar, ranging in 2012–2014 between 2.7% DM and 2.8% DM. The deepest sludge layers (in vertical profile) were characterized by the highest phosphorus content (samples from 2.5–3.0 m depth) at average value of 3.2% DM. The lowest phosphorus contents were recorded in sludge samples from the shallowest layer 0.0–0.5 m (sludge that was stabilized for the shortest time); value at this profile ranged from 1.9% DM in 2014 to 2.3% DM in 2012. Studies have confirmed that very long stabilization and decomposition of organic matter can retain phosphorus in the sludge.

The changes in the contents of each of the analyzed 16 PAHs in six layers of the vertical profile of STRB dehydrated and stabilized sludge conducted in 2012–2014 are presented in Table 1 as average values with standard deviation.

1.1

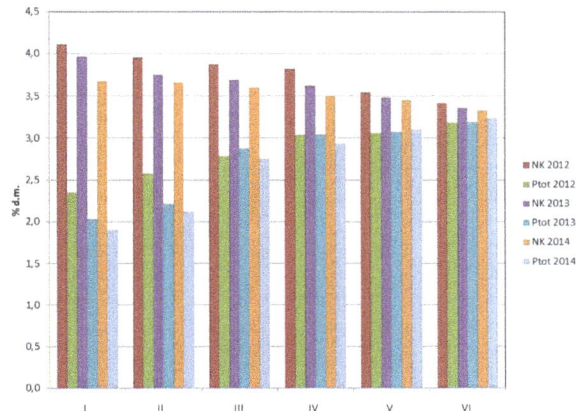

1.2

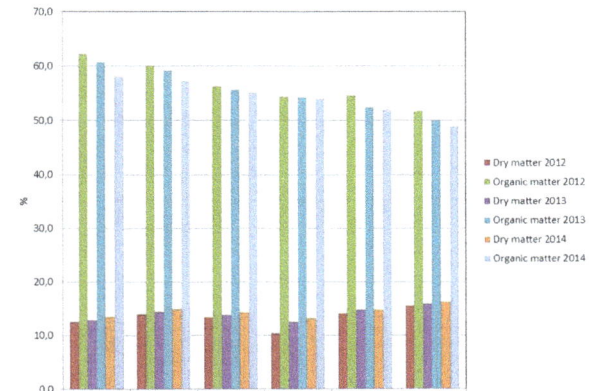

Figure 1. Physicochemical parameters in sludge in reed bed system in 2012, 2013 and 2014 year

Table 1. Quantitative PAHs changes (μg/kg d.m) in sludge stabilized in reed bed system in 2012, 2013 and 2014 year

Year	I	II	III	IV	V	VI	Average
Naphthalene							
2012	278.0±3.6	276.3±5.5	266.3±5.5	265.0±5.6	264.7±6.5	265.0±5.0	269.2±6.2
2013	234.7±10.5	243.0±10.8	220.3±6.0	253.7±2.5	260.3±5.5	258.3±7.5	245.1±15.5
2014	194.3±5.0	235.7±7.0	242.3±7.0	255.0±5.6	239.0±6.6	248.0±8.0	235.7±21.4
Acenaphtylene							
2012	331.7±8.7	320.3±7.6	308.7±6.1	305.7±5.0	277.3±6.5	254.7±5.1	299.7±28.6
2013	315.0±6.0	304.0±6.0	296.3±7.8	286.7±7.1	260.7±3.5	234.3±5.0	282.8±30.1
2014	302.0±7.5	314.0±6.0	277.3±7.0	278.7±1.5	263.3±7.5	235.7±5.0	278.5±27.9
Acenaphtene							
2012	164.3±3.1	177.0±6.6	251.7±7.6	299.3±6.4	284.7±2.1	204.3±3.5	230.2±56.6
2013	176.7±5.5	181.0±3.6	268.3±6.1	332.0±6.0	302.7±6.5	224.3±2.1	247.5±64.2
2014	164.3±6.8	192.7±6.0	234.3±4.5	355.3±7.0	366.3±6.0	270.3±8.0	263.9±83.4
Fluorene							
2012	78.7±2.1	79.0±2.0	84.0±1.7	92.3±1.5	87.3±2.1	63.3±2.5	80.8±10.0
2013	84.7±1.5	84.0±2.0	97.0±2.0	122.3±7.0	104.0±4.6	81.0±4.0	95.5±15.8
2014	76.7±1.5	81.7±1.5	97.3±1.5	140.7±0.6	131.7±1.5	88.7±1.5	102.8±26.9
Phenanthene							
2012	763.3±10.6	781.7±10.6	852.3±7.5	990.0±8.0	893.0±7.0	814.7±6.5	849.2±83.5
2013	646.7±12.6	672.3±2.1	701.7±11.7	830.7±10.5	821.0±10.5	708.3±8.5	730.1±77.4
2014	552.7±5.0	566.7±5.5	676.7±6.1	788.3±8.0	776.0±6.0	739.3±7.0	683.3±103.4
Antracene							
2012	203.7±3.1	198.3±6.0	133.3±6.1	98.3±0.6	79.0±2.0	82.3±2.1	132.5±56.5
2013	177.3±8.0	168.7±8.5	155.0±7.5	122.0±6.6	90.3±2.5	71.7±4.5	130.8±43.4
2014	172.3±6.5	155.3±6.0	144.7±5.5	109.7±4.5	102.7±2.5	83.7±2.5	128.1±34.4
Fluoranthene							
2012	1643.3±30.3	1541.7±18.5	1494.3±26.5	1457.0±18.7	912.3±7.5	818.3±12.2	1311.2±352.2
2013	1551.3±13.7	1409.0±20.5	1386.0±11.0	1321.7±20.0	1026.3±15.0	943.0±19.1	1272.9±237.0
2014	1421.3±16.5	1342.0±19.3	1316.3±15.0	1199.3±8.6	1105.0±6.0	933.7±7.4	1219.6±179.1
Pyrene							
2012	731.3±8.0	842.0±7.2	881.0±11.5	875.3±8.3	908.3±10.2	845.7±11.0	847.3±61.8
2013	656.3±15.5	702.7±10.7	872.3±12.7	884.0±15.5	914.3±13.0	863.0±11.0	815.4±107.7
2014	744.7±8.5	775.0±8.2	835.7±8.6	888.7±8.1	940.3±9.5	908.0±7.0	848.7±77.4
Benz[a]antracene							
2012	241.7±11.4	265.3±5.0	260.0±5.6	301.3±3.1	291.3±6.1	267.3±8.5	271.2±21.7
2013	232.0±11.1	256.3±11.5	244.0±15.5	296.0±10.5	303.7±14.0	277.0±12.1	268.2±28.8
2014	203.7±4.5	219.7±10.1	233.3±5.0	309.0±63.3	287.7±9.5	282.3±6.1	255.9±42.6
Chrysene							
2012	421.7±11.0	455.7±10.7	435.3±8.5	465.7±10.0	486.0±4.6	489.7±9.5	459.0±27.1
2013	403.0±9.5	454.0±14.1	463.3±13.1	477.3±12.7	495.3±13.5	485.0±13.5	463.0±32.9
2014	431.7±10.5	477.3±12.0	488.0±11.5	525.7±16.2	525.7±6.0	516.0±8.0	494.1±36.6
Benzo (b) fluoranthene							
2012	822.0±12.5	707.0±9.5	812.0±9.0	886.3±9.3	311.3±9.0	409.7±10.3	658.1±239.6
2013	853.7±12.5	801.3±11.6	842.7±15.0	854.0±14.1	421.3±17.0	447.3±12.5	703.4±209.5
2014	787.7±11.1	811.3±9.0	775.3±5.5	794.0±7.2	530.3±8.0	502.7±6.5	700.2±143.1
Benzo (k) fluoranthene							
2012	567.7±8.6	501.3±11.0	537.3±8.0	578.0±9.5	432.7±11.1	331.0±9.8	491.3±94.6
2013	622.3±16.6	608.0±11.1	564.7±12.5	558.0±9.0	467.7±12.0	384.3±8.0	534.2±91.2
2014	602.7±7.5	577.0±6.6	555.0±7.0	519.3±10.6	405.3±6.1	366.7±6.5	504.3±96.4
Benzo (a) pyrene							
2012	719.7±12.3	719.7±10.6	822.7±9.5	847.3±10.0	653.7±11.5	473.0±16.5	706.0±134.9
2013	747.3±8.5	754.0±12.1	878.0±10.1	887.3±10.6	619.7±16.3	520.7±14.6	734.5±143.8
2014	561.7±6.7	686.7±6.5	820.7±11.1	831.3±9.3	693.7±11.5	555.0±13.1	691.5±119.8

Table 1. cont.

Indeno(1.2.3-cd) phyrene							
2012	878.0±9.2	824.0±11.0	1128.7±19.6	1298.7±21.5	827.7±12.0	614.7±9.7	928.6±244.6
2013	655.3±12.0	734.3±13.5	1025.0±24.1	1016.3±12.6	933.7±15.8	712.7±11.0	846.2±164.5
2014	607.7±10.8	878.0±9.2	990.7±8.7	1011.0±9.5	869.3±5.5	764.0±9.2	853.4±150.3
Dibenzo (a. h) anthracene							
2012	41.7±0.6	37.3±1.5	52.7±2.5	60.3±0.6	28.7±0.6	25.0±1.0	40.9±13.7
2013	12.0±1.0	47.0±3.0	54.3±3.1	64.0±3.0	31.7±0.6	31.7±1.5	40.1±18.7
2014	11.0±1.0	37.3±2.5	61.7±2.1	56.7±2.1	27.7±1.5	29.3±0.6	37.3±19.1
Benzo (g. h. i) perylene							
2012	869.3±15.9	889.7±8.7	1213.3±16.5	1314.0±21.9	934.3±9.5	864.7±11.0	1014.2±197.4
2013	577.0±6.6	658.0±9.5	1114.0±16.4	1184.7±5.5	1005.0±6.0	779.3±12.2	886.3±250.7
2014	595.3±47.5	621.3±11.6	984.0±6.1	995.0±4.0	946.3±7.1	631.7±9.7	795.6±197.6
Total PAHs							
2012	8756.0±43.6	8616.3±26.1	9533.7±50.7	10134.7±14.0	7672.3±14.3	6823.3±49.9	8589.4±1205.7
2013	7945.3±17.6	8077.7±104.0	9183.0±30.3	9490.7±29.5	8057.7±35.6	7022.0±76.3	8296.1±901.6
2014	7429.7±42.5	7971.7±42.8	8733.3±27.0	9057.7±62.9	8210.3±19.0	7155.0±38.2	8092.9±733.3

± – standard deviation
(I – depth 0.0–0.5 m, II – depth 0.5–1.0 m, III – depth 1.0–1.5 m, IV – depth 1.5–2.0 m, V – depth 2.0–2.5 m, VI – depth 2.5–3.0 m)

The mean content of 16 PAHs sum in the sewage sludge was from 8092.9 µg/kg DM in 2014 to 8589.4 µg/kg DM in 2012. The lowest sum of 16 PAHs was recorded for the deepest sludge layer in the vertical profile (at a depth of 2.5–3.0 m) in 2012, which amounted to 6823.3 µg/kg DM. The highest level of 16 PAHs sum was found in the middle sludge layer of the vertical profile at a depth of 1.5–2.0 m in 2012, which was 10134.7 µg/kg DM. Among the 16 analyzed PAHs, the highest average content in the sewage sludge in STRB characterized fluoranthene, the amount of which equalled 1311.2 µg/kg DM, as well as benzo[g,h,i]perylene with the content of 1014.2 µg/kg DM. Out of the tested 16 PAHs, the lowest mean concentration in STRB sewage sludge was determined for dibenzo[a,h]anthracene, which amounted to 37.3 µg/kg DM.

The general trend of changes in the sum of 16 PAHs in sludge subjected to stabilization and dehydration in STRB in 2012–2014 was similar. The sludge layers from the shallowest to a depth of 1.5–2.0 m showed asteady increase in the sum of 16 PAHs. Then, in the two deepest layers located at the depths between 2.0 m and 3.0 m, a decrease in 16 PAHs sum was observed.

The studies revealed that all sludge samples contained 11 PAHs sum according to WD [2000] and 16 PAHs sum according to EPA above 6 mg/kg DM.

Figure 2 presents the changes in PAHs contents divided into 3,4,5,6-ring ones. During the sludge stabilization and dehydration process in STRB, the highest concentration was recorded for 4-ring PAHs, which ranged from 2421.0 µg/kg DM to 3104.7 µg/kg DM. The highest content was observed for 3-ring PAHs group, which amounted from 1268.0 µg/kg DM to 1785.7 µg/kg DM.

The sum of 16 PAHs in the lowest sludge layer (2.5–3.0 m depth) was always lower than 16 PAHs sum for sludge from 0.0–0.5 m depth. The highest difference was recorded in 2012, when the sum of 16 PAHs in the oldest sludge layer was lower by 22.1 % than the sum of 16 PAHs in the youngest (shallowest) layer of the sludge.

The research on the changes in the content of PAHs in sewage sludge stabilized and transformed in various ways by different researchers suggests the possibility of different behavior of these compounds during the process. A very high degree of 16 PAHs degradation sum, ranging from 89.8% to 98.9%, was obtained during the fermentation of sewage sludge in an autoclave for 60 days, as reported by Chang et al. [2003]. Bernal-Martinez et al. reported that about 50% decomposition of 16 PAHs sum occurred during fermentation of municipal sewage sludge [Bernal-Martinez 2005]. The use of anaerobic processes of biological decomposition of municipal sewage sludge has reduced the concentrations of PAHs from 45% to 55% [Boruszko 2013b]. According to Oleszczuk et al., the sewage sludge compositing process with various proportions of structure-forming materials can cause a decrease

2.1

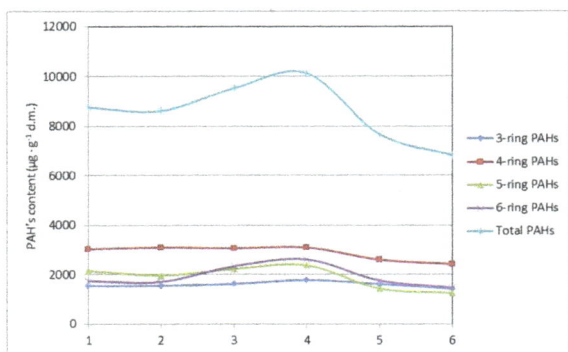

2.2

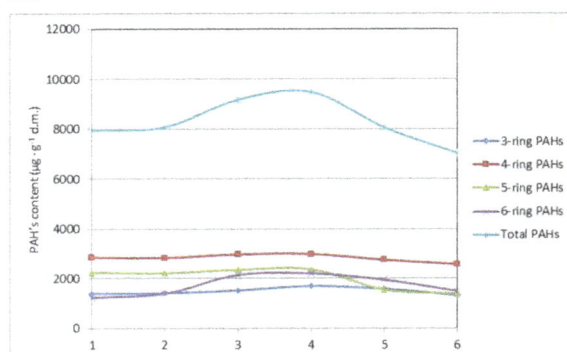

2.3

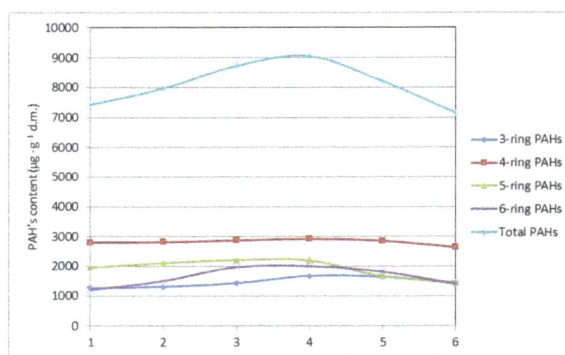

Figure 2. Changes of PAHs content in relations to number of rings in 2012, 2013 and 2014 year

in the sum content of 16 PAHs, even from 30% up to 70% [Oleszczuk 2004, 2009]. The authors also indicate the correlations between decomposition and degradation of individual PAHs as well as their sum vs. type and amount of available organic matter in the biomass [Oleszczuk 2006]. Other authors also report a high (about 50%) degradation of 16 PAHs during composting of sewage sludge [Li et al. 2008; Amir et al., 2005; Antizar-Ladislao et al. 2004].

Different results were achieved by Villar, who found that composting and sludge stabilization process increased the content of PAHs sum [Villar 2006]. Decomposition of hydrocarbons can be performed both by pure bacterial strains and mixed populations, for which the intermediate products are a substrate for other ones. Besides bacteria, bio-degradation can be carried out by fungi, actinomycetes, some cyanobacteria and algae [Kanaly et al. 2000; Kastner et al. 1999]. Due to the action mechanism, microorganisms can be divided into two groups: ones that use hydrocarbons as the only source of carbon and energy (e.g. bacteria: *Pseudomonas, Micrococcus, Alcaligens, and fungi: Candida, Fusarium, Aspergil-*

lus) and the ones that have the ability to carry out co-metabolic transformations [Perez et al. 2001].

Table 2 presents the average concentrations of selected heavy metals in the sludge stabilized in reed bed system. Th eaverage sum of heavy metals in the sludge ranged from 1607.2 mg/kg DM in 2014 to 1635.2 mg/kg DM in 2012. The highest content, among all analyzed heavy metals, was recorded for zinc with its mean concentration of 1301.1 mg/kg DM. Zinc is considered the most essential microelement for plant growth and its very high concentration (above 2500 mg/kg DM) becomes toxic [Pathak et al. 2009]. The lowest content among all tested heavy metals was showed by mercury, at average content of 1.3 mg/kg DM. The analysis of heavy metal contents in particular layers (depths) of sewage sludge reveals an increase in zinc and copper levels, along with their processing time in the reed lagoon. The highest contents of these metals were recorded at the depth of 1.5–2.0 m (layer IV), which amounted to 1412.3 mg Zn/kg DM and 247.0 mg Cu/kg DM. The opposite trend was observed in the case of lead and chromium, the overall content of which decreased with the

Table 2. The average concentrations of heavy metals in sludge stabilized in reed bed system (mg/kg d.m.)

Year	I	II	III	IV	V	VI	Average
			Pb				
2012	38.5±0.9	35.6±0.8	31.3±0.6	27.8±0.5	27.1±0.4	25.2±0.5	30.9±5.2
2013	35.1±0.9	34.7±0.9	32.5±0.7	26.8±0.6	26.2±0.4	24.2±0.8	29.9±4.7
2014	33.8±0.7	34.2±0.4	33.2±0.6	27.2±0.6	25.8±0.6	24.1±0.4	29.7±4.5
			Cd				
2012	3.5±0.1	3.4±0.1	3.3±0.1	3.2±0.1	3.1±0.0	2.6±0.1	3.2±0.3
2013	3.2±0.1	3.2±0.0	3.0±0.0	3.0±0.0	3.0±0.0	2.5±0.1	3.0±0.2
2014	3.1±0.0	3.1±0.0	2.9±0.1	2.5±0.1	2.4±0.0	2.4±0.0	2.7±0.3
			Cr				
2012	75.4±0.7	61.2±0.7	54.3±0.6	45.0±0.6	42.3±0.7	41.7±0.8	53.3±13.2
2013	71.1±0.6	62.2±0.9	53.2±0.5	46.0±0.6	43.0±0.6	42.4±0.5	53.0±11.6
2014	70.1±0.5	64.3±0.6	55.1±0.6	48.3±0.4	44.1±0.4	41.2±0.5	53.9±11.5
			Cu				
2012	156.3±3.1	183.0±3.0	223.7±4.2	247.0±6.6	236.7±2.5	224.3±3.5	211.8±34.8
2013	143.3±5.5	173.3±2.5	211.0±5.6	232.0±7.0	243.7±6.0	243.7±5.5	207.8±41.3
2014	154.0±5.6	178.3±1.5	223.3±5.5	211.7±63.0	251.3±6.5	254.3±5.0	212.2±39.9
			Ni				
2012	18.2±0.3	19.0±0.4	19.5±0.4	18.9±0.3	19.1±0.3	18.5±0.3	18.9±0.5
2013	19.6±0.7	19.8±0.7	19.1±0.4	19.8±0.6	19.5±0.5	19.5±0.4	19.6±0.3
2014	18.2±0.6	19.2±0.2	19.5±0.3	19.3±0.4	20.2±0.3	20.0±0.5	19.4±0.7
			Zn				
2012	1224.3±14.0	1316.0±14.5	1355.7±19.6	1399.0±12.5	1374.0±9.0	1225.3±14.2	1315.7±75.4
2013	1184.7±7.0	1276.0±11.0	1324.3±14.6	1412.3±10.5	1398.0±8.0	1244.3±5.0	1306.6±88.9
2014	1206.0±7.5	1254.7±9.1	1312.0±9.0	1396.3±7.0	1346.7±8.6	1213.3±10.6	1288.2±76.4
			Hg				
2012	1.5±0.1	1.5±0.1	1.3±0.1	1.3±0.1	1.3±0.1	1.2±0.1	1.4±0.1
2013	1.0±0.0	1.3±0.0	1.4±0.0	1.2±0.1	1.2±0.1	1.3±0.0	1.2±0.1
2014	1.1±0.0	1.2±0.0	1.2±0.0	1.2±0.0	1.2±0.0	1.2±0.0	1.2±0.1
			Total				
2012	1517.8±10.8	1619.8±10.9	1689.1±23.9	1742.3±5.6	1703.5±7.0	1538.8±10.7	1635.2±92.0
2013	1458.0±11.9	1570.5±7.9	1644.6±9.8	1741.1±18.0	1734.7±13.3	1577.9±12.2	1621.1±108.5
2014	1486.2±13.7	1555.0±9.3	1647.2±14.2	1706.5±56.6	1691.7±15.9	1556.6±7.1	1607.2±87.8

± – standard deviation
(I – depth 0.0–0.5 m, II – depth 0.5–1.0 m, III – depth 1.0–1.5 m, IV – depth 1.5–2.0 m, V – depth 2.0–2.5 m, VI – depth 2.5–3.0 m)

sludge layer depth in the reed lagoon. The highest average loss of the total content was observed in relation to chromium, the concentration of which fell by 26.4%.

Figure 3 presents the average concentrations of selected heavy metals and their speciation in the sludge stabilized in a reed bed system. All the analyzed heavy metals were related mainly to the most stable connections (fraction IV). The highest share of fraction IV in the total content was observed in the case of lead, which was on average 91.4%, and chromium – 90.3%. The lowest participation of fraction IV in relation to the total content was observed for zinc, and it amounted to an average of 48.2%. The highest share in the most mobile fraction I was observed in the case of

nickel, which amounted to an average of 10.9%. With regard to lead, cadmium and chromium, there was no presence of these metals compounds in fractions I and II, while their share in the fraction IV was high. Therefore, they are metals present in the most stable form. Moreover, other authors showed that in the sewage sludge stabilized by means of conventional methods, lead was the most bound, and the most abundant in the fraction IV [Alvarez et al. 2002; Scancar et al. 2000].

In fraction III (organic), the highest share was revealed by copper from 24.4% to 24.6%, zinc from 23.4% to 23.8%, as well as nickel from 21.4% to 22.9%. The analyzed sludge types showed the largest share in mobile fractions (I, II, III) by zinc from 50.8% to 51.8% and nickel

3.1

2012

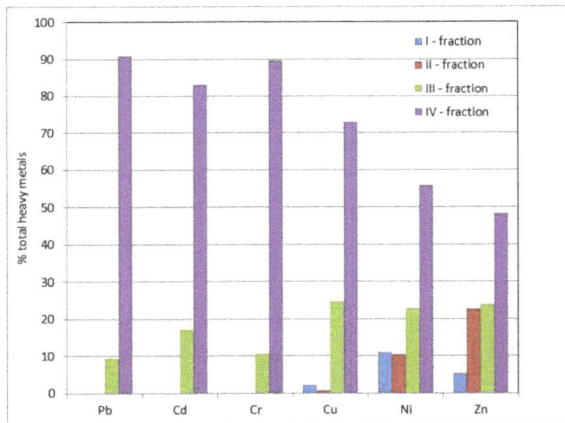

3.2

2013

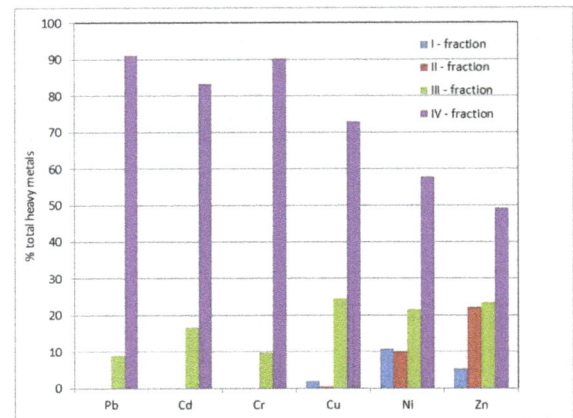

3.3

2014

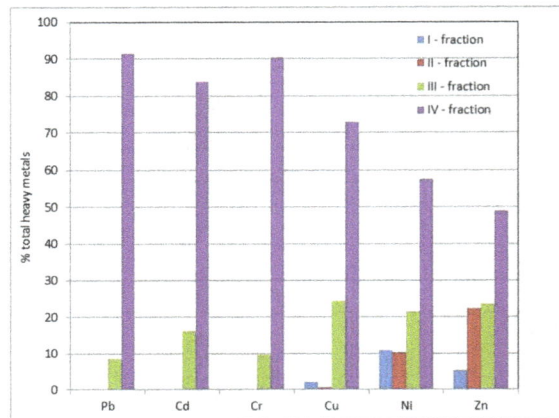

Figure 3. The average concetrations of heavy metals, their speciacion in sludge stabilized in reed bed system in 2012, 2013 and 2014 year

from 42.2% to 44.2%. In addition, the literature data indicate that zinc and nickel are present at largest quantities in mobile fractions (I, II, III) of stabilized sludge [Fuentes et al. 2008; Alvarez et al. 2002; Scancar et al. 2000].

The study on the changes in the physico-chemical properties of sludge dehydrated and stabilized in reed lagoons showed significant differences with respect to the working phase – closing of the lagoon and to the work time of the object (vertical profile).

Spearman's rank correlation coefficient was used to study the relationship between the quantitative parameters and to describe the strength of the correlation. Table 3 shows the results of statistical calculations of Spearman's rank correlation order for a series of tests during the conducted experiment. The determined PAHs contents in sludge types during the experiment were grouped dividing into the number of rings. All studied pa-

rameters (PAHs, heavy metals) were grouped in pairs with respect to the depth (sludge layers).

The correlation coefficients are significant at $p < 0.05$. The rank correlation coefficient takes values from the interval [-1;1]. The same rank values of tested variables testify to the existence of a positive correlation between them (X=Y=1), i.e. Y increases always altogether with X and vice versa. The opposite numbering suggests a negative correlation (when one parameter increases, the second decreases). A positive sign of the coefficient indicates the existence of a positive correlation (increase of one parameter causes an increase in the other). The closer the correlation coefficient is to unity, the stronger the correlative relationship. The calculations reveal that the strongest positive correlation (RS = 0.9750) characterizes the total phosphorus content in sludge with respect to the sludge layer (i.e. duration of its stabilization in STRB). Slightly lower posi-

Table 3. Correlation of order of Spearman ranks, indicated coefficients of correlation are essential p < 0.05000

	Depth	3-ring PAHs	4-ring PAHs	5-ring PAHs	6-ring PAHs	Tot PAHs	Pb	Cd	Cr	Cu	Ni	Zn	Hg	HM$_{tot}$	DM	OM	N$_k$	P$_{tot}$
Depth	1.0000	0.2853	-0.4922	-0.5236	0.1975	-0.2916	-0.9562	-0.7057	-0.9813	0.8971	0.3262	0.3229	-0.1600	0.4483	0.5361	-0.9562	-0.8621	0.9750
3-ring PAHs	0.2853	1.0000	0.5583	0.3457	0.8349	0.7110	-0.2198	0.0237	-0.3106	0.4729	0.1156	0.8576	0.1146	0.8493	-0.4943	-0.2260	0.0464	0.3519
4-ring PAHs	-0.4922	0.5583	1.0000	0.7668	0.6037	0.9009	0.5294	0.5369	0.4757	-0.2148	-0.1332	0.3953	0.4316	0.3251	-0.7337	0.4861	0.7090	-0.3849
5-ring PAHs	-0.5236	0.3457	0.7668	1.0000	0.5191	0.8225	0.4819	0.4223	0.5046	-0.3098	0.0465	0.3271	0.1642	0.2652	-0.7668	0.4613	0.6594	-0.4840
6-ring PAHs	0.1975	0.8349	0.6037	0.5191	1.0000	0.8328	-0.1496	0.0330	-0.1950	0.3934	0.0186	0.8514	0.2912	0.8349	-0.4551	-0.1785	0.0898	0.2384
Tot PAHs	-0.2916	0.7110	0.9009	0.8225	0.8328	1.0000	0.3127	0.3655	0.2859	-0.0299	-0.0186	0.6491	0.3779	0.5851	-0.7214	0.2755	0.5253	-0.2157
Pb	-0.9562	-0.2198	0.5294	0.4819	-0.1496	0.3127	1.0000	0.7754	0.9340	-0.8673	-0.3913	-0.2611	0.3366	-0.3932	-0.4964	0.9773	0.8906	-0.9298
Cd	-0.7057	0.0237	0.5369	0.4223	0.0330	0.3655	0.7754	1.0000	0.6464	-0.5475	-0.4081	-0.0196	0.5351	-0.1600	-0.5596	0.8240	0.8952	-0.6618
Cr	-0.9813	-0.3106	0.4757	0.5046	-0.1950	0.2859	0.9340	0.6464	1.0000	-0.8942	-0.2747	-0.3437	0.1373	-0.4634	-0.4902	0.9174	0.8142	-0.9567
Cu	0.8971	0.4729	-0.2148	-0.3098	0.3934	-0.0299	-0.8673	-0.5475	-0.8942	1.0000	0.4003	0.4739	-0.0444	0.6205	0.3665	-0.8673	-0.6939	0.9107
Ni	0.3262	0.1156	-0.1332	0.0465	0.0186	-0.0186	-0.3913	-0.4081	-0.2747	0.4003	1.0000	0.1631	-0.2882	0.2488	0.2850	-0.3645	-0.3531	0.3366
Zn	0.3229	0.8576	0.3953	0.3271	0.8514	0.6491	-0.2611	-0.0196	-0.3437	0.4739	0.1631	1.0000	0.1260	0.9690	-0.3271	-0.2755	-0.0712	0.3168
Hg	-0.1600	0.1146	0.4316	0.1642	0.2912	0.3779	0.3366	0.5351	0.1373	-0.0444	-0.2882	0.1260	1.0000	0.0764	-0.0754	0.3273	0.4244	-0.0733
HM$_{tot}$	0.4483	0.8493	0.3251	0.2652	0.8349	0.5851	-0.3932	-0.1600	-0.4634	0.6205	0.2488	0.9690	0.0764	1.0000	-0.2466	-0.4180	-0.2054	0.4448
DM	0.5361	-0.4943	-0.7337	-0.7668	-0.4551	-0.7214	-0.4964	-0.5596	-0.4902	0.3665	0.2850	-0.3271	-0.0754	-0.2466	1.0000	-0.5212	-0.7193	0.4654
OM	-0.9562	-0.2260	0.4861	0.4613	-0.1785	0.2755	0.9773	0.8240	0.9174	-0.8673	-0.3645	-0.2755	0.3273	-0.4180	-0.5212	1.0000	0.9174	-0.9236
N$_k$	-0.8621	0.0464	0.7090	0.6594	0.0898	0.5253	0.8906	0.8952	0.8142	-0.6939	-0.3531	-0.0712	0.4244	-0.2054	-0.7193	0.9174	1.0000	-0.7936
P$_{tot}$	0.9750	0.3519	-0.3849	-0.4840	0.2384	-0.2157	-0.9298	-0.6618	-0.9567	0.9107	0.3366	0.3168	-0.0733	0.4448	0.4654	-0.9236	-0.7936	1.0000

tive correlation (RS = 0.9009) was achieved for 4-ring PAHs in relation to the sum of 16 PAHs. High positive Spearman correlation coefficient (RS > 0.9) was also recorded for the total lead and chromium contents in relation to the organic matter. The strongest negative Spearman correlation (RS > 0.9) with respect to the sludge layer depth was observed for organic matter and total contents of lead and chromium.

Another statistical analysis of the achieved results was based on the use of two ANOVA tests: Kruskal-Wallis's and Friedman's. The Kruskal-Wallis ANOVA allowed to analyze the data as independent from each other (not related in pairs), while the Friedman's test allows to analyze data as dependent on one another (linked in pairs). The analyses were performed in two variants: comparing the data from three years (2012, 2013, 2014), (Table 4), and comparing the data according to 6 depths (Table 5). The red color highlights the statistically significant K-W test results, i.e. p<0.05. In addition, only for those results, the median test results are presented. It allows to precisely recognize which data groups are similar or different. The test results close to zero indicate the data pairs with strong statistical dependencies.

The results of statistical tests show a strong affinity for most of the tested parameters in the sewage sludge stabilized in STRB to the depth (sludge layer), and thus indirectly to the time of their dehydration and stabilization (Table 5), both as dependent and independent from each other. However, in relation to the three years of study (2012–2014), such affinity was not found

Table 4. Results of the statistic tests – year rank

Statistic test	3-ring PAHs	4-ring PAHs	5-ring PAHs	6-ring PAHs	Tot PAHs	Pb	Cd	Cr	Cu	Ni	Zn	Hg	HM$_{tot}$	DM	OM	N$_k$	P$_{tot}$
Friedman	0.0421	0.5134	0.3114	0.1146	0.5134	0.1146	0.0025	0.5134	0.3114	0.3114	0.0421	0.1146	0.2231	0.0025	0.0025	0.0025	0.5134
Kruskal-Wallis	0.3235	0.4758	0.63	0.63	0.7165	0.7961	0.0388	0.9599	0.9322	0.057	0.6918	0.0621	0.8948	0.3993	0.6227	0.2026	0.9599

Table 5. Results of the statistic tests – depth rank

Statistic test	3-ring PAHs	4-ring PAHs	5-ring PAHs	6-ring PAHs	Tot PAHs	Pb	Cd	Cr	Cu	Ni	Zn	Hg	HM$_{tot}$	DM	OM	N$_k$	P$_{tot}$
Friedman	0.0193	0.0412	0.0142	0.0193	0.0193	0.0121	0.0236	0.0104	0.0254	0.6126	0.0104	0.1938	0.0121	0.0121	0.0121	0.0104	0.0104
Kruskal-Wallis	0.0254	0.0521	0.0153	0.0129	0.0162	0.0071	0.1089	0.0058	0.0131	0.7418	0.0075	0.3908	0.0069	0.0106	0.0073	0.0223	0.0062
Median	0.0752	0.0752	0.0267	0.0029	0.0752	0.0029	0.0752	0.0029	0.0267	0.8491	0.0267	0.4579	0.0029	0.0267	0.0029	0.0267	0.0029

for most of the tested parameters. The results of these tests can suggest that the closure of the reed lagoon (end of the sludge supply), for a short time (one year before and one after the closure) had a significant effect only on the content of nitrogen as well as dry matter and organic matter in the deposited sludge.

CONCLUSIONS

Summing up the results and study results obtained from the long-term stabilization and dehydration of sewage sludge in the reed lagoon, the following conclusions can be drawn:

1. An increase in dry matter and decline of organic matter occurred in the stabilized sewage sludge.
2. The phosphorus concentration in sewage sludge increased along with the depth of the sludge layer. In contrast, the concentration of total nitrogen decreased with increasing sludge layer depth.
3. The concentrations of nitrogen and phosphorus in dry matter were relatively high.
4. The concentration of 16 PAHs sum was very high, which prevents the use of sludge in agriculture.
5. Long-term stabilization of sewage sludge in STRB reduces the sum of 16 PAHs, although their increase was observed in the first few years.
6. The total concentration of heavy metals allows the agricultural use of sewage sludge.
7. Heavy metals were usually bound to the residual fraction, which is the most stable.
8. The statistical analysis showed very strong correlations with the depth of the sludge layer.

Acknowledgments

The study was conducted as a research project S/WBiIŚ/3/2014 in Faculty of Building and Environmental Engineering of BUT and financed by Ministry of Science and Higher Education.

REFERENCES

1. Alvarez E.A., Mochon M.C., Sanchez J.C.J. & Rodriguez M.T., 2002 Heavy metal extractable forms in sludge from wastewater treatment plants. Chemosphere, 47, 765–775.

2. Amir S., Hafidi M., Merlina G., Hamdi H., Revel J.C. 2005 Fate of polycyclic aromatic hydrocarbons during composting of lagooning sewage sludge. Chemosphere, 58, 449–458.

3. Antizar-Ladislao B., Lopez-Real J.M., Beck A.J. 2004 Bioremediation of polycyclic aromatic hydrocarbons – contaminated waste using composting approaches. Critical Reviews in Environmental Science and Technology, 34, 249–289.

4. Aparicio I., Santos J.L., Alonso E., 2009 Limitation of the concetration of organic pollutants in sewage sludge for agricultural purposes: A case study in South Spain, Waste Management, 29, 217–228.

5. Bernal-Martinez A., Carrere H., Patureau D., Delgenes J.P. 2005 Combining anaerobic digestion and ozonation to remove PAH from urban sludge. Process Biochemistry, 40, 3244–3250.

6. Boruszko D. 2013a Fractionation of Heavy Metals in Sewage Sludge processed by Low-input methods. Annual Set The Environment Protection, 15 (2), 1787–1803.

7. Boruszko D. 2013b Impact low-cost processing methods on the contents of polycyclic aromatic hydrocarbons (PAHSs) in sewage sludge. Ecological Chemistry and Engineering, 20 (10), 1153–1161.

8. Boruszko D., Butarewicz A. 2015 Impact of effective microorganisms bacteria on low-input sewage sludge treatment. Environment Protection Engineering, 41 (4), 83–96.

9. Chang B.V., Chang S.W., Yuan S.Y. 2003 Anaerobic degradation of polycyclic aromatic hydrocarbon in sludge. Advances in Environmental Research, 7, 623–628.

10. Fuentes A., Llorens M., Saez J., Aguilar M.I., Ortuno J. & Meseguer V.F. 2008 Comparative study of six different sludges by sequential speciation of heavy metals. Bioresourse Technology, 99 (3), 517–525.

11. Gomez Palacios J.M., Ruiz de Apodaca A., Rebollo C., Azcarate J. 2002 European policy on biodegradable waste: a management perspective. Water Science Technology, 46, 311–318.

12. Kanaly R.A., Harayama S. 2000 Biodegradation ofhigh molecular weidght polycyclic aromatic hydrocarbons by bacteria. Journal of Bacteriology, 182, 2059–2069.

13. Kapanen A., Vikman M., Rajasarkka J., Virta M., Itavaara M. 2013, Biotest for environmental quality assessment of composed sewage sludge. Waste Management, 33, 1451–1460.

14. Kastner M., Streibich S., Beyer M., Richnow H.H., Fritsche W. 1999 Formation of bount residues during microbial degradation of [14C] anthracene in soil. Applied and Environmental Microbiology, 65, 1834–1842.

15. Kazi T.G., Jamali M.K., Kazi G.H., Arian M.B., Afridi H.I. & Siddiqui A. 2005 Evaluating the mobility of toxic metals in untreated industriak wastewater sludge using a BCR sequential extraction procedurę and a leaching test. Analytical and Bioanalytical Chemistry, 383, 297–304.

16. Kołecka K. & Obarska-Pempkowiak H. 2008 The quality of sewage sludge stabilized for a long time in reed baśni. Environment Protection Engineering, 34 (3), 13–20.

17. Langenkamp H., Part P. 2001 Organic contaminants in sewage sludge for agricultur use, European Commision Joint Research Centre Institute for Environment and Sustainability Soil and Waste Unit.

18. Li H., Wu W., Liu Y., Chen Y., Murray B. 2008, Effect of Composting on Polycyclic Aromatic Hydrocarbons Removal in Sewage Sludge. Water, Air Soil Pollution, 193, 259–267.

19. Matamoros V., Nguyen L. X., Arias C.A., Nielsen S., Laugen M.M. & Brix H. 2012 Musk fragrances, DEHP and heavy metals in a 20 year old sludge treatment reed bed system. Water Research, 46 (12), 3889–3896.

20. Nielsen S., 2005 Sludge reed bed facilitis: operation and problems. Water Science and Technology, 51 (9), 99–107.

21. Nielsen S., 2011 Sludge treatment reed bed facilities – organic load and operation problems. Water Science and Technology, 63 (5), 941–947.

22. Nielsen S., Peruzzi E., Macci C., Doni S. & Masciandaro G. 2012 Stabilisation and mineralization of sludge in reed bed system after 10–20 years of operation. 13th International IWA Specialist Group Conference on Wetland Systems for Water Pollution Control 25–30 November 2012. Perth, Australia.

23. Oleszczuk P., Baran S. 2004 The concentration of mild-extracted polycyclic aromatic hydrocarbons in sewage sludges. Journal of Environmental Science and Health, 11, 2799–2815.

24. Oleszczuk P. 2006 Influence of different bulking agents on the disapparance of PAHs during sewage sludge composting. Water, Air and Soil Pollution, 175, 15–32.

25. Oleszczuk P. 2009 Application of three methods used for the evaluation of policyclic aromatic hydrocarbons (PAHs) bioaccessibility for dewage sludge composting. Bioresource Technology, 100, 413–420.

26. Pathak A., Dastidar M.G. & Sreekrishnan T.R. 2009 Bioleaching of heavy metals from sewage sludge: a review. Journal of Environmental Management, 90 (8), 2343–2353.

27. Pérez S, La Farré M, García JM, Barceló D. 2001 Occurrence of polycyclic aromatic hydrocarbons in sewage sludge and their contribution to its toxicity in the ToxAlert® 100 bioassay. Chemosphere, 45 (6–7), 705–712.

28. Peruzzi E., Macci C., Doni S., Masciandaro G., Peruzzi P., Aiello M. & Ceccanti B. 2009 Phragmites australis for sewage sludge stabilization. Desalination, 246, 110–119.

29. Peruzzi E., Masciandaro G., Macci C., Doni S., Mora Rayelo S.G., Peruzzi P & Ceccanti B. 2011 heavy metal fractonation and organic mater stabilization in sewage sludge treatment wetlands. Ecological Engineering, 37 (5),771–778.

30. Scancar J., Milacic R. & Burica O. 2000 Total metal concentrations and partotioning of Cd, Cr, Cu, Fe, Ni, Pb and Zn in sewage sludge. The Science of the Total Environment, 250, 9–10.

31. Sidney S. 1956 Non-parametric statistics for the behavioral sciences. Nowy York, McGraw-Hill.

32. Smolik E. 2005 Polycyclic aromatic hydrocarbons (PAHs). Instytut Medycyny Pracy i Zdrowia Środowiskowego. [in Polish]

33. Standard Methods for the Examination of Water and Wastewater 2005 21th edn, American Public Health Association/American Water Works Association/Water Environment Federation, Washington DC, USA.

34. Sobczyk M. 2002 Statystyka, Wydawnictwo PWN. [in Polish]

35. Uggetti E., Ferrer I., Llorens E & Garcia J. 2010 sludge treatment wetlands a review on the state of the art. Biresouce Technology, 101 (9), 2905–2912.

36. Vacha R., Vyslouzilova M., Horvathova V., Cechmankova J. 2006 Recommended maximum contents of persistent organic pollutants in sewage sludge for application on agricultural soils. Plant Soil Environment, 52 (8), 362–367.

37. Villar P., Callejon M., Alonso E., Jimenez J. C., Guiraum A. 2006 Temporal evolution of polycyclic aromatic hydrocarbons (PAHs) in sludge from wastewater treatment plants: Comparison between PAHs nad Heavy metals. Chemosphere, 64, 535–541.

38. Working Document on Sludge, and Biowaste, 2010 Council of the European Community, Brussels, Belgie.

39. Working Document on Sludge, 3rd Draft, 2000 Council of the European Community, Brussels, Belgie.

40. Zwara W. & Obarska-Pempkowiak H. 2000 Polish experience with sewage sludge utilization in reed beds. Water Science and Technology, 41 (1), 65–68.

18

Ashes from Sewage Sludge and Bottom Sediments as a Source of Bioavailable Phosphorus

Tomasz Ciesielczuk[1*], Czesława Rosik Dulewska[2], Joanna Poluszyńska[3], Ewelina Ślęzak[3], Katarzyna Łuczak[1]

[1] Department of Land Protection, Opole University, ul. Oleska 22, 45-052 Opole, Poland

[2] Institute of Environmental Engineering of the Polish Academy of Sciences, ul. Skłodowskiej-Curie 4, 41–819 Zabrze, Poland

[3] Institute of Ceramics and Building Materials, ul. Oswiecimska 21, 45-641 Opole, Poland

[*] Corresponding author's e-mail: tciesielczuk@uni.opole.pl

ABSTRACT

Phosphorus is an element necessary for the growth of plants. As phosphate rock gets depleted, it becomes an increasingly scarce resource. Therefore, it seems necessary to implement simple methods of cheap and effective phosphorus recovery from waste. The ashes of municipal sewage sludge and bottom sediments constitute particularly valuable sources of phosphorus. However, these materials usually carry significant amounts of pollutants, including heavy metals. Optimization of ash phosphorus sequential extraction methods from a thermal conversion of sewage sludge and bottom sediments allows to select an effective and simple technology of phosphorus recovery, while maintaining low heavy metal pollution, which is one of the main restrictions in use of ashes. Determination of an amount of bioavailable phosphorus is therefore a basis for estimation of the possibility of using it from waste. Extraction using the Golterman method or shaking out with calcium lactate or Trougs reagent indicates that the ashes from sewage sludge and bottom sediments are rich sources of bioavailable phosphorus, which could find use under field conditions as a viable alternative to fertilizers containing fossil phosphorus.

Keywords: phosphorus, bioavailability, ash, sewage sludge, bottom sediments, extraction

INTRODUCTION

Phosphorus is an essential element in agriculture, and the yield depends, i.a. on the occurrence of its bioavailable forms. In the agricultural practice, it is widely used in the form of mineral and organic fertilizers.

Apart from manure, bird guano was considered in the last century as a source of phosphorus. Acquiring this bird manure (coming mainly from seabirds) which is extremely rich in nutrients, allowed for its introduction to the European market despite a high price (Szpak et al. 2012). After popularization of mineral fertilizers production, the exploitation of guano has become uneconomic. However, this necessitated phosphate mining in North America, Ukraine, Russia and West Africa.

About 90% of these minerals are converted into phosphate fertilizers, the use of which will grow at a rate of 1.5–3.6% per annum (Szaja 2013). However, taking into consideration the material price instability and its dwindling resources, new, alternative sources of this precious element are sought (van Vuuren et al. 2010).

The phosphorus-rich waste is also generated in large quantities of municipal sewage sludge, the amount of which in Poland in 2016 equaled 568.3 thousand Mg d.m., with the average phosphorus content of about 3%, which gives over 17 thousand Mg of phosphorus, annually (Bień 2012, Statistical Yearbook of Environmental Protection 2017).

The use of municipal sewage sludge is most often limited by a significant content of heavy metals, persistent organic pollutants and sometimes by the presence of living intestinal parasitic eggs (Poluszyńska and Ślęzak 2015). Actions were also taken to produce agricultural fertilizer, stabilizing sediments with ash from brown coal combustion. However, the problem of high content of heavy metals and polycyclic aromatic hydrocarbons (PAHs) that can migrate from the soil into the food chain, may limit its use in agriculture (Rosik-Dulewska et al. 2008, Rosik-Dulewska et al. 2016, Poluszyńska 2013).

The current trends of using solar energy for sludge drying create opportunities for using the dry sludge for thermal transformation, i.e. one with energy recovery (Bień 2012). The ash generated as a result of combustion of sewage sludge consists mainly of Si, Fe, Al, Ca and P, including significant amounts of heavy metals, 9 to 13 times higher in relation to their content in the sediments from which the ash was created. It excludes their use in agriculture (Guedes et al. 2016). However, a significant amount of phosphorus, which may reach even 11–13.4%, predestines it for recovery with prior heavy metal elimination (Donatello et al 2010, Guedes et al. 2014). The importance of the problem is emphasized by the European SU-SAN project aimed at optimizing the methods of nutrients recovery from sewage sludge, while at the same time minimizing the risk of heavy metals or persistent organic pollutants (Adam et al. 2009). The bottom sediments of water reservoirs and rivers are also of similar importance. Although they release significant amounts of heavy metals when used as fertilizers, their drying and combustion opens new possibilities for phosphorus recovery, especially in its bioavailable forms (Strzebońska et al. 2015).

MATERIALS AND METHODS

The research material involved the ashes from the thermal transformation of municipal sewage sludge and bottom sediments rich in organic matter. The sewage sludge came from municipal wastewater treatment plants without (OL) and with increased nutrient removal (OP) as well as the ashes from industrial incineration plant (OS). The bottom sediments (OD) came from the dam part of the "Jezioro Turawskie Duże" reservoir located on the Mała Panew river in the Opolskie voivodeship (Poland). OP, OL and OD samples were incinerated at 600°C for 3 hours, while the OS sample was created as a result of burning sewage sludge at 850°C in the municipal sludge incineration plant in Nowiny (Poland).

The following parameters were designated in the tested materials: pH, proper electrolytic conductivity by potentiometric methods and the content of organic matter by weight. The content of calcium, sodium and potassium were determined using the FES method and the BWB-XP apparatus.

In order to assess the quantitative share of bioavailable phosphorus forms in the studied ashes, six methods proposed in the literature were used. The first one includes distilled water extraction (1:20, w/v) in 22 h. After centrifugation, the supernatant was infused with 0.2 cm^3 0.12 M $MgCl_2$ and centrifuged again (Boen et al., 2013). The second method is extraction with 2% citric acid (CA) (Krupa-Żuczek et al. 2012). The third method involves extraction of phosphorus with the Trougs's reagent (TS) (0.002 N H_2SO_4 + 0.3% $(NH_4)_2SO_4$), which was buffered to pH 3.00. The mass-volume ratio ash:extractant was 1:200 at the shaking time of 30 min. The obtained extract was marked colorimetrically using the molybdate blue method (Chiu et al. 2006). The fourth method is the Egner-Rhiem (E-R) procedure using 0.04 M calcium lactate with lactic acid after buffering to pH 3.55 (Wójcik et al., 2014). In the preliminary tests, the control sample was the extractant itself. The fifth method was based on the extraction with sodium bicarbonate buffered to pH 8.5 according to Olsen (OLS), which is recommended for alkaline soils. The mass-volume ratio ash:extractant was 1:20 at the shaking time: 30 min. (Olsen et al. 1954).

For comparison, the Golterman methodology, developed for the bottom sediments of water reservoirs, was applied. This method consists of shaking the sample with extrinsic agents of varied strength (Table 1). Total phosphorus content (P-TOT) in the tested samples was determined with ICP-MS after microwave digestion in aqua regia.

RESULTS AND DISCUSSION

The tested ashes from sewage sludge incineration were alkaline and neutral in the case of ash from bottom sediments (Table 2), which is characteristic for this type of waste. The electrolytic conductivity was high and in the case of application as fertilizer it could pose a threat to plant seedlings (Mazur et al. 2013).

Table 1. Procedure for extracting tested ashes according to Golterman (Golterman 1996)

Stage and extraction time	Extractant	Fraction
G I 4 h	0.05M Ca-EDTA	Phosphorus bound with ferric, aluminium and manganese oxides and hydroxyoxides
G II 18 h	0.1M Na$_2$-EDTA	Phosphorus bound with carbonates
G III 2 h	0.5 M H$_2$SO$_4$	Phosphorus from soluble compounds with organic matter
G IV 2 h	2 M NaOH	Phosphorus bound with aluminosilicates and in organic matter resist on sulphuric acid digestion in stage G III

The content of nutrients (CaO, MgO, K$_2$O) was high and characteristic for the ashes from bottom sediments and sewage sludge incineration (Arnout and Nagels 2016, Donatello et al. 2010), which proves their high value, not only as a fertilizer, but also as a deacidifying agent for soil.

Sewage sludge and bottom sediments contain a significant amount of phosphorus-rich organic matter. If the content of heavy metals meets the legal requirements, these sediments can be used naturally (also as fertilizers) to improve the soil structure, increase its water absorption and as suppliers of many valuable elements for plants, including micronutrients (Rosik-Dulewska et al. 2016). Nevertheless, in many cases the sewage sludge, after initial drying, is subject to incineration in specialized incinerators. In some countries (Holland, Switzerland), the entire amount of generated sediments is burned (Herzel et al. 2016).

However, the ash obtained from sewage sludge may contain high concentrations of metals, especially zinc, copper, chromium and lead. Making the law more flexible and rapidly passing new legal acts would allow to improve the phosphorus recovery technology (energy and material consumption) on the one hand, and to establish economic incentives (taxes, subsidies) for the companies investing in phosphorus recovery installations on the other (Hukari et al. 2016). After incineration of sewage sludge or bottom sediments, while there is no longer any valuable organic matter, the amount of phosphorus increases to even a dozen or so percent, making this material a valuable source of this element, equally to phosphate fertilizers, and the attempts to wash out phosphorus from ash are carried out by strongly acidic extractants (Atineza-Martinez et al. 2014, Ebbers et al. 2015).

In the ashes from sewage sludge, the bioavailable phosphorus is present in relatively large quantities and constitutes on average over 30% of the total phosphorus, which creates considerable opportunities for using this material as an agricultural fertilizer (Kruger and Adam 2015). Although the process of sludge incineration demands significant investment costs, it eliminates the organic toxic compounds found in sediments, and at the same time, by mineralization of organic matter, it "enriches" the obtained material with phosphorus. Finally, in the ashes from sewage sludge, the total phosphorus can exceed 16.0% (Guedes et al 2016). In the municipal sewage sludge research, the total phosphorus content was recorded in the range 2.97–6.64% (Poluszyńska and Ślęzak 2015), while in the sludge ash examined here, these values range between 13.8 and 18.4%. Only in bottom sediments, the phosphorus content was lower (4.6%) and comparable with the data obtained by the other authors (Bezak-Mazur and

Table. 2. General characteristics of tested ashes OP, OL, OS and OD (n = 3)

Description	OP	OL	OS	OD
pH $_{H2O}$ [-]	7.78–7.83	8.48–8.52	9.78–9.85	6.90–7.01
EC [mS/cm]	3.77(0.11)	2.98(0.08)	2.26(0.04)	2.33(0.13)
TC [%]	0.84(0.16)	1.13(0.15)	0.17(0.08)	0.91(0.01)
CaO [g/kg d.m.]	109.3	16.68	20.59	1.973
MgO [g/kg d.w.]	12.2	14.8	18.6	2.49
Na$_2$O [g/kg d.w.]	3.419	6.280	3.665	0.355
K$_2$O [g/kg d.w.]	5.364	8.247	10.85	2.301

Note: Standard deviation (SD) value in brackets.

Stoińska 2013, Havukainen and others 2016). Many efficient methods of phosphorus recovery have been developed thus far, but they do not give the right perspective of bioavailable forms content, because of the use of aggressive extractants and high temperature, which also makes the forms practically unavailable to plants (Petzet et al. 2012, Shiba and Ntuli 2016, Weigand et al. 2013, Wzorek 2008).

PRELIMINARY TESTS

In the conducted tests, there was a strong correlation between the amount of "bioavailable" phosphorus and the pH of an extraction solution. While using the TS method to extract the OP samples, acidification of the solution with H_2SO_4 with pH 3.00 (recommended in the method) to 0.88, increased the amount of marked phosphorus by 184%. The correlation ratio of the pH value and the amount of eluted phosphorus was 0.997. The method, used to extract highly alkaline biomass ash (after adding the extractant to the sample, a reaction of 7–8 is obtained instead of the expected 3–4), after acidification from pH 7.31 to 0.55, the amount of phosphorus obtained increased by 2731%, and the correlation coefficient in this case was 0.999. This confirms the effect of H_2SO_4 as an extractant capable of decomposing even hardly soluble phosphorus-containing minerals (Tujaka et al. 2006). The need to correct the reaction by adding a strong extractant was confirmed by other authors (Wzorek et al. 2006). Additional acidification facilitates the dissolution of calcium-bound phosphates, as well as partially iron-bound, which, because of a significant amount of

such forms in the ash, can significantly affect the recovery rate (Bednarek and Reszka 2007).

In our own research, we have also tested the influence pertaining to the time of contact between the extractant and the ash sample, on changing the pH value of the extraction solution. For comparison, the data obtained for oak biomass ashes burned at 600°C (Fig. 1) was also introduced. The obtained results indicate a small increase in the pH value in

The case of OP and OL samples as well as pH change by 2 degrees in the case of biomass ash (BM test). The reaction of the extractant is of fundamental importance for the amount of phosphorus leached from a given material; therefore, special attention should be paid to the possibility of obtaining low results, especially in the case of strongly alkaline substances (Golterman 1996).

COMPARISON OF EXTRACTION METHODS

The ashes from the analyzed sewage sludge contained between 13.8% and 18.4% P_2O_5, which is also confirmed by the results of the research carried out by other authors, who assessed it as very rich in this element (Havukainen et al. 2016). Only the ash from bottom sediments was characterized by a much lower amount of phosphorus, i.e. 0.46% P_2O_5. It is still a valuable amount as far as the usage as a fertilizer is concerned.

On the basis of the preliminary tests and available literature, it was assumed that the amount of bioavailable phosphorus obtained in the experiments with OP, OL, OS and OD samples shall depend on the type of the extractant used. The phos-

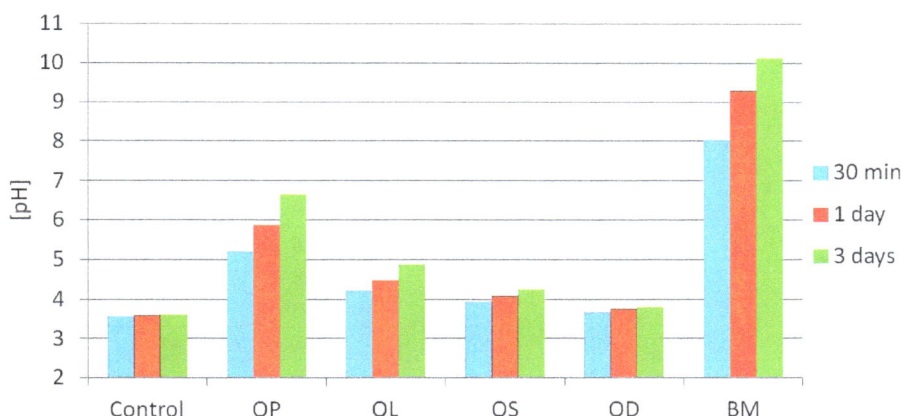

Figure 1. Changes in the pH value in the process of bioavailable phosphorus extraction using the E-R method and 0.04 M calcium lactate as control

phates contained in the tested samples (despite the long contact time), dissolved in water only to a small extent – WE method (Table 1). The phosphorus levels eluted in this way were very low and only slightly exceeded 0.04% of total phosphorus. The highest efficiency of this method (WE) was obtained for the OD ash. Such low amounts washed out with water are not confirmed by the results of Bednarek and Reszka, obtained in a pot experiment with limed soil fertilized with phosphorus. However, it should be noted that in the experiment, the ashes from sediments were not used (Bednarek and Reszka 2007). The citric acid method was much more effective, since it obtained between 4.5 and 6.4% of the total amount of phosphorus marked in the samples.

Using the Trougs and Egner-Rhiem methods, i.e. using extraction liquids with a similar pH, the results were obtained with a predominance (up to 2x) in favor of the E-R method (Table 3). Their efficiency was at the level of 1.2–3.3% of the total phosphorus content in the case of sewage sludge ashes. Higher amounts exceeding 6 and 12% of total phosphorus were recorded only in dermal sediment samples (OD samples) for TS and E-R respectively. Due to the fact that acidified calcium lactate HCl elutes mainly phosphorus compounded with calcium and iron, the presence of these compounds will have a significant impact on the obtained effects (Bednarek and Reszka 2007). Considering the influence of the extractant reaction on the amount of leached phosphorus (part: Preliminary tests), it should be concluded that the values obtained by W, CA, TS and ER, due to the high ash reaction, are understated and the actual amount of bioavailable forms is higher.

Olsen extraction of phosphorus used for acidic and neutral soils may give erroneous results, but in the case of alkaline soils, it simulates the conditions present during the water-soil contact very well. Therefore, in the ash from sediments characterized by relatively high pH, this method should determine the actual amount of phosphorus available for plants. The amount of phosphorus determined according to the Olsen's methodology in the eluates from OP, OL and OS samples is 2–8 times lower than in the E-R method, which is justified by the higher pH of the extractant, and thus the reduced solubility of phosphorus compounds. The best results were noted for the OD samples which obtained almost 4.5% of the total phosphorus content. The fly ashes from coal combustion are characterized by a reaction above 10 and containing 1.17 g P_2O_5/kg d.m. showed a similar (3.37%) amount of bioavailable phosphorus as in the OD samples. Moreover, it has been shown that a significant content of calcium reduces the amount of bioavailable phosphorus up to 10 times (Seshardi et al. 2013).

Using the Golterman sequential extraction method, the highest amounts of bioavailable phosphorus were marked in the OS sample (over 74.6% of the total content) and slightly lower in the OP and OL samples (74.1 and 69.8% respectively). The least bioavailable phosphorus was obtained in OD samples. Although this method is meant to be used with bottom sediments, the ashes obtained after combustion of sediments have significantly different properties (alkaline nature, lack of organic matter) than the sediment. Due to the efficiency and simplicity of the CA method, it seems that it should be recommended for sew-

Table 3. Content of bioavailable and total phosphorus in the examined ashes [g P_2O_5/kg d.m.] and [%] of the total phosphorus content (n = 3)

Description	OP		OL		OS		OD	
	[g P_2O_5/kg d.w.]	[%]	[g P_2O_5/kg d.w.]	[%]	[g P_2O_5/kg d.w.]	[%]	[g P_2O_5/kg d.w.]	[%]
WE	0.006	0.004	0.011	0.008	0.002	0.001	0.002	0.043
CA	9.840	5.906	6.225	4.501	11.865	6.441	0.296	6.407
TS	1.998	1.199	3.516	2.542	3.749	2.035	0.283	6.126
E-R	3.717	2.231	4.535	3.279	4.424	2.402	0.578	12.51
OLS	1.477	0.887	0.820	0.593	0.558	0.303	0.252	5.45
G I	5.126	3.077	4.072	2.944	8.636	4.688	0.184	3.983
G II	38.05	22.84	28.85	20.86	32.07	17.41	0.176	3.810
G III	78.23	46.96	62.52	45.20	94.24	51.16	1.78	38.53
G IV	1.98	1.188	1.167	0.844	2.519	1.368	0.832	18.01
G Σ	123.4	74.06	96.60	69.85	137.5	74.63	2.972	64.33
P-TOT	166.6	100	138.3	100	184.2	100	4.621	100

age sludge ashes, with possible correction after adding the extractant to the sample. It is true that while using the Golterman method, much higher contents were obtained, but due to the time-consuming nature and considerable workload, its use may be limited in favor of the CA method. The use of the Olsen method is justified when assessing fertilizing properties, and thus allow for a better estimation of the amount of bioavailable phosphate introduced into the soil, in case the materials are used for soil deacidification.

The ashes from municipal sewage sludge and bottom sediments are an interesting, easily available and rich source of phosphorus. The total content in the tested samples was 4.621 g P_2O_5/kg d.m. in the bottom sediment and between 138.3 and 184.2 g P_2O_5/kg d.m. in the sewage sludge (Table 3). However, it should be noted that as far as the use as a fertilizer is concerned, the crucial thing is the amount of bioavailable phosphorus, i.e. the part of phosphorous compounds present in a soluble form in soil, thus becoming a potential source of phosphorus for plants, while an indicated amount of total phosphorus content is here for the information purposes.

CONCLUSIONS

The recovery of phosphorus from the ashes obtained from the thermal processing of municipal sewage sludge and bottom sediments seems to be very effective, which confirms their usefulness as a source of phosphorus in agricultural production. The determination of assimilable phosphorus content was largely dependent on the method used. In the case of the tested ash samples, the Golterman method proved to be the most effective, although due to the required workload, citric acid (CA) extraction seems to be the preferred method. The distilled water extraction method gave unsatisfactory results in all cases, which confirms its unsuitability in determining phosphorus content in highly alkaline materials.

REFERENCES

1. Adam C., Peplinski B., Michaelis M., Kley G., Simon F.-G. 2009. Thermochemical treatment of sewage sludge ashes for phosphorus recovery. Waste Management 29: 1122–1128.

2. Arnout S., Nagels E. Modelling thermal phosphorus recovery from sewage sludge Ash 2016. CALPHAD: ComputerCouplingofPhaseDiagramsandThermochemistry55, 26–31 http://dx.doi.org/10.1016/j.calphad. 2016.06.008

3. Atienza-Martınez M., Gea G., Arauzo J., Kersten S.R.A., Maarten A., Kootstra J.2014. Phosphorus recovery from sewage sludge char Ash. Biomass and bioenergy 65, 42–50. doi.org/10.1016/j.biombioe.2014.03.058

4. Bednarek W., ReszkaR. 2007. The effect of liming and fertilization with various nitrogen forms on the content

5. of available forms and mineral fractions of phosphorus in the soil. Annales Universitatis Mariae Curie – Skłodowska Lublin – Polonia Vol. Lxii (2) Sectio E 234–242 (in Polish)

6. Bezak-Mazur E., Stoińska R. 2013. Speciation of phosphorus in wastewater sediments from selected wastewater treatment plant. Ecol Chem Eng A. 20(4–5): 503–514. DOI: 10.2428/ecea.2013.20(04)047

7. Bień J.D. 2012. Utilisation of Sewage Sludge in Poland by Thermal Method. Inżynieria i Ochrona Środowiska 15(4): 439–449 (in Polish)

8. Bøen A., Haraldsen T.K., Krogstad T. 2013. Large differences in soil phosphorus solubility after the application of compost and biosolids at high rates. Acta Agriculturae Scandinavica, Section B – Soil & Plant Science. DOI: 10.1080/09064710.2013.801508

9. Chiu S.W., Gao T., Chan,C.S.S., Ho C.K.R. 2009. Removal of spilled petroleum in industrial soils by spent compost of mushroom Pleurotus pulmonarius. Chemosphere 75 837–842

10. Ciesielczuk T., Rosik–Dulewska Cz., Kusza G. 2016. Extraction of phosphorus from sewage sludge ash and sewage sludge – problem analysis. Polish Journal for Sustainable Development 20, 21–28, DOI: 10.15584/pjsd.2016.20.3 (in Polish)

11. Czechowska-Kosacka A. 2016. Phosphorus Speciation Forms in Sewage Sludge from Selected Wastewater Treatment Plants. Annual Set The Environment Protection 18: 158–168

12. Ebbers B., Ottosen L.M., Jensen P.E. 2015. Comparison of two different electrodialytic cells for separation of phosphorus and heavy metals from sewage sludge ash. Chemosphere 125: 122–129.

13. Golterman H.L. Hydrobiologia. 1996. 335(1);87–95. DOI: 10.1007/BF00013687

14. Guedes P., Couto N., Ottosen L.M., Ribeiro A.B. 2014. Phosphorus recovery from sewage sludge ash through an electrodialytic process. Waste Management 34: 886–892.

15. Guedes P., Couto N., Ottosen L.M., Kirkelund G.M., Mateus E., Ribeiro A.B. 2016. Valorisation of ferric sewage sludge ashes: Potential as a phosphorus source. Waste Management 52: 193–201.

16. Havukainen J., Nguyen M.T., Hermann L., Horttanainen M.. Mikkilä M. Deviatkin I., Linnanen L. 2016. Potential of phosphorus recovery from sewage sludge and manure ash by thermochemical treatment. Waste Management 49, 221–229.

17. Herzel H., Krüger O., Hermann L., Adam Ch. 2016. Sewage sludge ash – A promising secondary phosphorus source for fertilizer production. Science of the Total Environment 542, 1136–1143.

18. Hukari S., Hermann L., Nättorp A. 2016. From wastewater to fertilisers – Technical overview and critical review. Science of the Total Environment 542, 1127–1135.

19. Kruger O., Adam Ch. 2015. Recovery potential of German sewage sludge Ash. Waste Management 45, 400–406.

20. Krupa-Żuczek K., Podraza Z., Wzorek Z. 2012 Extraction of phosphorus from sewage sludge ash and sewage sludge – problem analysis. Chemistry. Technical Transactions 16, 65–70 (in Polish)

21. Mazur Z., Radziemska M., Tomaszewska Z., Świątkowski Ł. 2013. Effect of sodium chloride salinization on the seed germination of selected vegetable plants. Scientific Review – Engineering and Environmental Sciences 62, 444–453.

22. Olsen SR, Cole CV, Watanabe FS, Dean LA (1954) Estimation of available phosphorus in soils by extraction with sodium bicarbonate. USDA Circular Nr 939, US Gov Print Office, Washington, DC, pp 1–19.

23. Petzet S., Peplinski B., Cornel P. 2012. On wet chemical phosphorus recovery from sewage sludge ash by acidic or alkaline leaching and an optimized combination of both. water research 46, 3769–3780.

24. Poluszyńska J. 2013. Assessment of contamination possibility of soil by polycyclic aromatic hydrocarbons (PAHs) contained in the fly ash from power boilers. Scientific Works of Institute of Ceramics and Building Materials 12, 60–77 (in Polish)

25. Poluszyńska J., Ślęzak E. 2015. Phosphorus from municipal sewage sludge. Scientific Works of Institute of Ceramics and Building Materials 22, 44–55 (in Polish)

26. Rosik-Dulewska Cz., Karwaczyńska U., Głowala K., Robak J. 2008. Elution of heavy metals from granulates produced from municipal sevage deposits and fly-ash of hard and brown in the aspect of recycling for fertilization purposes. Archives of Environment Protection 34(2), 63–71.

27. Rosik-Dulewska Cz., Nocoń K., Karwaczyńska U. 2016 Wytwarzanie granulatu z komunalnych osadów ściekowych i popiołów lotnych w celu ich przyrodniczego (nawozowego) odzysku (in Polish).

Instytut Podstaw Inżynierii Środowiska Polskiej Akademii Nauk, Prace i Studia 87, pp. 187.

28. Seshadri B., Bolan N., Choppala G., Naidu R. 2013. Differential effect of coal combustion products on the bioavailability of phosphorus between inorganic and organic nutrient sources. Journal of Hazardous Materials 261 (2013) 817– 825, http://dx.doi.org/10.1016/j.jhazmat.2013.04.051.

29. Shiba N.C., Ntuli F. 2016. Extraction and precipitation of phosphorus from sewage sludge. Waste Management http://dx.doi.org/10.1016/j.wasman.2016.07.031.

30. Statistical Yearbook of Environmental Protection 2017. Central Statistical Office Warszawa 2017.

31. Strzebońska M., Kostka A., Helios-Rybicka E., Jarosz-Krzemińska E. 2015. Effect of Flooding on Heavy Metals Contamination of Vistula Floodplain Sediments in Cracow; Historical Mining and Smelting as the Most Important Source of Pollution. Polish J. of Environ. Stud., 34, (3), 1317–1326, doi 10.15244/pjoes/33202

32. Szaja A. 2013. Phosphorus Recovery from Sewage Sludge via Pyrolysis. The Annual Set the Environment Protection 15, 361–370.

33. Szpak P., Millaire J.F., White Ch.D., Longstaffe F.J. 2012. Influence of seabird guano and camelid dung fertilization on the nitrogenisotopic composition of field-grown maize (Zea mays). Journal of Archaeological Science 39, 3721–3740.

34. Tujaka A., Gosek S., Gałązka R. 2006. Estimation of Hedley's fractionation method applicability to the determination of changes in phosphorus fractions in soil Polish Journal of Agronomy 2011, 6, 52–57 (in Polish)

35. van Vuuren D.P., Bouwman A.F., Beusen, A.H.W. 2010. Phosphorus demand for the 1970–2100 period: A scenario analysis of resource depletion. Global Environmental Change 20(3), 428–439. doi:10.1016/j.gloenvcha.2010.04.004

36. Weigand H., Bertau M., Hübner W., Bohndick F., Bruckert A. 2013. RecoPhos: Full-scale fertilizer production from sewage sludge Ash. Waste Management 33, 540–544.

37. Wzorek Z. 2008. The pfosphorus compounds recovery from thermally treated waste and its use as substitute of natural phosphorus raw materials. Monogrph 356 Kraków 2008. (in Polish)

38. Wzorek Z., Jodko M., Gorazda K., Rzepecki T. 2006. Extraction of phosphorus compounds from ashes from thermal processing of sewage sludge. Journal of Loss Prevention in the Process Industries 19, 39–50.

19

Application of SS-VF Bed for the Treatment of High Concentrated Reject Water from Autothermal Termophilic Aerobic Sewage Sludge Digestion

Wojciech Dąbrowski[1*], Paweł Malinowski[1], Beata Karolinczak[2]

[1] Bialystok University of Technology, Faculty of Building and Environmental Engineering, Wiejska 45E, 15-351 Białystok, Poland

[2] Warsaw University of Technology, Faculty of Building Service, Hydro and Environmental Engineering, Nowowiejska 20, 00-653 Warsaw, Poland

* Corresponding author's e-mail: dabrow@pb.edu.pl

ABSTRACT

The autothermal termophilic aerobic digestion (ATAD) technology is used in the municipal and industrial wastewater treatment plants (WWTPs) with personal equivalent up to 30.000. The process provides a high level of sewage sludge stabilization and its hygienization. The main operation problems are caused by the high concentration of nitrogen and phosphorus in the reject water from sewage sludge treatment and air purification (odor removal). Reject water usually is returned to the main sewage treatment, which has a negative impact, especially on the sewage treatment systems based on the sequence batch reactors (SBR). Applying high-performance and expensive separate reject water treatment methods such as SHARON, Anammox or CANON in small facilities is not justified economically. The article presents the research results concerning the effectiveness of applying subsurface vertical flow constructed wetlands (SS VF) for reject water treatment from the ATAD process. An innovative type of bed filling was used, which was produced from waste (ash from a heat and power plant). The efficiency of reject water treatment during the research period was on average at 45.6% for ammonia nitrogen, 32.3% for total phosphorus and 85.1% for BOD_5. Applying SS VF beds for separate reject water treatment might ensure a stable and effective functioning of municipal WWTPs by decreasing the load of biological part of a WWTP.

Keywords: autothermal thermophilic aerobic digestion (ATAD), reject water, vertical flow constructed wetland (SS VF)

INTRODUCTION

The main tendencies in the development of municipal and industrial wastewater treatment plants (WWTPs) are connected with sewage sludge management. It is caused by the necessity of decreasing the content of organic substance, eliminating pathogenic microorganisms and limiting the volume and mass of sludge. Applying new technologies for treating the sludge from wastewater treatment plants (WWTPs) is commonly related to the problem of reject water, which is generated during thickening and final dewatering of sewage sludge.

Returning the reject water to the main sewage treatment line is a common solution. In the facilities applying aerobic digestion, the reject water does not cause any major exploitation problems; however, in large facilities with anaerobic digestion, its impact is substantial [Janus and Van der Roast 1997, Gajewska & Obarska-Pempkowiak 2013, Dąbrowski et al. 2017].

Nowadays, the sewage sludge stabilization technology which involves the process of autothermal termophilic aerobic sewage sludge digestion (ATAD) is more popular in Poland [Bartkowska 2017]. The process can be applied both in the municipal and industrial WWTPs with personal equivalent (PE) up to 30.000 or flow capacity is up to 20.000 m³/day [Bartkowska & Dzienis 2007, Bartkowska 2016]. In these WWTPs, using the anaerobic sewage sludge stabilization which allows for biogas and finally electric and heat energy production is not justified economically. The most essential advantages of the process are a very high level of sewage sludge stabilization and its hygienization, which can be safely recycled to the natural environment and used, for instance, as fertilizer [Layden et al. 2007, Schugen et al. 2011]. Among the issues connected with exploitation, there is a high concentration of biogenic compounds in the reject water produced during the sewage sludge dewatering and additionally from the air treatment carried out by means of a scrubber. The composition of reject water in a WWTP applying the ATAD process is similar to the objects utilizing sewage sludge under anaerobic conditions (digestion) [Dąbrowski 2014].

The reject water from ATAD process might have an exceptionally negative impact on the sewage treatment systems based on sequence batch reactors (SBR), especially when there is no retention and averaging tank. It is possible to apply separate reject water treatment. The most commonly known process is called Single Reactor System for High-Rate Ammonia Removal Over Nitrite (SHARON) [van Kempen et al. 2001]. This method, although highly effective, is nevertheless connected with high investment and exploitation costs. Alternatively, a low-cost constructed wetlands method can be used. It is currently applied for treating not only household or municipal wastewater, but also septage as well as various types of industrial sewage and reject water with high concentration of pollutants [Carvalho et al. 2017, Karolinczak & Dąbrowski 2017, Liang et al. 2017, Tan et al. 2017].

The research explores the possibility of separate high efficiency treatment of the reject water generated during the ATAD process with subsurface vertical flow constructed wetland (SS-VF) filled with innovative lightweight sintered aggregate Certyd. The research focuses on decreasing the concentration of nitrogen and phosphorus, as well as organic substances. According to Science Direct, no studies on ATAD reject water treatment in constructed wetlands have been carried out thus far (date of search: 15th January 2017, keywords: ATAD, reject water, constructed wetlands). Due to the lack of experiments concerning that process, the necessity for conducting

the research with such goals was confirmed. The obtained results can be applied by the engineers designing full scale installations.

MATERIALS AND METHODS

Characteristic of municipal WWTP

The municipal WWTP in Wysokie Mazowieckie has been operating since 2016, the project flow equals 1344 m³d⁻¹ and PE – 11200. The sewage flow in 2017 varied from 1100 to 1500 m³day⁻¹, BOD_5 in raw sewage changed from 140 to 990 mg O_2 l⁻¹.

It is a typical plant with mechanical, biological and chemical treatment. The screen and grid chamber prepare sewage for the biological treatment in two sequencing batch reactors (SBRs) aerated with Biogest Hyper Classic mixers and decanters. There is no averaging tank applied in the sewage line. The chemical precipitation of phosphorus occurs simultaneously with the treatment process. The excess sewage sludge is treated with the ATAD process. Two chambers are supplied with sludge after thickening with centrifuge; the final dewatering is realized with a filter press. Stabilization occurs in two sequential reactors. The sewage sludge from the first ATAD chamber is sent to the second chamber in portions using a sludge transfer pump after removing the portion of stabilized sludge from the second reactor. The temperature in the second reactor is over 56°C and the retention time is around 6 days. Figure 1 presents a general view of the municipal WWTP in Wysokie Mazowieckie with a scheme of sewage sludge treatment. Finally, sewage sludge is used as fertilizer. Due to the nuisance caused by the odor released from ATAD reactors (mainly high concentration of ammonia nitrogen), preliminary air treatment is necessary before it can be discharged into the atmosphere. The main elements of air purification installation are flow scrubber and NEUTRALOX device. Photoionization is an effective odor treatment process applied for all odors coming from the sewage and sludge treatment. The process is based on the effective interaction between UV-light and a catalytic converter.

The problem with a high concentration of nutrients in the reject water was observed in the WWTPs applying the ATAD process. The reject water (sewage sludge dewatering, thickening

Fig. 1. A general view of municipal WWTP in Wysokie Mazowieckie with a scheme of sewage sludge treatment (1-centrifuge for thickening, 2-ATAD chambers, 3-filterpress for dewatering, 4-scroober) [Dąbrowski and Zdziarski]

and scroober) is returned to the main sewage line without any separate treatment. The concentration of ammonia nitrogen and phosphorus in the reject water is similar as in the WWTPs applying anaerobic sewage sludge stabilization [Dąbrowski et al. 2017]. Taking into consideration the amount of generated reject water and its content, it should undergo primary treatment to remove the nitrogen compounds, phosphorus and organic matter before it enters the main technological line in a WWTP [Gui et al. 2009].

Research installation

The research installation based on the constructed wetlands technology was designed by

the authors using the earlier experience with reject water treatment in dairy and municipal WWTPs [Kadlec & Wallace 2009, Karolinczak & Dąbrowski 2017, Dąbrowski et al. 2107]. The main element is a SS-VF bed with the height of 0.8 m. In addition, the installation includes a retention tank, outflow and control wells. Figure 2 depicts the research installation in the summer 2017 and microscopic view of bed filling called *Certyd*.

In the SS-VF research installation, passive aeration systems were applied. The bed was built with innovative filling – *Certyd*. Lightweight sintered aggregate (LSA) *Certyd* is a ceramic and porous material. It is produced in accordance with the LSA technology featuring an innovative sintering process in a rotary furnace. It is obtained

Fig. 2. ATAD reject water treatment installation and microscopic view of *Certyd* [Dąbrowski and Wasilewska]

by a thermal processing of ash. The product conforms to the EN-13055:2016–07 standard as a lightweight aggregate. It has the declaration of National Institute of Hygiene no. HR/B/86/2015 [LSA LLC]. The bed is filled with three layers of *Certyd* characterized by varying granulation (2–9 mm). The beds were planted with specially prepared reed (*Phragmites australis*) obtained from the Zambrów municipal WWTP.

Sampling and scope of the determination

The study was carried out from August to September 2017. The SS-VF bed was supplied from the retention tank. The daily hydraulic load was 0.1 $m^3 m^{-2}$. The samples were collected in 12 research series (influent to SS-VF and effluent). The air temperature during the experiment varied from 8° to 12°C. The basic physical and chemical analyses were performed: biochemical oxygen demand BOD_5, chemical oxygen demand COD, total organic carbon TOC, total Kjeldahl nitrogen TKN, ammonia nitrogen $N-NH_4^+$, nitrate nitrogen (V) $N-NO_3^-$, nitrite nitrogen (III) $N-NO_2^-$ and total phosphorus TP. Unit loads in influent and effluent were used to calculate the removed load. Determinations were conducted in a certified laboratory, in accordance with the procedures set out in the Regulation of the Environmental Protection Minister from 18th November 2014 and in line with the American Public Health Association (2005). Box and whiskers plots have been used as a graphical interpretation of the statistical analysis of influent, effluent and removed loads.

RESULTS AND DISCUSSION

Table 1 presents the basic operation parameters of the WWTP in 2017, concerning the actual quantity of sewage sludge and reject water generated during the sludge treatment.

Table 1. Basic parameters of the municipal WWTP in Wysokie Mazowieckie in 2017

Parameter	Unit	Value
Sewage flow	$m^3 d^{-1}$	1300
P.E of sewage	-	9035
Sewage sludge quantity	tons of dry mass year^{-1}	80.1
Reject water flow	$m^3 d^{-1}$	120
P.E. of reject water	-	660

The amount of reject water in the analyzed WWTP utilizing ATAD was 9% of the total amount of raw sewage. According to the research by Ryzińska [2006], in 40 municipal WWTPs with anaerobic stabilization, the quantity of the reject water was ranged from 2.7 to 7%. According to Janus & van der Roest [1997], the quantity of reject water in Dutch WWTPs was below 2%. Thus, it can be concluded that applying ATAD is connected with the generation of greater quantity of reject water in comparison with the plants applying common stabilization methods (aerobic or digestion).

Figure 3 compares the pollutants load in raw sewage and the reject water generated in ATAD process in WWTP.

The load of pollutants contained within the reject water depends on many factors, among which the most important are the type of treated sewage and stabilization methods. The analysis of the loads in WWTP with ATAD shows that in the case of organic and some non-organic compounds indicators, expressed as BOD_5 and COD, a higher load of pollutants is present in the sewage than in the reject water. The TN and TP loads are similar in the sewage and reject water from the ATAD process. However, the ammonia nitrogen load in this reject water exceeds its load in the municipal sewage. In the municipal WWTPs applying the typical stabilization processes, the load of total nitrogen in the reject water does not exceed 25% of its load in raw sewage [Janus & van der Roest 1997, Meyer & Wilderer 2004, Fux *et al.* 2006]. According to Ryzińska [2006] and Rosenwinkel *et al.*[2009], the share of biogenic compounds load might consitute from 10 to 30% of load in raw sewage, whereas the share of ammonia nitrogen varied from 1 to 50%. The treatment systems for the reject water from the ATAD process ought to be designed with great consideration for the ammonia nitrogen load. Table 2 presents the characteristics of reject water from ATAD process before and after treatment with SS VF bed during the research period and removed pollutants load.

Comparing the achieved results of the research on the reject water from the ATAD process with the results present in literature, it can be observed that the nitrogen and phosphorus compounds have similar concentrations. According to Borowski [2000], the average concentration of ammonia nitrogen in the reject water from the ATAD process varied from 290 to 715 mg N l^{-1}, total nitrogen from 975 to 1569 mg N l^{-1}, and total

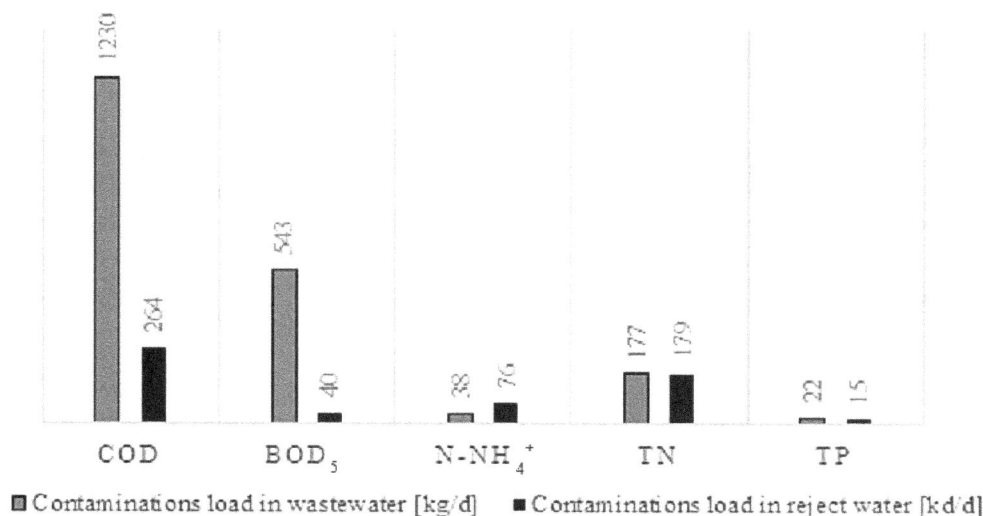

Figure 3. Comparison of pollutants load in raw sewage and the reject water generated in ATAD process in WWTP

Table 2. The characteristics of reject water from ATAD process before and after treatment (n=12) in SS VF bed and removed pollutants load

Parameter	Inflow (mg l^{-1})	Outflow (mg l^{-1})	Removed load (g m^{-2} d^{-1})
BOD$_5$	330±47	48±6	28.2±4.9
COD	2200±212	1100±169	110±17.0
TOC	711±82	299±39	41.2±6.8
TKN	1487±166	619±72	86.8±14.5
N-NH$_4^+$	636±76	347±55	28.9±3.9
N-NO$_3^-$	4.0±0.8	49.7±6.6	-
N-NO$_2^-$	0.6±0.1	0.1±0.02	-
TP	128.5±16.7	86.3±7.0	4.2±1.1

Mean ± Standard deviation

phosphorus from 217 to 327 mg P l^{-1}. The results of his research, conducted with stabilization time of 4, 6 and 8 days, point to the conclusion that the ammonia nitrogen and total nitrogen concentrations in the reject water decrease when the sludge stabilization time is extended. In increased temperature and high pH, ammonia nitrogen occurs as ammonia and is released into the atmosphere. Ammonia nitrogen cannot undergo biological oxidation into nitrite and nitrate as nitrification does not occur during the ATAD process. Moreover, the research carried out by the same author showed that the increased concentration of phosphorus compounds in reject water was not as great as in the case of nitrogen. Extension of the stabilization time caused an increased release of phosphorus from sludge, but it also supported phosphorus assimilation and storage by thermophilic microorganisms. Transferring great amounts of nitrogen

and phosphorus to reject water during the ATAD process must be taken into consideration while designing such installations in WWTPs.

The concentration of pollutants in the reject water generated during the ATAD process is similar to the one observed in the reject water from the dewatering of sludge after the digestion process. The ammonia nitrogen concentration, which is of great importance in this research, is similar to the one reported by other authors. According to Hans & van der Roest [1997], it reaches 657±56 mg N-NH$_4^+$ l^{-1} and according to Fux *et al.* [2003] it changes from 450 mg N-NH$_4^+$ l^{-1} to 750 mg N-NH$_4^+$ l^{-1}. On the other hand, according to Sperczyńska [2016], the average pollutants concentrations are: ammonia nitrogen 844 mg N-NH$_4^+$ l^{-1}, TKN 891 mg l^{-1}, phosphorus 125 mg P-PO$_4^{3-}$, chemical compounds expressed as ChZT 592 mg O$_2$ l^{-1}.

The low value of BOD$_5$/COD points to low biodegradability of the reject water from the ATAD process. It is lower than in the case of the reject water from the aerobic sewage sludge stabilization in dairy WWTPs, which ranged from 0.35 to 0.66 [Dąbrowski *et al.* 2017]. What is more, it is also lower than in the case of the reject water from the anaerobic sludge stabilization in municipal WWTPs, which – according to various authors – varied from 0.2 to 0.54 [Fux *et al.* 2006, Gajewska & Obarska – Pempkowiak 2013]. According to Klimiuk *et al.*[2007], the BOD$_5$/COD and BOD$_5$/TN ratios decrease in progress with the decomposition process. Their low values point to the lack of necessity of high efficiency treatment with the activated sludge system or a trickling filter. Moreover, the content of easily biodegradable organics is insufficient; hence, the conventional path of nitrogen removal cannot be applied. New microbiological processes, such as SHARON, Anammox, or CANON can be used for nitrogen removal [Ryzińska 2006]. Constructed wetlands might also be applied for that purpose. According to data given by Kadlec & Wallace [2009], the BOD_5/TKN ratio value below 1 is necessary for ensuring effective nitrification process in constructed wetlands. Figure 4 presents the graphical interpretation of the statistical analysis of influent, effluent and removed loads.

The calculated treatment efficiency was on average at: BOD_5 85.1%, COD 50.0%, TOC 57.9%, TKN 58.4%, $N-NH_4^+$ 45.6%, TN 55.2%, TP 32.3%. Similar values were obtained in the case of treating the reject water from digested sludge dewatering [Gajewska & Obarska-Pempkowiak 2013].

Generally, the effectiveness of removing organic compounds expressed as BOD_5 is similar to that observed in the case of high concentration household sewage or septage under comparable climatic conditions. Slightly inferior

results were obtained for COD, which is caused by lower biodegradability of reject water from the ATAD process (ratio BOD_5/COD = 0.04). In the case of total and ammonia nitrogen, high treatment efficiency was observed mainly due to high concentration of nitrogen in the reject water. The phosphorus removal effect was more stable than in the case of household sewage and septage treatment [Karolinczak & Dąbrowski 2017, Obarska-Pempkowiak et al. 2015].

The effectiveness of removing ammonia nitrogen in a CW bed of 45% is much lower than the effectiveness of its removal using advanced methods. In the SHARON method, ammonia nitrogen is removed with over the effectiveness of 90% [van Kempen et al. 2001]. Taking into consideration a substantial cost of installation and high exploitation demand, its application is not justified in most facilities.

An alternative method of reject water treatment might be the one involving the chemical process of precipitation. The aim of the research conducted by Sperczyńska [2016] was to evaluate the effectiveness of removing phosphorus

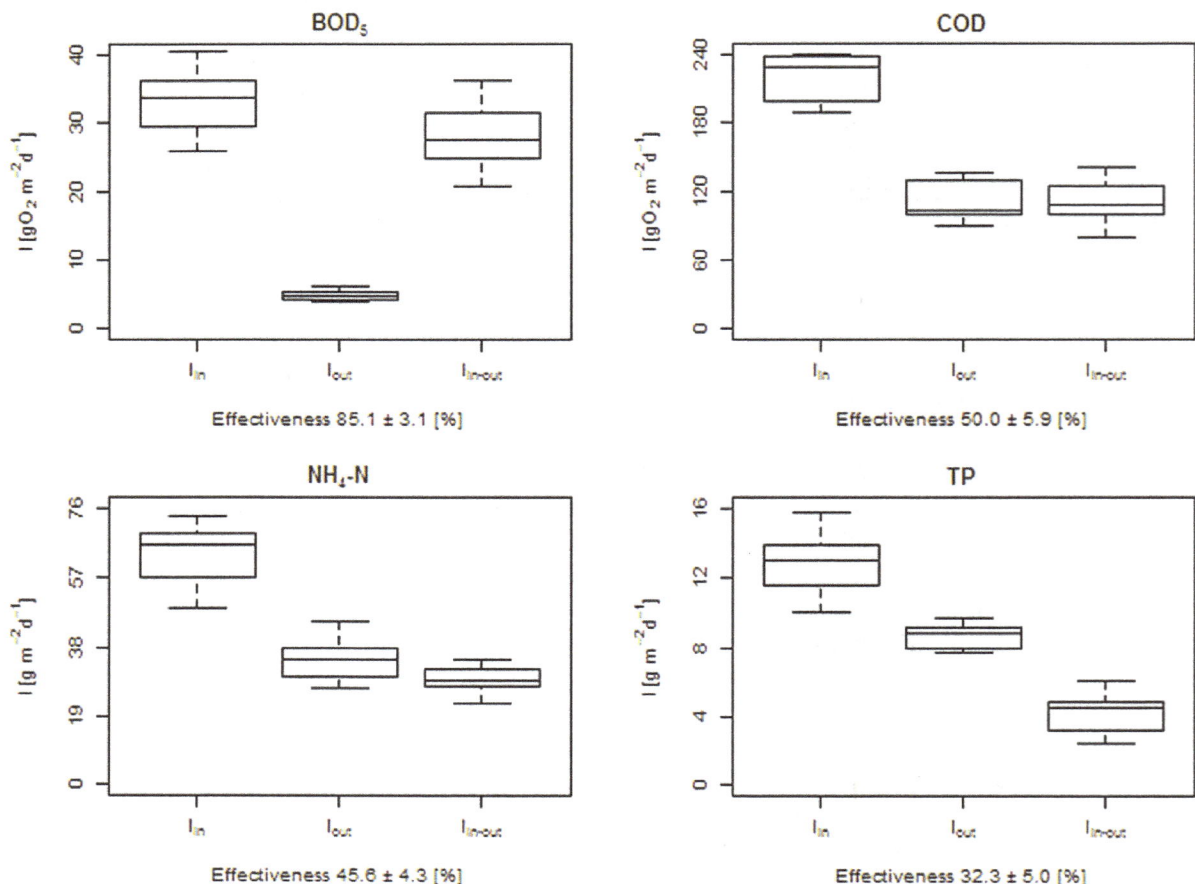

Figure 4. Graphical interpretation of the statistical analysis

compounds from post-fermentation reject water using pre-hydrolyzed polialuminium chlorides and aluminium sulfate. The removal effectiveness changed from 51 to 96%, while the removal effectiveness of organic compounds measured as COD increased from 19 to 47%, depending on the kind and dose of coagulant. In the authors' own research, the average effectiveness of removing phosphorus in a CW bed was lower and reached 32%, while the effectiveness of removing organic compounds and some non-organic elements measured as COD was higher.

The problem of minimizing the negative impact of reject water requires applying economically justified methods. They should be a compromise between highly effective and cost biological methods and a total omission of the reject water pre-treatment process.

CONCLUSIONS

The ATAD technology is commonly used in municipal and industrial WWTPs with PE less than 30.000. In such an object, applying anaerobic sewage sludge stabilization is not justified. The main problem related to the ATAD exploitation is a high concentration of nitrogen and phosphorus in the reject water from sewage sludge treatment and air purification (odor removal). The issue concerns mainly the highly loaded systems operating with SBRs without averaging tanks.

In the conducted research it was proven that applying separate reject water treatment with SS VF beds might ensure stable and effective functioning of municipal WWTPs by decreasing the load of biological part of WWTP. The efficiency of reject water treatment during the research period was on average 45.6% for ammonia nitrogen, 32.3% for total phosphorus and 85.1% for BOD_5. An innovative type of filling was used in SS VF beds, which was produced from waste (ash from a heat and power plant). Its application decreases the negative impact on the environment connected with exploitation of extracted aggregate (sand and gravel) and decreases the amount of waste.

Acknowledgements

The study was conducted as a research project S/WBiIŚ/3/2014 in Faculty of Building and Environmental Engineering of BUT and financed by Ministry of Science and Higher Education of Poland.

REFERENCES

1. American Public Health Association (APHA) 2005.Standard Methods for Examination of Water and Wastewater. 21st edition. American Public Health Association, Washington.

2. Bartkowska I., Dzienis L. 2007. Technical and economic aspects of autothermal thermophilic aerobic digestion exemplified by sewage treatment plant in Giżycko. Env. Prot. Eng. 33(2), 17–24.

3. Bartkowska I. 2016. Alternative application of ATAD process for sewage sludge stabilization, Gaz Woda i Technika Sanitarna, 90(11), 413–417, doi: 10.15199/17.2016.11.4 (in Polish)

4. Bartkowska I. 2017. Autothermal Thermohilic Aerobic Digestion. Seidel-Przywecki, Warsaw, (in Polish).

5. Borowski S. 2000. Aerobic thermophilic sewage sludge stabilization, Ochrona Środowiska 4(79), 21–25 (in Polish).

6. Carvalho P.N. , Arias A.C., Brix H. 2017. Constructed Wetlands for Water Treatment: New Developments. Water 9, 397, doi: 10.3390/w9060397.

7. Dąbrowski W., Karolinczak B., Gajewska M., Wojciechowska E. 2017. Application of subsurface vertical flow constructed wetlands to reject water treatment in dairy wastewater treatment plant. Environmental Technology 38(2), 175–182. http://doi.org/10.1080/09593330.2016.1262459.

8. Dąbrowski W. 2014. A study of the digestion process of sewage sludge from dairy WWTP to determine the composition and load of reject water, Water Pract. Technol. 9(1), 71–78.

9. Fux Ch., Valten S., Carozzi V., Solley D., Keller J. 2006. Efficient and stable nitrification and denitrification of ammonium-rich sludge dewatering liquor using SBR with continuous loading, Water Research 40 (14), 2765–2775.

10. Gajewska M., Obarska-Pempkowiak H. 2013. Multistage treatment wetland for treatment of reject waters from digested sludge dewatering, Water Sci. Technol., 68(6), 1223–1232.

11. Guo C.H., Kuang S.L., Stabnikov V., Ivanov V. 2009. The removal of phosphorus from wastewater using anoxic reduction of iron ore in the rotating reactor. Biochem. Eng. J. 46, 223–226.

12. Janus H.M., van der Roest H.F. 1997. Do not reject the idea of treating reject water, Water Sci. Technol. 35(10), 27–34.

13. Kadlec, R.H, Wallace S.D., 2009. Treatment Wetlands, 2nd ed.; CRC Press: Boca Raton, FL, USA.

14. Karolinczak B., Dąbrowski W. 2017. Effectiveness of septage pre-treatment in vertical flow constructed wetland. Water Science and Technology, 76.9, 2544–2553, doi: 10.2166/wst.2017.398.

15. Kempen R., Mulder J.W., Uijetrlinde C.A., Loosdrecht M.C.M. 2001. Overview: full scale experience of the SHARON process for treatment rejection water of digested sludge dewatering, Water Science and Technology 44(1), 145–152.

16. Klimiuk E., Kulikowska D., Koc-Jurczyk J. 2007. Biological removal of organics and nitrogen from landfill leachates – a review. In: Management of Pollutant Emission from Landfills and Sludge (M. Pawłowska & L. Pawłowski, eds.), Taylor and Francis Group, London, 187–204.

17. Liang Y., Zhu H., Bañuelos G., Yan B., Zhou Q., Yu X., Cheng X. 2017. Constructed wetlands for saline wastewater treatment: A review, Ecological Engineering, 98, 275–285.

18. Layden N.M., Mavinic D.S., Kelly H.G, Moles R., Bartlett J. 2007. Autothermal thermophilic aerobic digestion (ATAD) – Part I: Review of origins, design, and process operation, Journal of Environmental Engineering and Science, 6(6), 665–678, https://doi.org/10.1139/S07–015

19. Meyer S.S., Wilderer A., 2004. Reject Water: Treating of Process Water in Large Wastewater Treatment Plants in Germany – A Case study, Journal of Environmental Science and Health, 39(7), 1645-1654.

20. Obarska – Pempkowiak H., Gajewska M., Wojciechowska E., Kołecka K. 2015. Sewage gardens – constructed wetlands for single family households, Environmental Protection Engineering, 41(4), doi: 10.5277/epe150406.

21. Regulations of the Minister of Environment from 18th of November 2014 on conditions to be met for disposal of treated sewage into water and soil and concerning substances harmful to the environment (Dz.U. 2014. no. 1800) (in Polish)

22. Rosenwinkel K.H., Beier M., Phan L. C., Hartwig P. 2009. Conventional and advanced Technologies for biological nitrogen removal in Europe, Water Practice & Technology, 4(1), doi: 10.2166/WTP.2009.014

23. Ryzińska J. 2006. Problems of reject water and possibility of its treatment in Poland, Gaz, Woda i Technika Sanitarna, 7–8, 58–62 (in Polish).

24. Shugen L., Nanwen Z., Loretta Y.L. 2011. The one-stage autothermal thermophilic aerobic digestion for sewage sludge treatment, Chemical Engineering Journal, 174 (2–3), 564–570, https://doi.org/10.1016/j.cej.2011.09.043

25. Sperczyńska E. 2016. Usuwanie fosforanów z pofermentacyjnych cieczy osadowych. Ecological Engineering 46, 196–201, doi: 10.12912/23920629/63262, (in Polish).

26. Tan Y.Y., Tang F. E., Ho C.L.I., Jong V.S.W. 2017. Dewatering and treatment of septage using vertical flow constructed wetlands, Technologies 5(70), doi: 10.3390/technologies5040070

27. Website of LSA sp. z o.o. http://lsa.biz.pl/

20

Application of an Innovative Ultrasound Disintegrator for Sewage Sludge Conditioning Before Methane Fermentation

Marcin Zieliński[1*], Marcin Dębowski[1], Mirosław Krzemieniewski[1],
Paulina Rusanowska[1], Magdalena Zielińska[2], Agnieszka Cydzik-Kwiatkowska[2],
Agata Głowacka-Gil[1]

[1] Department of Environment Engineering, Faculty of Environmental Sciences, University of Warmia and Mazury in Olsztyn, ul. Warszawska 117, 10-720 Olsztyn, Poland

[2] Department of Environmental Biotechnology, Faculty of Environmental Sciences, University of Warmia and Mazury in Olsztyn, ul. Słoneczna 45G, 10-709 Olsztyn, Poland

[*] Corresponding author's e-mail: marcin.zielinski@uwm.edu.pl

ABSTRACT

Ultrasonic disintegration is one of the most interesting technologies among all known and described technologies for sewage sludge pre-treatment before the process of methane fermentation. This study was aimed at determining the effects of an innovative ultrasonic string disintegrator used for sewage sludge pre-treatment on the effectiveness of methane fermentation process. In this experiment, we used a device for disintegration of organic substrates, including sewage sludge, with the use ultrasonic waves. Its technical solution is protected by a patent no. P. 391477 – *Device for destruction of tissue and cell structures of organic substrate*. The volume of biogas produced ranged from 0.194 ± 0.089 dm^3/g o.d.m. at loading of 5.0 g o.d.m./dm^3 and power of 50 W to 0.315 ± 0.087 dm^3/g o.d.m. at loading of 4.0 g o.d.m./dm^3 and ultrasounds power of 125 W. The study demonstrated a positive effect of sewage sludge sonication on the percentage content of methane in biogas. Sewage sludge exposure to 125 W ultrasounds increased methane content in biogas to 68.3 ± 2.5 % at tank loading of 3.0 g o.d.m./dm^3.

Keywords: sewage sludge, ultrasound disintegrator, methane fermentation, biogas

INTRODUCTION

Ultrasonic disintegration is one of the most interesting technologies among all known and described technologies for sewage sludge pre-treatment before the process of methane fermentation. The main reasons behind studies on the ultrasonic field and application of installations based on its use in the existing facilities include very wide applicability and technological versatility of the sonication process [Quiroga et al. 2014].

This physical factor may induce deep physicochemical changes in the sonicated sludge that are highly desired from the viewpoint of sewage sludge processing. A special role is ascribed in this case to disintegration of sewage sludge bacteria which are next subjected to the process of methane fermentation. Damage of cell structures of microorganisms secreted in settling tanks of a wastewater treatment plant enables biodegradation of organic biomass in the process of sludge stabilization with anaerobic methods. Once the cell wall of microorganisms is damaged, cytoplasm and cellular enzymes are released, while the substrates released in this way, in either dissolved or colloidal form, are immediately available for biological degradation by anaerobic bacteria [Negral et al. 2013]. To ensure the maximal release of organic matter from dead cells of microorganisms, disintegration aims to result in particles smaller than 10 μm. In many research works and in many facilities operating in the technical scale, contents of organic substances in the dissolved phase of the super-sludge liquid (expressed as COD, BOD$_5$, TOC) were demonstrated to increase immediately after sonication [Negral et al. 2013, Martínez-Moral et Tena Focused 2013]. This phenomenon is indicative of immediate sonolysis of solid substances contained in the sludge bulk and of the release of the cellular material from dead microorganisms, and finally points to intensification of the hydrolytic stage of the fermentation process. This allows for significant reduction in time needed for anaerobic stabilization of the sludge and for increasing efficiency of producing gaseous metabolites of anaerobic bacteria [Alagöz et al. 2018].

Although the first installations based on the use of ultrasounds for excess sludge pre-treatment via disintegration are already operating in municipal wastewater treatment plants, the acoustic and mechanical disintegration technology of sewage sludge is still in the experimental phase. Studies in progress are focused on both the phenomena occurring in the sonicated sludge and their reference to the effects of disintegration as well as on the search for effective methods and devices for ultrasonic disintegration [Dhar et al. 2012].

This study was aimed at determining the effects of an innovative ultrasonic string disintegrator used for sewage sludge pre-treatment on the effectiveness of methane fermentation process.

METHODS

Study design

The experiment was divided into two stages. The first involved analyses of the effect of ultrasounds on sewage sludge parameter significant for methane fermentation, i.e. content of organic carbon in the dissolved form (TOC). The second stage was aimed at determining the course, efficiency and characteristics of the methane fermentation process. The experiment was conducted under laboratory conditions, in 5 series differing in the power of ultrasounds applied to the sewage sludge (variant 1 – 0 W, variant 2 – 25 W, variant 3 – 50 W, variant 4 – 100 W, variant 5 – 125 W). In each variant, exposure time to ultrasounds was 600s.

To determine methane fermentation efficiency, model anaerobic reactors (by WTW company) were used in the study. They consisted of reaction tanks with an active volume of 0.5 dm^3 coupled tightly with devices measuring and registering changes in partial pressure in the measuring tank induced by biogas production. In each experimental variant, 200 cm^3 of anaerobic sludge (inoculum) were introduced into the reaction tank which was then filled with the assumed volumes of pre-disintegrated sewage sludge Anaerobic conditions were provided inside the exploited fermentation tanks by blowing the installation through with nitrogen before measurements to remove atmospheric air from respirometers. The initial organic matter load ranged from 3.0 g o.d.m./dm^3 to 5.0 g o.d.m./dm^3 depending on variant. A complete measuring set was put into a thermostat cabinet with hysteresis not exceeding ± 0.5°C. Measurements were conducted at a temperature of 35°C for 20 days, whereas pressure values in the reaction tank were registered every 24 h.

MATERIALS

Excessive sewage sludge used in the experiment originated from the Municipal Wastewater Treatment Plant "Łyna" in Olsztyn. It is a bio-mechanical wastewater treatment plant, with the mean flow capacity of 56,135 m^3/d (max 63,310 m^3/d). The plant operates based on a three-stage treatment process with biological removal of biogenes. Sludge was obtained from gravity belt thickener intended for sludge thickening before introducing it to fermentation chamber. Characteristics of excess sewage sludge used in the study was provided in Table 1.

Table 1. Characteristics of sewage sludge used in the experiment

Parameter	Unit	Min. value	Max. value	Mean value
Hydration	[%]	97.25	96.92	97.08
Dry matter	[g/dm3]	27.490	28.840	28.165
Mineral substances	[g/dm3]	5.950	7.760	6.855
Volatile substances	[g/dm3]	19.570	20.560	20.065
TOC	[mg/dm3]	2696.3	2949.0	2822.6
Acidity	[pH]	5.78	6.49	6.13
Clostridium perfringens	[CFU/g d. m.]	6.7×10^4	7.9×10^4	7.3×10^4
Coli group bacteria	[MPN/g d. m.]	7.1×10^6	7.9×10^6	7.5×10^6
Fecal coliform bacteria	[MPN/g d. m.]	2.6×10^6	3.8×10^6	3.2×10^6

Experimental stand

In this experiment, we used a device for disintegration of organic substrates, including sewage sludge, with the use ultrasonic waves. Its technical solution is protected by a patent no. P. 391477 – *Device for destruction of tissue and cell structures of organic substrate*. This device has the form of a cylindrical tank with strings mounted inside that transfer ultrasound vibrations (Figure 1). The flow in the reactor is from bottom up. Sewage sludge flowing through the reactor is exposed to the ultrasounds along the entire active length of the reactors. Ultrasounds cause abrupt condensation and expansion of the medium they act upon. Cavitation "bubbles" are formed which damage the structure of the organic substrate fed to the reactor by their abrupt collapse (implosion).

The device for destruction of tissue and cell structures of the organic substrate consists of a cylindrical tank with an inlet channel (2) and an outlet channel (3) on a side wall, and is equipped in a disintegrating ultrasound generator. Inside the tank (1), strings (4) are fixed to two discs (5 and 6) located in the upper and bottom part of the tank (1). The top disc (5) is connected to the disintegrating ultrasound generator (7), and the bottom disc (6) is connected to the instrument (8) that is used to pull the strings (4).

The device consists of undo elements coupled with each other, namely: main tank made of stainless steel (1–3 mm thick) which is mounted on a 0.2 m high stand and has an inlet in the form of a cylinder (ϕ80 mm, height 50mm). On the bottom of the cylinder there is a disc which fixes ϕ50 strings with thickness of 10 mm. An inlet pipe (50 mm in diameter, 100 mm in length) is mounted on tank side which has a tip allowing to connect it (using a quick connector) with a pressing pipe. From the upper part, the cylinder is connected with a channel (50 mm in diameter and 1200 mm in length). Strings (1 mm in diameter) are mounted in the channel in 0.6 mm spacing, with external strings fixed 3 mm away from channel walls. Three ball valves (ϕ 10mm) for sample collection are mounted inside the channel along its whole length (in equal spacing). Corpus channel is closed with a top cylinder (ϕ 80mm). The top cylinder (50 mm thick) which has a side inlet (made in the same way as the inlet at the bottom) should be mounted so as the inlet tip was in the position of a 90° elbow to be fixed with a quick connector. A disc with fixed strings is mounted in the top part of the cylinder. Its position may be modified to change string tension using one or three turnbuckles. The central part of the disc is coupled with the disintegrating ultrasound generator operating with a frequency of 30 kHz and power from 50 to 300 W, whose casing is coupled with the top cylinder.

Analytical methods

The scope of conducted analyses included determinations of concentrations of: dry matter (d.m.), organic dry matter (o.d.m.), mineral dry matter (m.d.m.), total carbon (TC), and total organic carbon (TOC), as well as pH measurements. Concentrations of dry matter, organic dry matter and mineral dry matter in sewage sludge were determined with the gravimetric method. Concentrations of TC and TOC were assayed in biomass samples dried at 105°C. Analyses were carried out with Flash 2000 elementary particle analyzer by Thermo company. Sewage sludge pH was determined by weighing 10 g of a homogenized air-dry sample into a 100-mL beaker, adding 50 mL of distilled water to the beaker and sample mixing, followed by apparatus calibration and pH measurement.

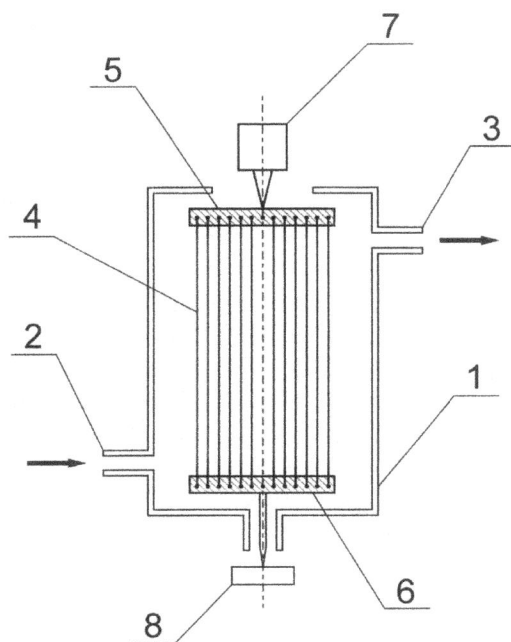

Figure 1. Device for damage of tissue and cell structure of organic substrate with ultrasounds

The presence of *Coli* group bacteria and fecal coliforms was analyzed acc. to the Polish Standard PN-EN-ISO 9308–1:2002(U). Samples of sludge were seeded and incubated onto an isolation medium – a lauryl sulfate broth, to determine gas production capability. *Coli* group bacteria were incubated at 37°C whereas fecal coliforms at 44.5°C. The samples determined for the presence of *Salmonella* genus bacteria were proliferated onto Muller-Kauffman medium, and next the bacteria were isolated onto the solid differentiating and selective S-S medium. Biochemical analyses were conducted with API 20E tests by BioMerieux company. Once bacteria from the genus *Salmonella* were detected in the samples, they were subjected to serological tests to confirm their capability for agglutination with the slide method using HM serum by IMMUNOLAB. To confirm the presence of spore forming bacteria from the species *Clostridium perfringens,* sewage sludge was analyzed acc. to the Polish Standard PN-EN-ISO 2646–1:2002. Samples of sludge were seeded and incubated on the Wilson-Blair medium in anaerostats using AnaeroGen sachets by OXOID, to generate anaerobic conditions. Bacteria around which a black precipitate of iron sulfide has developed were counted.

The biogas potential of substrate at different organic compounds loading was tested. The study was carried out in the methane potential analysis tool (AMPTS II Bioprocess Control). This device was used to measure the quantity of biogas produced. Reactors of the volume of 500 mL were connected to the multifunctional agitation system. Mixing in the reactor run for 30 seconds each 10 minutes. Rotating speed was 100 rpm. The 200 mL of digestate was introduced to the reactors. Anaerobic conditions were achieved by continuous flushing of pure nitrogen through the sludge. Methane fermentation was carried out under mesophilic conditions at 37°C for 40 days. The experiments were performed in triplicates.

Samples of biogas were collected from respirometers for qualitative gas analysis using a needle and a gas-proof syringe. Respirometers were equipped in valves with rubber stoppers enabling gas-proof sample collection. A single collected sample contained 20 cm^3 of biogas, the composition and percentage content of which were analyzed with a GC Agillent 7890 A gas chromatograph. In addition, quality of the produced biogas was evaluated with a GMF 430 analyzer by Gas Data company. Biogas was determined for percentage contents of: methane CH_4 and carbon dioxide CO_2. The volume of biogas produced during respirometric analyses was computed based on the equation of perfect gas. The extent of changes in pressure determined inside the measuring chamber allowed calculating the volume of produced biogas expressed per normal conditions.

Statistical analysis

Statistical analysis of the results was made in STATISTICA package 10.0 PL. All physicochemical analyses were carried out in three replications. The hypothesis on distribution of each analyzed variable was verified with the Shapiro-Wilk W test. One-way analysis of variance (ANOVA) was conducted to determine the significance of differences between mean values. Homogeneity of variance in groups was checked with Levene's test, whereas HSD Tukey's test was used to determine the significance of differences between the analyzed variables. Differences were found significant at $p = 0.05$.

RESULTS

The conducted experiment allows concluding that the use of ultrasonic waves contributed to a significant increase ($p=0,05$) in TOC concentration in the effluent after vacuum filtration of the disintegrated sewage sludge. The study showed a direct impact of the applied power of the physical factor on TOC concentration. The highest TOC concentration in the effluent, accounting for 3933 ± 112 mg/dm^3, was determined in the experimental variant with ultrasound waves having the power of 100 W and was by nearly 40 % higher compared to the non-sonicated sludge. Increasing disintegrator power had no significant effect ($p=0,05$) on the technological effects related to TOC concentration in the effluent, which in this variant equaled to 3970 ± 270 mg/dm^3. The use of lower powers resulted in a decreased concentration of organic matter in the dissolved phase of the sludge (Figure 2).

It was proved that the volume and composition of biogas produced during respirometric tests was directly dependent on the power of ultrasounds. In turn, efficiency of methane fermentation was not significantly ($p=0,05$) affected by the initial load of organic compounds fed to reactors. Irrespective of the variant of tanks load-

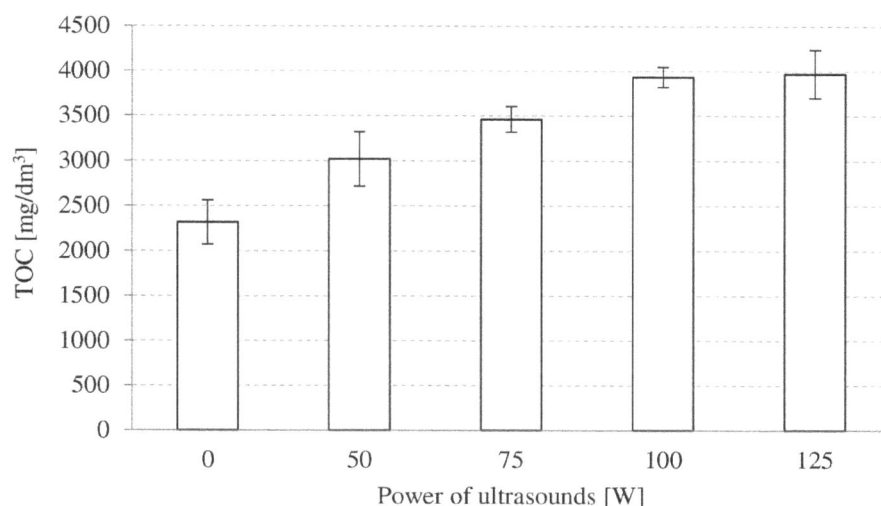

Figure 2. TOC concentration in the effluent after vacuum filtration in relation to used power of ultrasounds

ing with organic matter, the volume of produced biogas increased along with an increasing power of ultrasound waves. The volume of biogas produced ranged from 0.194 ± 0.089 dm^3/g o.d.m. at loading of 5.0 g o.d.m./dm^3 and power of 50 W to 0.315 ± 0.087 dm^3/g o.d.m. at loading of 4.0 g o.d.m./dm^3 and ultrasounds power of 125 W. The study demonstrated a positive effect of sewage sludge sonication on the percentage content of methane in biogas. Fermentation of non-sonicated sludge allowed producing from $50.1\pm2.1\%$ to $56.0\pm1.8\%$ of methane in biogas depending on respirometric tanks loading with organic compounds. The use of ultrasounds with the power of 125 W increased methane content in biogas to $68.3\pm2.5\%$, in the variant with organic compounds load of 3.0 g o.d.m./dm^3. Results of experiments were presented in Tables 2 and 3.

DISCUSSION

Ultrasonic disintegration aiding the methane fermentation process is mainly aimed at production of biogas as a carrier of renewable energy. The total volume of biogas produced from sonicated excess sludge increases by 30–50% (or even by 100% under laboratory conditions) compared to non-sonicated sludge, which was demonstrated by many authors [Chang et al. 2011]. In our study, the increase in biogas volume was from 12 to 50% depending on the experimental variant.

The first outcome of ultrasonic waves is disintegration and dispersion of flocks with no damage caused to cells of microorganisms – effects of this type may be monitored with methods of optical or laser microscopy, but also through determination of organic matter content in the dissolved phase. At longer exposure to ultrasound waves or at their higher energy, when cell walls of microorganisms are disrupted and intracellular material is released to the liquid phase, the number of microorganisms decreases [Rochebrochard et al. 2011]. Depending on its origin, the sludge contains microorganisms with various resistance of their cell walls. These walls can be broken only at a high load of energy imposed on a sonicated medium, i.e. either by a long exposure to ultrasounds emission (significantly longer than that needed to damage particle structure of the sludge), or by a short impulse of a very high power [Braguglia et al. 2011, Feng et al. 2009]. Our study demonstrated that the applied disintegrator enabled destruction of cellular structures of the excess sludge and that this phenomenon resulted in increased by 40% TOC concentration in the dissolved phase compared to the non-sonicated sludge.

It was concluded that sewage sludge should be disintegrated with ultrasounds generated at relatively low frequencies, ranging from 16 kHz t 50 kHz, however cavitation excited by acoustic wave intercepted by the sludge requires at least 1.0 W/cm^2 and higher intensity of the acoustic field [Rochebrochard et al. 2013, Grönroos et al. 2005]. For practical applications, especially for estimating costs of the sonication process in wastewater treatment plants, we postulate the calculated index of energy consumption for ultrasound disintegration be at 0.5 kWh/kg d.m. sludge [Wang et al. 2018]. In our study, the frequency of ultrasounds was at 30 kHz.

Table 2. Effect conditioning method on the qualitative characteristics of biogas

Parameter		Carbon content in gaseous phase	CO$_2$ content in gaseous phase	CH$_4$ content in gaseous phase	CO$_2$	CH$_4$
Unit		mmol/ g o.d.m.	mmol/ g o.d.m.	mmol/ g o.d.m.	%	%
Power [W]	Load [g o.d.m./dm³]			Value±		
crude	3.0	0.335±0.003	0.147±0.009	0.187±0.016	44.0±1.2	56.0±1.8
	4.0	0.354±0.008	0.170±0.011	0.184±0.019	48.1±1.9	51.9±1.9
	5.0	0.317±0.005	0.158±0.012	0.159±0.015	49.9±2.0	50.1±2.1
50	3.0	0.373±0.011	0.149±0.013	0.224±0.028	39.9±1.8	60.1±1.6
	4.0	0.393±0.009	0.146±0.009	0.247±0.022	39.2±2.1	60.8±1.8
	5.0	0.351±0.008	0.159±0.011	0.192±0.017	45.3±1.9	54.7±1.1
75	3.0	0.373±0.012	0.139±0.020	0.234±0.019	37.2±2.5	62.8±1.6
	4.0	0.393±0.011	0.137±0.015	0.256±0.023	34.9±2.4	65.1±1.8
	5.0	0.351±0.011	0.151±0.016	0.201±0.011	42.9±1.9	57.1±2.0
100	3.0	0.345±0.016	0.160±0.009	0.185±0.017	36.4±3.1	63.6±2.6
	4.0	0.357±0.007	0.155±0.009	0.202±0.009	33.4±3.6	66.6±2.8
	5.0	0.317±0.008	0.153±0.016	0.164±0.018	41.3±1.7	58.7±1.7
125	3.0	0.356±0.005	0.139±0.012	0.217±0.017	37.1±2.9	68.3±2.5
	4.0	0.361±0.008	0.128±0.007	0.233±0.014	35.4±3.1	64.6±2.7
	5.0	0.326±0.009	0.151±0.011	0.155±0.021	40.4±2.9	59.6±1.6

Table 3. Effect of conditioning method on the volume of biogas produced during methane fermentation

Parameter		Biogas volume	Biogas volume	Biogas volume
Unit		dm³/g d.m.	dm³/g o.d.m.	dm³
Power [W]	Load [g o.d.m./dm³]		Value	
crude	3.0	0.280±0.059	0.188±0.066	0.414±0.111
	4.0	0.310±0.092	0.208±0.067	0.701±0.201
	5.0	0.260±0.71	0.174±0.045	1.154±0.178
50	3.0	0.330±0.023	0.221±0.055	0.488±0.099
	4.0	0.370±0.040	0.248±0.076	0.836±0.176
	5.0	0.290±0.039	0.194±0.089	1.288±0.199
75	3.0	0.390±0.099	0.261±0.058	0.577±0.166
	4.0	0.420±0.021	0.281±0.044	0.949±0.189
	5.0	0.330±0.043	0.221±0.050	1.465±0.123
100	3.0	0.360±0.036	0.241±0.090	0.533±0.171
	4.0	0.430±0.028	0.288±0.098	0.972±0.132
	5.0	0.340±0.022	0.228±0.121	1.510±0.245
125	3.0	0.420±0.062	0.281±0.090	0.622±0.165
	4.0	0.470±0.031	0.315±0.087	1.062±0.156
	5.0	0.400±0.056	0.268±0.098	1.776±0.233

Changes in the medium induced by the effects of the ultrasonic field may vary and depend not only on the aforementioned acoustic values but also on the physicochemical properties of the medium subjected to sonication: viscosity of liquid, presence of electrolytes and polyelectrolytes, macrostructure and character of the suspension, temperature of the medium, and many others; and therefore may be influenced by many factors acting in the same time [Yaowei, et al. 2018].

The achieved effect of disintegration is largely affected by physical conditions of the process and by parameters of the reactor used for disintegration. Each system will have its own optimal technical conditions of ultrasound disintegration which apart from emission parameters include geometrical conditions of wave propagation: shape and size of the tank, emitter (or emitters) location in the sonicated area and many others [Li et al. 2017].

Ultrasound disintegrators for sewage sludge pre-treatment have so far been produced as individual items by specialized groups in scientific centers or research-and-development units. Their quality and effectiveness are, therefore, largely dependent on the knowledge and skills of their constructors in the fields of acoustics and electronics [de Moortel et al. 2017]. Devices for ultrasound conditioning of sewage sludge before the fermentation process applied so far in municipal wastewater treatment plants represent the type of tubular or tank flow-through reactors. In majority of cases, these are innovative facilities adjusted in their shape, size and power to conditions of a specific installation. Most of devices installed so far have been operating in a frequency range of 20–40 kHz, with feeding electric power ranging from 300 to 2000 W.

CONCLUSIONS

1. The experiment was conducted with an innovative device for damage of tissue and cell structures of an organic substrate which consisted of a cylindrical tank equipped in a disintegrating ultrasound generator. Its characteristic trait are strings fixed to two discs mounted inside the tank at its bottom and top sides, wherein the top disc is connected with the disintegrating ultrasound generator and the bottom disc – with a device for string tension control. It is an untypical, so far underexploited technical solution.

2. The applied device contributed to a significant increase in TOC concentration in the effluent after vacuum filtration of the disintegrated sewage sludge. The highest TOC concentration in the effluent, accounting for 3933±112 mg/dm^3, was determined in the experimental variant with ultrasound waves having the power of 100 W and was by nearly 40 % higher compared to the non-sonicated sludge.

3. It was proved that the volume and composition of biogas produced during respirometric tests was directly dependent on the power of ultrasounds. The volume of biogas produced ranged from 0.194±0.089 dm^3/g o.d.m. at loading of 5.0 g o.d.m./dm^3 and power of 50 W to 0.315±0.087 dm^3/g o.d.m. at loading of 4.0 g o.d.m./dm^3 and ultrasounds power of 125 W. The study demonstrated a positive effect of sewage sludge sonication on the percentage content of methane in biogas. Sewage sludge exposure to 125 W ultrasounds increased methane content in biogas to 68.3±2.5% at tank loading of 3.0 g o.d.m./dm^3.

Acknowledgements

The solution of ultrasound disintegrator is promoted under the project *Research Coordination for a Low-Cost Biomethane Production at Small and Medium Scale Applications* (Record Biomap) Horizon 2020 research and innovation programme (Grant Agreement 691911).

The research was conducted under the Project No. 18.610.008–300 entitled "Improving methods of wastewater treatment and sludge disposal" from the University of Warmia and Mazury in Olsztyn.

REFERENCES

1. Alagöz B.A., Yenigün O., Erdinçler A. 2018. Ultrasound assisted biogas production from co-digestion of wastewater sludges and agricultural wastes: Comparison with microwave pre-treatment. Ultrasonics Sonochemistry, 40 (B), 193–200.

2. Braguglia C.M., Gianico A., Mininni G. 2011. Laboratory-scale ultrasound pre-treated digestion of sludge: Heat and energy balance. Bioresource Technology, 102 (16), 7567–7573.

3. Chang T-C., You S-J., Damodar R.,A., Chen Y-Y. 2011. Ultrasound pre-treatment step for performance enhancement in an aerobic sludge digestion process. Journal of the Taiwan Institute of Chemical Engineers, 42 (5), 801–808.

4. de Moortel N., den Broeck Rob., Degrève J., Dewil R. 2017. Comparing glow discharge plasma and ultrasound treatment for improving aerobic respiration of activated sludge. Water Research, 122, 207–215.

5. Dhar B.,R., Nakhla G., Ray M.,B. 2012. Techno-economic evaluation of ultrasound and thermal pretreatments for enhanced anaerobic digestion of municipal waste activated sludge. Waste Management, 32 (3), 542–549.

6. Feng X., Deng J., Lei H., Bai T., Fan Q., Li Z. 2009. Dewaterability of waste activated sludge with ultrasound conditioning. Bioresource Technology, 100 (3), 1074–1081.

7. Grönroos A., Kyllönen H., Korpijärvi K., Pirkonen P., Paavola T., Jokela J., Rintala J. 2005. Ultrasound assisted method to increase soluble chemical oxygen demand (SCOD) of sewage sludge for digestion. Ultrasonics Sonochemistry, 12 (1–2), 115–120.

8. K., Yaowei, Ning X., Liang J., Zou H., Sun J., Cai H., Lin M., Li R., Zhang Y. 2018. Sludge treatment by integrated ultrasound-Fenton process: Characterization of sludge organic matter and its impact on PAHs removal. Journal of Hazardous Materials, 343, 191–199.

9. Li X., Guo S., Peng Y., He Y., Wang S., Li L., Zhao M. 2017. Anaerobic digestion using ultrasound as pretreatment approach: Changes in waste activated sludge, anaerobic digestion performances and digestive microbial populations. Biochemical Engineering Journal, In Press, Accepted Manuscript, Available online 22 November DOI: https://doi.org/10.1016/j.bej.2017.11.009.

10. Martínez-Moral M.T., Tena.Focused M.T. 2013. Ultrasound solid–liquid extraction of perfluorinated compounds from sewage sludge. Talanta, 109 (15), 197–202.

11. Negral L., Marañón E., Fernández-Nava Y., Castrillón L. 2013. Short term evolution of soluble COD and ammonium in pre-treated sewage sludge by ultrasound and inverted phase fermentation. Chemical Engineering and Processing: Process Intensification, 69, 44–51.

12. Quiroga G., Castrillón L., Fernández-Nava Y., Marañón E., Negral L., Rodríguez-Iglesias J., Ormaechea P. 2014. Effect of ultrasound pre-treatment in the anaerobic co-digestion of cattle manure with food waste and sludge. Bioresource Technology, 154, 74–79.

13. Rochebrochard S., Naffrechoux E., Drogui P., Mercier G., Blais J-F 2013. Low frequency ultrasound-assisted leaching of sewage sludge for toxic metal removal, dewatering and fertilizing properties preservation. Ultrasonics Sonochemistry, 20 (1), 109–117.

14. Wang T., Zhang D., Sun Y., Zhou S., Li L., Shao J. 2018. Using low frequency and intensity ultrasound to enhance start-up and operation performance of Anammox process inoculated with the conventional sludge. Ultrasonics Sonochemistry, 42, 283–292.

21

Analysis of Bakery Sewage Treatment Process Options based on COD Fraction Changes

Joanna Struk-Sokolowska[1], Justyna Tkaczuk[1]

[1] Bialystok University of Technology, Faculty of Civil and Environmental Engineering, Department of Technology and Systems in Environmental Engineering, ul. Wiejska 45E, 15-351 Bialystok, Poland

[*] Corresponding author's e-mail: j.struk@pb.edu.pl

ABSTRACT

Municipal WWTPs often receive industrial wastewater including the bakery sewage. The effluent of the bakery industry has a high biological oxygen demand (BOD). In addition to high BOD, this wastewater contains high chemical oxygen demand (COD), total Kjeldahl nitrogen (TKN) and is characterized by a dark color. The effect of bakery wastewater contribution on the COD fraction changes in the municipal sewage is presented in this paper. The study was conducted in July 2016 in a WWTP located in Lipsk, East-North Poland. The sewage receiver is the Biebrza River. The volume contribution of bakery wastewater is 10%. The analytical results were used to compute the percentage value contribution of individual COD fractions in wastewater. During the study, the following fractions were identified: S_S – COD of readily-biodegradable dissolved organic matter, S_I – COD of non-biodegradable dissolved organic matter, X_S – COD of slowly-biodegradable non-soluble organic matter, X_I – COD of non-biodegradable non-soluble organic matter. The method used for the COD fraction determination in wastewater was developed based on the ATV 131P guidelines (ATV-DVWK-A131P). The aim of the study was to determine the effect of bakery wastewater contribution on the COD fraction changes during the technical scale biological wastewater treatment with an activated-sludge process. The percentage contributions of individual COD fractions in wastewater were compared with the shares in the wastewater from other food industries (dairy, olive mill, tomato, sugar beet, potato processing, winery). In raw wastewater, the X_S fraction was dominant 44.2%. S_S fraction was 38.8%. In raw wastewater, the S_I, X_I fractions ranged from 2.3 to 14.8%. In the effluent the S_S fraction was not noted, which is indicative of microorganisms consumption. The WWTP effluents mostly (43.4%) contained slowly-biodegradable non-soluble organic matter (X_S). Non-biodegradable dissolved organic matter (S_I fraction) had a high share of 42.3%.

Keywords: wastewater, municipal, bakery, food industry, COD fractions, activated sludge process

INTRODUCTION

Industrial pollution is still a major concern for sewage treatment and despite its significance, sound and systematic pollution control efforts are very poorly documented. The character and treatability of the industrial wastewater is highly variable and specific for each industrial activity. The issues related to the activated sludge treatment, such as biodegradability based characterization, modeling, assessment of stoichiometric and kinetic parameters and design, as well as the issues of industrial pollution control are not completely solved [Orhon et al. 2009]. The bakery industry is one of the world's major food industries and varies widely in terms of production scale and process. Traditionally, bakery products may be categorized as bread and bread roll products, pastry products (e.g., pies and pasties), and specialty products (e.g., cake, biscuits, donuts, and specialty breads). Flour, yeast, salt, water and oil/fat are the basic ingredients, as well as bread improvers – flour treatment agents like vitamin C (ascorbic acid) and preservatives in the commercial bakery production process [Chen et al. 2006].

It is possible to co-treat bakery and municipal wastewater to achieve the desired C:N:P ratio in order to carry out highly effective denitrification and biological phosphorus removal. The analysis of the current state of knowledge showed that there is a lack of data on the changes in quantitative and qualitative parameters of the municipal wastewater with a specified share of bakery wastewater during large scale treatment with an activated-sludge process. The results from fractionation of organic compounds by means of COD in this type of sewage are not available. The analysis of changes of particular fractions is also not available, making it impossible to control the purification processes in detail and obtain high efficiency of organic compounds and biogenic substance removal.

The aim of the study was to determine the effect of bakery wastewater contribution on the COD fraction changes during large scale biological wastewater treatment with the activated-sludge process. The results of the research on the effectiveness of municipal sewage treatment with a proportion of bakery wastewater and determined COD fractions can be used for: modification of the wastewater treatment process, modernization of bakery sewage pre-treatment plant, liquidation of the sewage treatment plant and discharge of bakery wastewater into the municipal sewage treatment plants.

Characteristics of wastewater from bakery industry

The bakery industry is one of the largest water users in Europe and the United States [Chen et al. 2006]. More than half of the water is discharged as wastewater. The wastewater in bakeries is primarily generated from the cleaning operations, including equipment cleaning and floor washing. It can be characterized as high loading, fluctuating flow containing oil and grease. The ratio of water consumed to products is about 10 in common food industry, much higher than that of 5 in the chemical industry and 2 in the paper and textiles industry. Normally, half of the water is used in the process, while the remainder is used for washing purposes, e.g., of equipment, floor, and containers [Chen et al. 2006]. Bakeries generate large amounts of acidic wastewater which contains flour, fat and sugar [bakerybazar.com]; they also use large amounts of butter, flour, shortening, eggs and fillings of various types. The bakery wastewater is biodegradable; hence, the biological treatment can be effective in reducing BOD. This type of wastewater is normally treated by physical and chemical processes and the various treatment possibilities are presented in Figure 1.

Large commercial bakeries producing cakes, pies, cookies, biscuits, brownies, rolls and a variety of other desserts generate wastewater with loads of contaminants which are too high for discharge to a municipal sewerage system without pretreatment. The flow diagram of bakery wastewater treatment plant is presented in Figure 2.

Pretreatment of the wastewater from the bakery industry is necessary before discharge to a municipal sewer system. The process flow diagram of the bakery wastewater pretreatment system is presented in Figure 3.

A very efficient way to accomplish this is through the installation of a Dissolved Air Flotation (DAF) system. These systems are the optimal solution for the removal of such materials as fats, oil and grease. A DAF system utilizes the gravity separation technique that uses air bubbles in the wastewater holding tank to attach to the insoluble materials and float them to the surface. Certain materials used in the baking process are heavier than water and thus require the addition of flocculants, which – when introduced – cause the materials to join in clusters light enough to float to the surface. These contaminants then accumulate on the surface of the DAF tank as sludge and are scraped off and removed from the wastewater.

Pretreatment using a DAF in commercial bakeries allows for a reduction of FOG (fats, oil and grease) by 99% and Total Suspended Solids (TSS) by 97%. The requirement for flocculants as well as pH adjustment of the water for effective FOG removal can be fully automated [ecologixsystems.com].

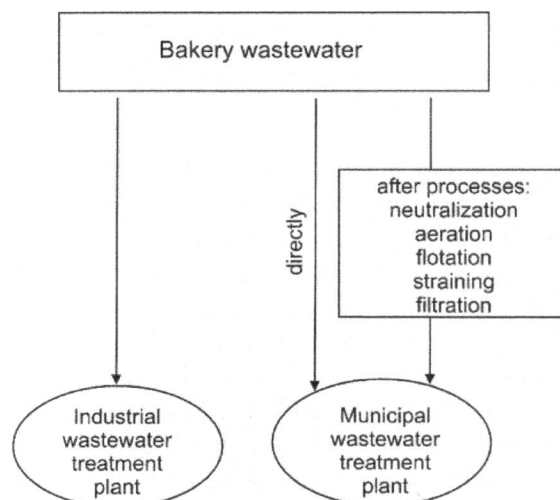

Figure 1. The possibilities of bakery wastewater treatment and co-treatment

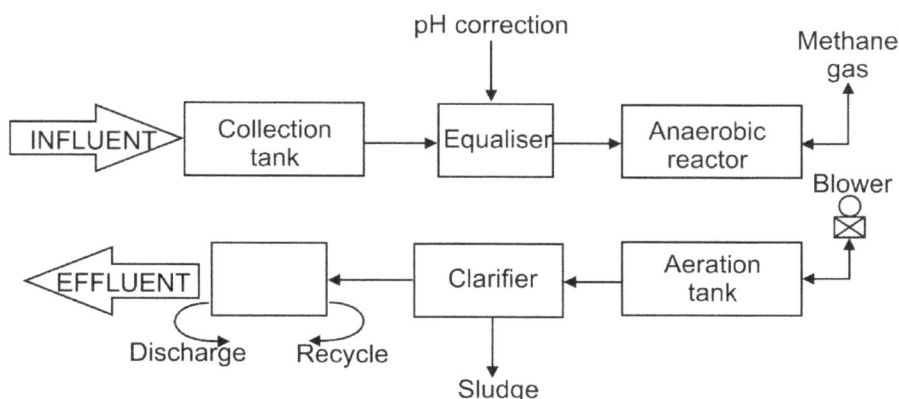

Figure 2. Flow diagram of bakery wastewater treatment plant [bakerybazar.com]

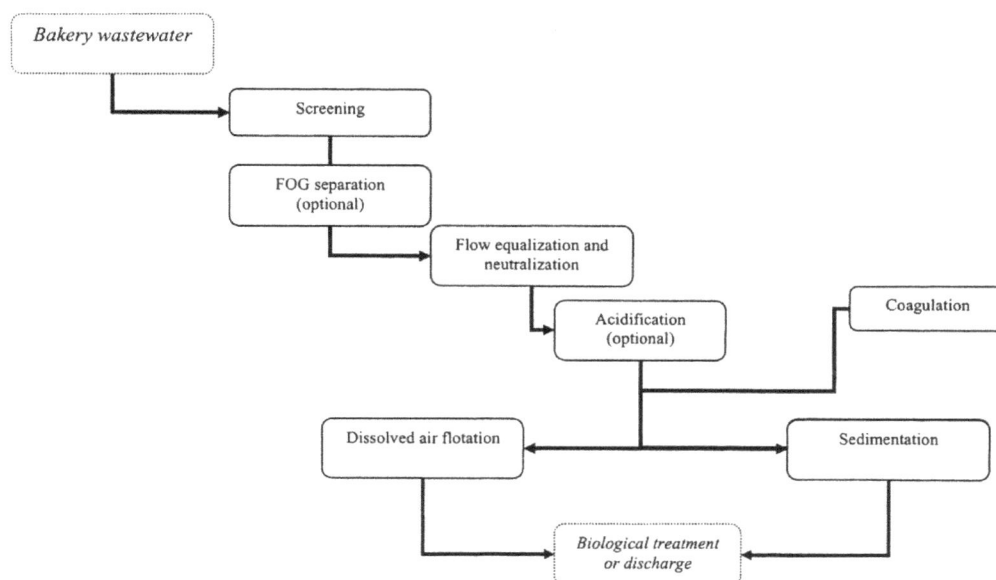

Figure 3. Bakery wastewater pretreatment system process flow diagram [Chen et al. 2006]

The bakery wastewater contains products which are made from the strains of yeast selected for their special qualities relating to the needs of the baking industry. The bakery effluent containing yeast is a major source of pollution as it has a high biological oxygen demand (BOD), high chemical oxygen demand (COD), total Kjeldahl nitrogen (TKN), dark color, and nonbiodegradable organic pollutants.

The biological treatment of wastewater eliminates important components of the organics in the wastewater. However, the biochemical decomposition by conventional treatment method may not be enough to obtain a complete color reduction. Although there is no effluent limitation set for color in the regulations, it can be necessary to use a more effective process for the reduction of color [Catalkaya and Sengul 2006].

The research on trends in the recovery of phosphorus in the bioavailable forms from wastewater has proven that other than in municipal WWTPs, struvite (a superior recovered P product in terms of plant availability) precipitation has recently been investigated in a broad variety of wastewater streams from the bakery production [Uysal et al. 2014; Melia et al. 2017].

Most bakery industries are of small or medium size, often located in densely populated areas, which makes environmental problems more critical. Nevertheless, the conventional "end-of-pipe" treatment philosophy has its restrictions in dealing with these problems, as it only addresses the result of inefficient and wasteful production processes, and should be considered only as a final option [Chen et al. 2006]. Table 1 shows level of BOD, suspended solids and fats, oils and grease in the bakery wastewater.

Table 1. Bakery wastewater characteristics

Parameter	Type of bakery							
	Unspecified			Variety	Cake			Bread
pH		6.0–10.0	4.7–5.1		5.6	4.7–8.4	5.0–12.3	6.9–7.8
COD		3000.0	7000.0	7500.0–10000.0	1600.0		180.0–1500.0	
BOD$_5$	1160.0–8200.0	1000.0	3200.0	4000.0–7000.0		2240.0–8500.0	119.6–807.3	155.0–620.0
TSS		1500.0	6000.0				12.4–1940.0	
TP			6.8				0.3–14.7	
TKN			36.0				44.4–280.0*	
FOG	1070.0–4490.0	100.0	820.0		630.0	400.0–1200.0		60.0–68.0
Ref.	[4]	[2]	[7]	[9]	[4]	[4]	[10]	[4]

FOG - fats, oil and grease; unit: ph -; COD, BOD$_5$ – mgO$_2$·dm^{-3}; TSS, TP, TN, TKN, FOG - mg·dm^{-3}; * - N$_{Tot}$.

Characteristic of organic compounds in wastewater based on COD fractions

COD fractionation is one of the most significant achievements in the modeling of the activated sludge systems and an indispensable part of wastewater characteristics [Sadecka 2010]. Calculation of COD fractions enables discrimination of organic contaminants based on the particle size and susceptibility to biochemical degradation including the readily- and hardly-biodegradable substrates [Struk-Sokolowska et al. 2016]. The biodegradable fractions (S_S and X_S) are mainly important at designing the systems for the biological removal of nitrogen and phosphorus. The

soluble fraction (S_S) consists of the substances directly available to microorganisms. The fraction of suspensions (X_S), prior to be used by microorganisms, has to be decomposed by extra-cellular enzymes. Non-biodegradable fractions (S_I and X_I) are not subject to any conversion (Fig. 4, 5). In a system for sewage treatment, the concentration of soluble fraction (S_I) in the inlet is equal to that in the outlet, while the suspended fraction (X_I) can remain in the system due to building into the microorganism biomass or be subject to mechanical processes [Sadecka et al. 2011, Smyk et al. 2015, Struk-Sokolowska 2016]. The percentage contribution of individual COD fractions in the municipal wastewater is shown in Figure 4.

4.1)

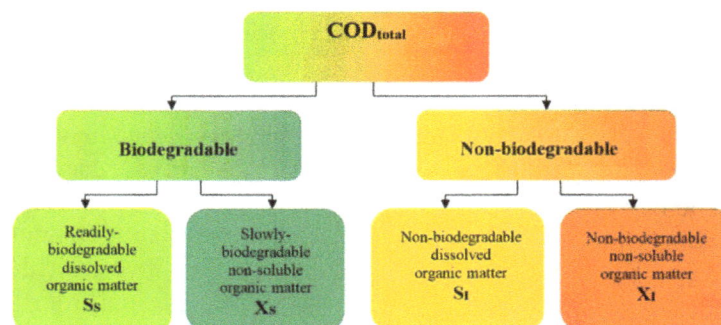

4.2)

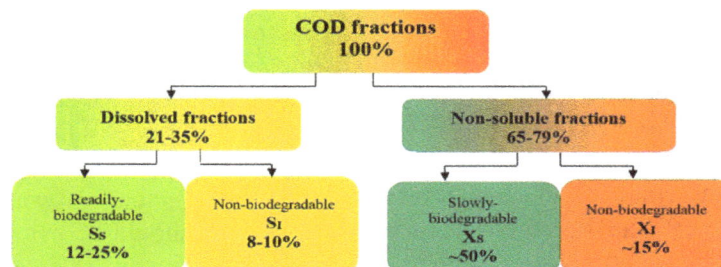

Figure 4. COD fractions and the percentage contribution of individual COD fractions in wastewater [Sadecka 2010, Fudala-Książek 2011, Płuciennik-Koropczuk 2011, Sadecka et al. 2011, Struk-Sokolowska 2015, 2016]

RBCOD (Readily Biodegradable COD), such as volatile fatty acids, is readily degraded by microbial metabolism. SBCOD (Slowly Biodegradable COD), composed of particulate organic matter, is degraded slowly by a series of microbial actions like adsorption, hydrolysis, and metabolism (Figure 5). NBDSCOD (Non-Biodegradable Soluble COD) refractory in biodegradation is contained in industrial discharges. Aromatic compounds used in various industries are typical examples of NBDSCOD. NBDPCOD is also non-biodegradable, but is removed easily by sedimentation in a conventional WWTP [Choi et al. 2017]. A schematic diagram of the chemical oxygen demand (COD) fractions and their fates in a biological wastewater treatment plant is shown in Figure 5.

MATERIALS AND METHODS

Research methodology is presented in Figure 6. The Bakery and Confectionery Plant in Lipsk, produces rye bread, mixed bread, wheat and semi-confectionary baked dough products. The daily production is about 500 mixed breads (400 kg), 60 rye breads (48 kg), 600 wheat rolls (60 kg), 300 shortbreads, doughnuts, croissants,

buns and challahs (36 kg). Yeasts are used for the production of 80% of the portfolio and rye sourdough is used for up to 20% of all products. The bakery also uses butter, margarine, shortening as well as sugar, honey, syrup, marmalade, milk, cream, oil and eggs for their products. The wastewater from this bakery is discharged directly to the municipal wastewater treatment plant located in Lipsk.

Investigation regarding the co-treatment of municipal sewage and bakery wastewater was conducted in the wastewater treatment plant in Lipsk, Eastern Poland. In this WWTP, the volume contribution of bakery wastewater is 10% and the sewage receiver is the Biebrza River. The maximum daily flow capacity of the mechanical-biological treatment plant in Lipsk is 430 m^3, the actual volume of influent wastewater in 2016 amounted to 250 $m^3 \cdot d^{-1}$ (PE = 2500). This plant is fed with the household sewage from local town and villages. The bio-treatment process is conducted with an activated-sludge process and includes: anaerobic stage of dephosphatation, aerobic stage of organic carbon compounds oxidation and nitrification, anoxic stage of denitrification, sedimentation of activated sludge. The aerobic conditions change in the range from 0.2 to 3.0 $gO_2 \cdot m^{-3}$. Parameters of raw wastewater: BOD_5 480.0–680.0 $mgO_2 \cdot dm^{-3}$,

5.1)

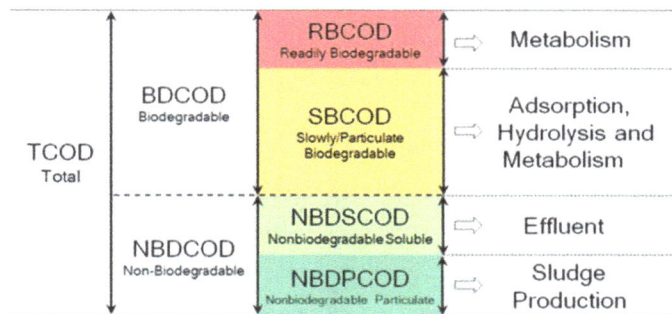

5.2)

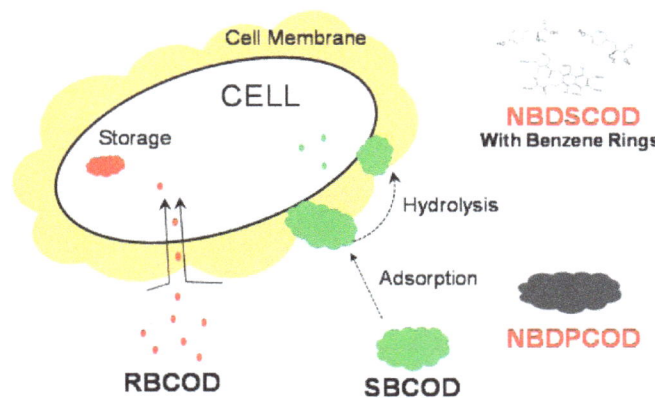

Figure 5. Schematic diagram of the chemical oxygen demand (COD) fractions and their fates in a biological wastewater treatment plant [Choi et al. 2017]

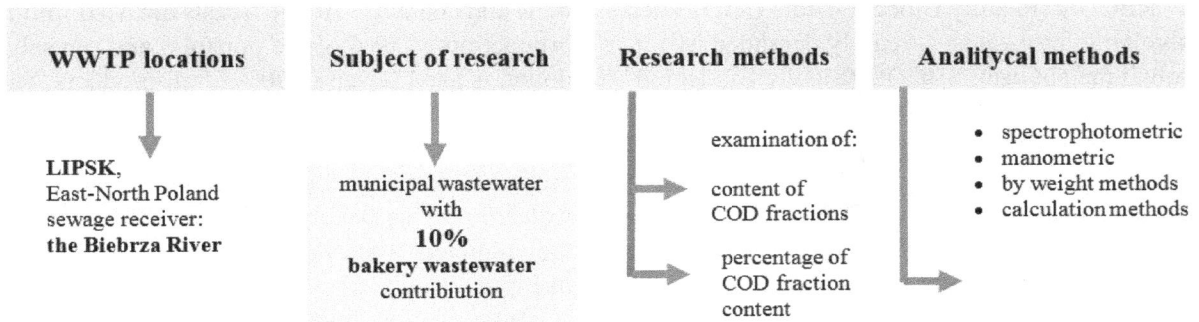

Figure 6. Research methodology

COD 626.0–998.0 mgO$_2$·dm^{-3}, pH 6.4–8.3, TSS 390.0–500.0 mg·dm^{-3}, TN 112.0–150.0 mgN·dm^{-3}, TP 24.0–48.0 mgP·dm^{-3}.

The samples of wastewater for physicochemical analyses were collected in July 2016. Four analytical series were conducted in the wastewater treatment plant receiving bakery wastewater. The list of process line points from which the analyzed samples of wastewater were collected is provided in Table 2.

The analytical results were used to compute the values and percentage contribution of indi-

vidual COD fractions in wastewater. The method used for COD fractions determination in wastewater was developed based on the ATV 131P guidelines. The list of formulas used for calculations is provided in Table 3.

In order to establish S$_{COD}$, the COD of crude wastewater was determined after centrifugation (RMP = 3000 rpm; t = 10 min), decantation, and filtration through a membrane filter made of glass fiber, with the pore size of 0.45 μm. The analysis of S$_I$ consisted of determining COD of the treated wastewater sample after centrifuga-

Table 2. List of process line points from which samples of wastewater were collected

Sample	Treatment process stage	Stage type
1	influent	raw wastewater
2	preliminary treated wastewater (after sand separator)	pre-treatment
3	preliminary treated wastewater (after grit chamber)	
4	wastewater collected from pumping station	
5	wastewater from chamber of dephosphatation	anaerobic
6	wastewater from chamber of denitrification	anoxic
7	wastewater from chamber of nitrification	aerobic
8	wastewater after sedimentation in secondary settling tank	effluent

Table 3. List of formulas used for calculations

Symbol	Formula	Name of organic matter
S$_{COD}$	$S_{COD} = S_S + S_I$	dissolved organic contaminants
S$_S$	$S_S = S_{COD} - S_I$	COD of readily-biodegradable dissolved organic matter
S$_I$	$S_I = S_{COD} - S_S$	COD of non-biodegradable dissolved organic matter
X$_S$	$X_S = BOD_T - S_S$	COD of slowly biodegradable insoluble organic matter
X$_I$	$X_I = A \cdot X_{COD}$ $X_I = X_{COD} - X_S$	COD of non-biodegradable insoluble organic matter
BOD$_T$	$BOD_T = BOD_5/0{,}6$	total biochemical oxygen demand
BOD$_t$	$BOD_t = BOD_T \cdot (1 - 10^{-k \cdot t})$	BOD after time t
X$_{COD}$	$X_{COD} = X_S + X_I$ $X_{COD} = X_S/0{,}75$	total concentration of molecular organic matter

A – coefficient ranged from 0.2 to 0.35, depending on the type of wastewater or relatively on wastewater retention time in the primary settling tank; A was assumed at A=0.25.

The obtained values of S$_S$, S$_I$, X$_S$, X$_I$, and total COD enabled the calculation of individual COD fractions percentage.

tion, decantation, and filtration. The total BOD was computed based on experimentally established BOD_5 of wastewater. The values of kinetics of the first phase of BOD, described with the first order reaction equation, k- reaction rate constant; for the substances contained in the household wastewater range from 0.1 to 0.3 d^{-1}, and the mean value is assumed at 0.23 d^{-1}. The total concentration of molecular organic matter was determined based on a dependency provided in the ATV-131 guidelines.

RESULTS

The values and percentage of COD fractions after subsequent steps of treatment in WWTP in Lipsk are shown in Table 4. The average values and percentage of COD fractions after subsequent steps of treatment in dependence of total COD are shown in Table 5.

In addition to the amount of organic compounds belonging to COD fraction, the proportions between fractions also are relevant to the wastewater treatment technologies. The organic compounds determined by the biodegradable fractions (S_S; X_S), will be used with the highest efficiency during the mechanical and biological wastewater treatment, while the organic compounds concentration expressed by hardly or non-biodegradable fractions is subject to the smallest variation. As a result, the percentage of fractions in total COD is changed.

DISCUSSION

The percentage of individual COD fractions in the municipal wastewater with bakery wastewater contribution was compared with the COD fractions shares in the wastewater from other industries (dairy, olive mill, tomato, sugar beet, potato processing, winery, paper mill) presented in literature [Rodriguez et al. 2007, Chiavola et al. 2014, Struk-Sokolowska 2014, 2015, Choi et al. 2017]. The comparison of the COD fractions in raw industrial or municipal wastewater with industrial sewage contribution and split per industry is shown in Table 6.

Comparing the percentage of individual COD fractions in municipal wastewater with 10% bakery wastewater contribution to the shares in municipal wastewater with 10% dairy wastewater contribution [Struk-Sokolowska 2014, 2015] proves close similarities. It may therefore be assumed that the co-treatment of bakery and municipal wastewater will occur with high efficiency using a technology which is well recognized and described in literature i.e. a sequencing batch reactor (SBR).

The research carried out in Italy by Chiavola et al. [2014] shows that the share of biodegradable organic matter in the sewage from olive mill was 66.6%, out of which the dissolved $S_S = 29.2\%$, and non-soluble $X_S = 37.4\%$, while non-biodegradable organic matter was 33.4%, including dissolved $S_I = 9.9\%$ and non-soluble $X_I = 23.5\%$. The municipal sewage with bak-

Table 4. Values and percentage of COD fractions after subsequent steps of treatment

Sample	Unit	COD fractions			
		S_S	S_I	X_S	X_I
1	$mgO_2 \cdot dm^{-3}$	368.9–411.3	18.7–28.1	355.4–564.4	118.5–188.1
	%	32.1–45.5	2.1–2.4	39.3–49.1	13.1–16.4
2	$mgO_2 \cdot dm^{-3}$	427.3–453.9	18.7–28.1	206.0–679.4	68.7–226.5
	%	32.7–59.3	2.0–2.6	28.6–49.0	9.5–16.3
3	$mgO_2 \cdot dm^{-3}$	392.9–643.3	18.7–28.1	123.4–507.1	41.1–169.0
	%	35.8–77.8	2.3–2.6	14.9–46.2	5.0–15.4
4	$mgO_2 \cdot dm^{-3}$	103.3–276.9	18.7–28.1	46.7–423.1	15.6–141.0
	%	31.9–56.0	3.2–10.1	25.3–48.7	8.5–16.2
5	$mgO_2 \cdot dm^{-3}$	8.9–52.3	18.7–28.1	11.0–71.1	3.7–23.7
	%	6.8–61.0	21.3–21.8	12.8–53.9	4.3–18.0
6	$mgO_2 \cdot dm^{-3}$	6.9–23.3	18.7–28.1	33.1–33.4	11.0–11.1
	%	8.7–26.9	21.6–35.5	38.6–41.8	12.8–13.9
7	$mgO_2 \cdot dm^{-3}$	4.9–21.3	18.7–28.1	11.8–37.0	3.9–12.3
	%	10.1–23.8	20.9–57.7	24.2–41.4	8.0–13.8
8	$mgO_2 \cdot dm^{-3}$	<0.1	18.7–28.1	20.0–28.3	6.7–9.4
	%	<0.1	33.2–51.3	36.5–50.2	12.2–16.7

Table 5. The average values and percentage of COD fractions after subsequent steps of treatment in dependence of total COD

Sample	Unit	COD fractions				total COD $mgO_2 \cdot dm^{-3}$
		S_S	S_I	X_S	X_I	
1	$mgO_2 \cdot dm^{-3}$	390.1	23.4	459.9	153.3	903.9–1149.5
	%	38.8	2.3	44.2	14.8	
2	$mgO_2 \cdot dm^{-3}$	440.6	23.4	442.7	105.1	720.7–1387.9
	%	46.0	2.3	38.8	12.9	
3	$mgO_2 \cdot dm^{-3}$	518.1	23.4	315.3	105.1	826.5–1097.1
	%	56.8	2.5	30.6	10.2	
4	$mgO_2 \cdot dm^{-3}$	190.1	23.4	234.9	78.3	184.3–869.1
	%	44.0	6.7	37.0	12.4	
5	$mgO_2 \cdot dm^{-3}$	30.6	23.4	41.1	13.7	85.7–131.8
	%	33.9	21.6	33.4	11.2	
6	$mgO_2 \cdot dm^{-3}$	15.1	23.4	33.3	11.1	86.5–79.1
	%	17.8	28.6	40.2	13.4	
7	$mgO_2 \cdot dm^{-3}$	13.1	23.4	24.4	8.1	48.7–89.3
	%	17.0	39.3	32.8	10.9	
8	$mgO_2 \cdot dm^{-3}$	<0.1	23.4	24.2	8.1	54.8–56.4
	%	<0.1	42.3	43.4	14.5	

Table 6. Comparison of the COD fractions in raw industrial or municipal wastewater (with industrial sewage contribution) depending on the industry

Industry	COD fractions, %				$COD_{tct'}$ $mgO_2 \cdot dm^{-3}$	Reference
	S_S	X_S	S_I	X_I		
Bakery (10%)	38.8	44.2	2.3	14.8	626.0–998.0	this paper
Dairy (10%)	32.5	47.8	2.7	17.0	863.7	Struk-Sokolowska, 2014
Dairy (10%)	38.8	45.5	1.0	15.2	1079.0	Struk-Sokolowska, 2015
Olive mill	29.2	37.4	9.9	23.5	60000.0	Chiavola et al., 2014
Tomato processing	*	*	**18.4		*	Rodriguez et al., 2007
Sugar beet processing	*	*	**18.9		*	Rodriguez et al., 2007
Potato processing	*	*	**28.8		*	Rodriguez et al., 2007
Winery	*	*	**20.9		*	Rodriguez et al., 2007
Paper mill (25%)	4.2	43.1	39.5	13.2	110.7	Choi et al., 2017
Paper mill (75%)	0.6	34.0	59.8	5.5	112.0	Choi et al., 2017

* – no data available; ** – sum of fractions S_I and X_I

ery wastewater contribution has a higher share of biodegradable organic matter (S_S and X_S), with smaller share of non-biodegradable (inert) fraction S_I and X_I, and is highly susceptible to biological treatment processes.

In the research presented by Rodriguez et al. [2007] the share of non-biodegradable (inert) fractions S_I and X_I in wastewater from tomato, sugar beet, potato and winery processing were 18.4, 18.9, 28.8 and 20.9%, respectively. In the municipal wastewater with 10% bakery wastewater, the contribution shares of non-biodegradable organic matter (as a sum of fractions S_I and X_I) were lower and amounted to 17.1%.

The research conducted by Chen et al. [2017] proved that the municipal wastewater with 25 or 75% contribution of paper mill sewage is char-

acterized by a different composition. The biggest differences were attributed to biodegradable dissolved organic matter (S_S faction) levels. In the municipal wastewater containing sewage from the paper mill, the (S_S) fraction was 4.2 or 0.6%, whereas in the municipal wastewater with bakery wastewater, it was up to 38.8%. This confirms that the wastewater from the paper mill is definitely less susceptible to the biological treatment.

CONCLUSIONS

1. The bakery sewage from small and medium-sized bakeries can be discharged to the municipal wastewater treatment plant and treated there successfully. The presence of bakery

wastewater in a stream of municipal sewage does not cause operational problems at the facility but positively influences the biodegradable organic matter formation.

2. The results indicated that the municipal wastewater containing bakery sewage is susceptible to biological degradation as evidenced by the ratio of $COD/BOD_5 < 1.4$.

3. In the raw municipal wastewater with bakery sewage contribution, the dominant percentage was recorded for the fractions of biodegradable contaminants (S_S and X_S) reaching about 83% of the total COD of sewage.

4. During the full-scale biological wastewater treatment with an activated-sludge process, the amount of readily-biodegradable dissolved organic matter (S_S) and slowly-biodegradable non-soluble organic matter (X_S) decreased, which confirms high treatment efficiency.

5. During the pre-treatment (sand separator, grit chamber), a decrease of slowly-biodegradable non-soluble organic matter (X_S) was observed, which confirms the high efficiency of mechanical treatment.

6. The highest decrease of readily-biodegradable dissolved organic matter (S_S) was noted in the wastewater from dephosphatation and denitrification chambers.

7. A dose of bakery wastewater significantly raised the denitrification capacity at the WWTP, because of the significant increase of readily-biodegradable wastewater substrates.

8. Low value of non-biodegradable non-soluble organic matter (X_I) in the effluent indicates a high efficiency of sedimentation in the secondary settling tank.

Acknowledgments

The research has been carried out in the framework of project No. S/WBiIS/3/2014 and financed from the funds for science MNiSW.

REFERENCES

1. ATV-DVWK-A131P 2000. Dimensioning of single stage activated sludge plant. WA rules and standards – German Association for Water. Attachment: COD balance sheet, Seidel-Przywecki (in Polish).

2. Bakery Industry, Wastewater treatment in bakeries, http://www.bakerybazar.com /2009/10/waste-water-treatment-in-bakeries_05.html, access: 17.01.2018.

3. Catalkaya E.C., Sengul F. 2006. Application of Box–Wilson experimental design method for the photodegradation of bakery's yeast industry with UV/H2O2 and UV/H2O2/Fe(II) process, Journal of Hazardous Materials B128, 201–207.

4. Chen J. P., Yang L., Bai R., Hung Y. T. 2006. Bakery Waste Treatment. Taylor&Francis Group. LLC.

5. Chiavola A., Farabegoli G., Antonetti F. 2014. Biological treatment of olive mill wastewater in a sequencing batch reactor. Biochem. Engi. Journ., 85, 71–78.

6. Choi Y-T., Beak S-R., Kim J-I., Choi J-W., Hur J., Lee T-U., Park C-J., Lee B. J. 2017. Characteristics and Biodegradability of Wastewater Organic Matter in Municipal Wastewater Treatment Plants Collecting Domestic Wastewater and Industrial Discharge, Water, 9, 409.

7. Ecologix Environmental Systems, Commercial Bakery Wastewater Treatment, http://www.ecologixsystems.com/industry-bakery.php, access: 17.01.2018.

8. Fudala-Ksiazek S. 2011. The impact of the discharge of landfill leachate on the efficiency of municipal wastewater treatment plant. PhD thesis, Gdansk University of Technology (in Polish).

9. Gąsiorowski H. 2004. Wastewater treatment for bakeries and confectionary plants. Przegląd Piekarski i Cukierniczy, 03/2004 (in Polish).

10. Krzanowski S., Wałęga A., Paśmionka I. 2008. Treatment of sewage from selected food manufacturing plants, Infrastructure and Ecology of Rural Areas, Polish Academy of Science, Kracow.

11. Melia P. M., Cundy A. B., Sohi S. P., Hooda P. S., Busquets R. 2017. Trends in the recovery of phosphorus in bioavailable forms from wastewater, Chemosphere, 186, 381–395.

12. Orhon D., Babuna F. G., Karahan O. 2009. Industrial Wastewater Treatment by Activated Sludge, IWA Publishing.

13. Płuciennik-Koropczuk E. 2011. COD fractions as a measure of the quality of wastewater. PhD thesis, University of Zielona Góra, (in Polish).

14. Rodriguez L., Villasenor J., Fernandez F. J. 2007. Use of agro-food wastewater for the optimisation of the denitrification process. Wat. Scie. Tech., 55, 10, 63–70.

15. Sadecka Z. 2010. Fundamentals of biological wastewater treatment, Warsaw, Wyd. Seidel-Przywecki (in Polish).

16. Sadecka Z., Płuciennik-Koropczuk E. 2010. COD fractions in kinematic models. In: Wastewater treatment and sludge treatment. Zielona Gora, Printing House of University of Zielona Góra, 39–48.

17. Sadecka Z., Płuciennik-Koropczuk E., Sieciechowicz A. 2011. COD fractions in wastewater kinematic models. Forum Eksploatatora, 54(3), 72–77 (in Polish).

18. Smyk J., Ignatowicz K., Struk-Sokolowska J. 2015. COD fractions changes during sewage treatment with constructed wetland, Journal of Ecological Engineering, 16(3), 43–48.

19. Struk-Sokołowska J. 2014. Speciation of organic matter by COD in wastewater on the chosen example. Interdysc. Zagadn. w Inż. i Ochr. Środ., Ofic. Wydawn. Pol. Wroc., 4, 807–820 (in Polish).

20. Struk-Sokolowska J. 2015. COD fraction changes in the process of municipal and dairy wastewater treatment in SBR reactors. PhD thesis, Bialystok University of Technology (in Polish).

21. Struk-Sokolowska J., Wiater J., Rodziewicz J. 2016. Wastewater organic compounds characterization on the basis of COD fractions. Gaz, Woda i Technika Sanitarna, 1(3), 14–20 (in Polish).

22. Uysal A., Demir S., Sayilgan E., Eraslan F., Kucukyumuk Z. 2014. Optimization of struvite fertilizer formation from baker's yeast wastewater: growth and nutrition of maize and tomato plants, Environ. Sci. Pollut. Res. 21, 3264–3274.

22

Analysis of Selected Technical and Technological Parameters of the Sewage Sludge Stability Process

Izabela Bartkowska[1], Dariusz Wawrentowicz[1]

[1] Bialystok University of Technology, Department of Technology and Environmental Engineering Systems, ul. Wiejska 45 A, 15-351 Bialystok, Poland

[*] Corresponding author's e-mail: i.bartkowska@pb.edu.pl

ABSTRACT

The article presents the results of the analysis, which was carried out in 2015–2017 on the example of autothermal installation of thermophilic sludge stabilization (ATAD) in a sewage treatment plant in Giżycko. The installation was created in 2003 as the first of its kind and still remains operational. The purpose of the conducted research was to assess the suitability of the analyzed technological parameters as tools that can be used by operators to determine the actual possibilities of changing the operating conditions of the installation or to develop an optimization strategy to reduce the energy demand. The dry mass content and organic dry mass content was used as the assessment indicators. In the analysed period, the sludge from the process contained from 47.47% to 60.80% of organic matter in the dry mass of the sludge. The organic dry matter decrease due to the process was also calculated, and it ranged from 26.4% to 48.7%. The amount of sludge undergoing the process and the amount of electricity consumption were also analysed. On this basis, the energy consumption indicators in the ATAD process were calculated.

Keywords: autothermal thermophilic stabilization of sewage sludge, dry mass of sewage sludge, organic substances in sewage sludge, sewage sludge hygienization, use of sewage sludge in agriculture.

INTRODUCTION

Sewage sludge is subject to the Directive of the European Parliament and of the Council of 19 November 2008 2008/98/EC on waste (the so-called Waste Framework Directive). According to the above-mentioned directive, the sludge as a waste is subordinated to a specific hierarchy of treatment procedure. First of all, waste generation should be prevented. Furthermore, the sludge should be prepared for re-use, subjected to recycling, or other methods of recovery/recycling, and eventually rendered harmless. Prevention of sewage sludge generation is, however, not possible because it is a kind of waste that cannot be avoided. Therefore, the other priorities are important in the waste processing hierarchy, i.e. preparation for re-use or final disposal or rendering the waste completely harmless. Its re-use, however, is possible only if the sewage sludge is stabilized and sanitary-safe.

The stabilization of sewage sludge can be achieved through the use of biological, chemical or thermal methods. The biological methods are used most commonly, in particular methane fermentation and oxygen stabilization. The biological methods also include composting. However, a few designers or specialists distinguish autothermal thermophilic sludge stabilization (ATAD) among the biological methods.

There are many known methods for the stabilization and hygienization of sewage sludge. There are even more technologies based on them and devices used to implement them. Therefore, a designer faces a difficult task to choose the right one, which will allow him to obtain a product that meets the formal and legal requirements, but at the same time fulfils the expectations of investors and users. The method of final disposal of sludge may be of key importance. The most popular choice in Europe is the thermal processing of sewage sludge. However, under domestic conditions its natural use is undoubtedly the most preferred option, especially for a certain size of sewage treatment plants [Podedworna and Heidrich 2010]. The autothermal thermophilic stabilization may be an ideal solution. This process, is unfortunately still not very well-known in Poland; however, it enables to obtain organic fertilizer instead of waste in sewage treatment plants.

The research, which is the subject of this article, concerns a sewage treatment plant in which the ATAD installation has been in operation since 2003. It is a pioneer installation, which was established in Poland and has been running for the longest time. The characteristics of the ATAD installation are presented by Bartkowska [Bartkowska 2017].

The operation of devices, machines or installations is still underestimated in society. Perhaps, this is due to the fact that the concept of operation is an interdisciplinary issue. It includes the organizational, technical, ecological, economic and social issues related to the activity and operation of people and machines. In engineering terms, it can be defined as a set of activities including planning, using, servicing, diagnosing, storing and others, aiming at the safe use of installation/devices and extending the period of its/their operation. The analysis of technical and technological parameters of the implemented process can also serve this purpose.

The research, the results of which were presented in the article, was aimed at analysing the effectiveness of ATAD installations after a period of 12 years of operation. The purpose of the conducted research was to assess the suitability of the analysed technological parameters as tools that can be used by operators to determine the actual possibilities of changing the operating conditions of the installation or to develop an optimization strategy in order to reduce the energy demand. The cognitive objective of this publication is still to supplement the knowledge about the subject process as a way of acquiring the biomass that can be used as a soil conditioner. Thus, the presentation of the ATAD process as a method that effectively contributes to solving problems with the disposal of constantly increasing amounts of sewage sludge.

MATERIALS AND METHODS

The research concerned the pre-compacted sludge before the ATAD process and the sludge after the process, and which was dehydrated. The stabilized sewage sludge, according to the decision of the Minister of Agriculture and Rural Development, since 2008 has been placed on the market as an organic fertilizer.

The municipal sewage sludge processed in a two- or three-stage autothermal termophilic stabilization plant was analysed. The total residence time of sludge in the installation amounts to 8.2 full days. The capacity of ATAD installation is 90 m^3/d [Bartkowska 2017]. The excessive sludge removed from the sewage line is subjected to stabilization alternatively in two places: from the recycle stream after secondary settling tanks or from the last aerated chamber of the biological reactor.

The analysis includes the results of research carried out in years 2015–2017. The sewage sludge samples for testing were collected in accordance with PN-EN ISO 5667–13: 2011. A representative sample was obtained by combining and thoroughly mixing a certain number of samples, taken at the same time from different places of the same installation, in accordance with the Regulation of the Minister of Environment of 6 February 2015 on municipal sewage sludge.

The analysis involved the dry mass content and organic substances in the concentrated and dehydrated sludge. All determinations were made in accordance with the reference test methods indicated in the Regulation of the Minister of Environment of 6 February 2015 on municipal sewage sludge.

RESULTS AND DISCUSSION

The test cycle included the analysis of the compacted sludge, which is fed to the ATAD installation. The basic parameters determining the proper operation are dry matter and organic dry matter. The results obtained during the study were subjected to a statistical analysis. Table 1 presents the numerical characteristics of the distribution of dry matter content and organic substances.

As can be seen from the data, the dry matter content in the sludge fed to the ATAD installation fluctuated within the limits of 4.37% to 6.72%, and its average value was 5.38%. Within the period covered by the research, an average of 77.12% of organic matter content was found in the dry mass of the sludge. The share of organic matter in the dry mass of crude sludge ranged from 73.16% to 82.48%.

The proper course of the autothermal stabilization process requires the supply of sewage sludge with the appropriate dry mass content, which ensures the supply of the appropriate amount of substrates to microorganisms, owing to which the distribution of organic substances

Table 1. Numerical characteristics of the analysed distributions of dry mass content (%) and organic substances (%) in compacted sludge

Distribution measure	The values of the distribution measures	
	dry mass	of organic substances
Amount	238	238
Arithmetic mean	5.38	77.12
Maximum	6.72	82.48
Minimum	4.37	73.16
Median	5.34	77.06
Variance	0.211	2.787
Standard deviation	0.459	1.669
Percentile 10%	4.82	75.00
Percentile 90%	6.03	79.23

is possible, provided the process is supplied with the proper amount of oxygen. The initial sludge concentration should reach the value of up to 5% of dry mass, resulting in a higher unit organic content, which should be from 65% to 70% of dry mass and should not be lesser than 40.0 g/l, expressed in COD. In the studied period, however, the ATAD installation operation regime ensured even higher values.

In the initial period of operation of this installation (in years 2003–2005), the dry mass content in the compacted sludge ranged from 2.1% to 8.4%, whereas organic substances constituted 70.05% to 86.16% of the entire volume [Bartkowska and Wawrentowicz 2011].

In other wastewater treatment plants that operate the ATAD installation, the dry mass content in compacted sludge ranged from 3.4% to 5.8%, 1.1% to 8.5% and from 2.28% to 8.46%. The organic matter content in the dry mass of this sludge was 35.5–74.3%, respectively and 35.51–73.78% [Bartkowska 2017].

In the study cycle presented in this paper, the content of dry mass and organic substances in the sludge after the ATAD process and dehydration was also controlled. The obtained results were subjected to a statistical analysis. Table 2 presents the numerical characteristics of the distribution of dry mass content and organic substances.

The sludge generated within the period under analysis contained from 15.08% to 24.38% of dry mass, on average 19.15%. The organic substance content was on average 57.13% and ranged from 47.47% to 60.80%.

In the initial period of operation of this installation, the content of organic matter in the dry mass of the sludge ranged from 63.2% to 68.8%.

In the next cycle of tests, the organic matter content in the dry mass of sewage sludge after the ATAD process and dehydration was on average 62.1% [Bartkowska and Dzienis 2007].

In other wastewater treatment plants in the country that operate the ATAD plant, the content of organic substances in the treated sewage sludge varies from 30% to 60% in the dry mass of the sludge.

In Poland, there is no legally regulated concept of stabilized sewage sludge. However, according to the German guidelines, the sludge, which contains 55% to 60% of organic matter in its dry mass, is known as the stabilized sludge. With regard to technical devices, the degree of sludge mineralization can be determined using the percentage loss of dry organic matter. The more so, because there is the concept of the so-called technical limit of stabilization, which is assumed at the level of 38–40% reduction of initial organic dry mass [Borowski and Szopa 2007, Movahedian et al. 2005].

On the basis of the analysed research results, the percentage decrease of organic matter content in the dry mass of sludge was calculated. During the period of research, it ranged from 26.4% to 48.7%. Similar values were obtained in other studies. For example, the reduction of organic matter content in the dry mass in the sludge recorded in Ireland was from 28.5% to even 53.8% [Layden 2007]. The loss of organic matter in the dry mass of sludge, according to the research by Zhelev et al. [2008], was 25÷56%. In other sources, the content of organic substances dropped within 32÷51% [Zupančič and Roš 2008, Song and Hu 2005]. In subsequent studies with an

Table 2. Numerical characteristics of the analysed distributions of dry mass content (%) and organic substances (%) in the compacted sludge

Distribution measure	The values of the distribution measures	
	dry mass	of organic substances
Amount	238	238
Arithmetic mean	19.15	57.13
Maximum	24.38	60.80
Minimum	15.08	47.47
Median	19.34	57.11
Variance	3.107	6.814
Standard deviation	1.762	2.610
Percentile 10%	16.84	53.86
Percentile 90%	21.14	60.46

extended retention time of up to 23 days, the average organic content dropped by 45% [Shugen et al. 2012]. While conducting tests at the time of sludge retention similar to the technical conditions (6–10 days), the organic matter content was not higher than 33.2% [Shugen et al 2013]. The quoted results are the result of the research carried out at a laboratory scale in a single reactor.

The summary of the amount of sludge brought to the ATAD installation was analyzed as well. In the analysed period, the monthly processing ranged from 1175 m³ to 3644 m³. A detailed summary of the amount of sludge is shown in Figure 1.

The sludge is supplied for ATAD installation once per day. In the period under study, this average quantity was 84.54 m³/d; however, it changed from the minimum value of 37.90 m³/d to the maximum value of 117.55 m³/d. The total amount

of sludge that underwent the ATAD process was 87,518,60 m³. This sludge was completely used for agricultural purposes as an organic fertilizer after dehydration.

The thermophilic population of microorganisms is characterized by a high rate of oxygen absorption The process requires constant aeration of sludge and mixing in the reactors. This is associated with a marked increase in the energy consumption. Figure 2 shows the energy consumption in ATAD reactors.

In the studied period, the average amount of energy consumed was 61,186.79 kWh in a month. The least energy was used in April 2016 and it was 47,640.00 kWh, and the most in October 2016, when it was 78,780 kWh. In the analysed period, the lower consumption of electricity fell on the months from March to July each year. However,

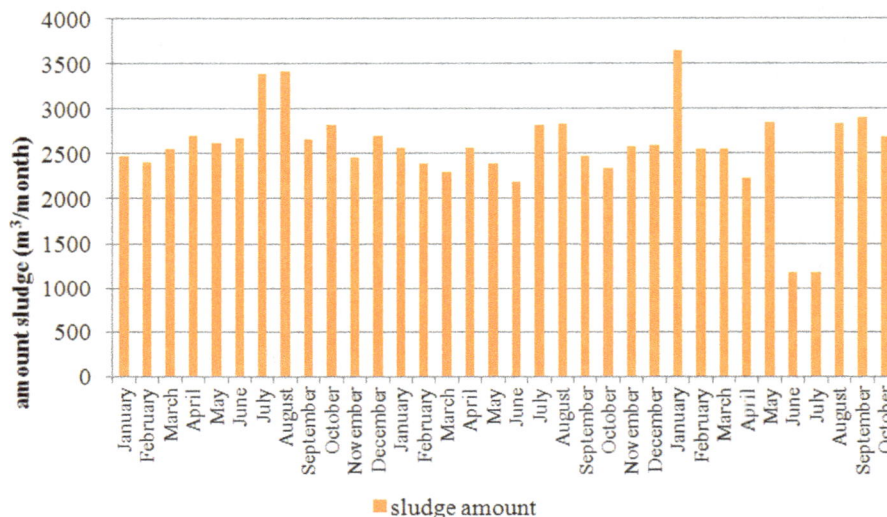

Fig. 1 Monthly quantities of sludge processed from January 2015 to October 2017

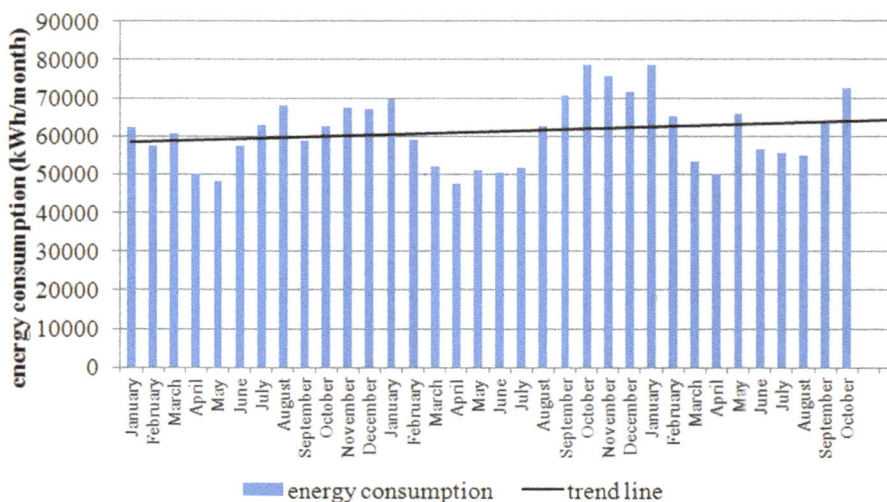

Fig. 2. Monthly values of energy consumption in the ATAD installation from January 2015 to October 2017

the trend line shown in the picture clearly indicates a growing energy consumption.

The knowledge on the energy balance of a sewage treatment plant allows to assess the energy consumption of individual processes using technical indicators that specify the amount of electricity consumption. Taking into account the monthly energy consumption, volume and mass of the generated sludge and the energy consumption indicators were calculated. The values of the calculated ratios are presented in Table 3.

Acquiring information on energy consumption for selected processes in a sewage treatment plant is relatively difficult. The existing metering usually allows to determine the total energy consumption. In certain wastewater treatment plants, however, it is possible to record the energy consumption for individual processes or devices. In other waste-water treatment plants in the country with the the sludge management based on the ATAD process, the calculated energy consumption rates were 13.3 kWh/m^3 for the processed sludge, and in relation to 1 kg of dry matter of sludge 0.44 kWh/kg of dry mass. The research carried out in the next sewage treatment plant enabled to calculate the energy consumption indicators in the years 2013–2015 in the range from 23.26 kWh/m^3 to 33.32 kWh/m^3. However, per kilogram of dry matter from 0.23 kWh/kg of dry mass to 0.28 kWh/kg of dry mass was achieved in the same period of time [Bartkowska 2017].

Rojas et al. [2010] state that the energy consumption indicators in ATAD installations operating in Spain and Ireland are 9 – 15 kWh /m^3 of processed sludge or 0.3 – 0.5 kWh/kg of dry mass. This means that under domestic conditions, the energy consumption with respect to 1 m^3 of processed sludge is higher, and in relation to one kilogram of dry matter it is lower or takes the lower values from the given range. Therefore, our installations can be considered more energy-consuming, since the expression of the amount of sludge in 1 m^3 is unambiguous. However, the approach to assuming the amount of sludge in kilograms of dry matter, is clearly different. Skilful operation can also affect the energy balance.

Unfortunately, there is little information on the energy consumption of the ATAD process in the scientific literature, apart from the remarksthat it is significant.

Other technologies of sludge stabilization also contribute to the increase of energy consumption in sewage treatment plants. On the basis of the energy consumption in the treatment plant for the process of oxygen stabilization of sludge, the discussed indicators range from 6.44 kWh/m^3 to 13.25 kWh/m^3 and from 2.96 kWh/kg of dry mass to 6.02 kWh/kg of dry mass [Dąbrowski et al. 2017].

CONCLUSIONS

The research conducted in the existing ATAD installations confirmed the effectiveness of this process in acquiring stabilized sewage sludge and it being safe in terms of hygiene and sanitation. The process enables to obtain sludge that does not rot and is devoid of pathogenic microorganisms, parasites and fungi. The ATAD installation requires a small area, and the process is carried out in closed tanks.

The results discussed in the article refer to the installation that has been in operation since 2003. Almost a 15-year operation period allows to realistically assess the advantages and disadvantages of this technology.

From the practical point of view, the analysis of the degree of stabilization of the sludge should be made using the methods which are simple to conduct and possible on-site at the sewage treatment plant. This gives the opportunity to react quickly and achieve greater efficiency of the installation. For this reason, the study has focused on the content of dry matter and organic substances. These two technological parameters determine the effectiveness of the process. As a measure of effectiveness, it is necessary to indicate the decrease in organic matter content in the dry mass of the sludge.

Exploitation of the installation is a set of purposeful organizational, technical and economic

Table 3. The values of energy consumption indicators in the ATAD installation in 2015–2017

Indicator	Energy consumption indicator		
	mean	minimal value	maximal value
Energy consumption index for 1 m^3 of processed sludge (kWh /m^3)	24.72	18.34	48.00
Energy consumption indicator for 1 kg of dry matter of sludge (kWh /kg of dry mass)	0.13	0.09	0.24

activities of people concerning their technical object and mutual relations occurring between them from the moment the object is commisioned for use in accordance with the intended purpose, up to the time of liquidation. The length of this period can be extended following certain rules most often given by the manufacturer.

The dry mass content constitutes a very important technological parameter of the sludge. The thickening of sludge to the required value of 4–5% ensures the supply of an appropriate amount of substrates to microorganisms, owing to which the distribution of organic substances is possible, provided, however, the right amount of oxygen is supplied. Maintaining the dry mass content at this level in the sludge supplied to the first level reactor will help avoid many operational problems. Higher content of dry mass may cause a faster wear of moving parts of aeration and mixing devices or foam controllers. It can also cause an increased energy consumption, which can be seen on the example of the installation. It was found that only in 20% of the analysed test results, the dry mass content was up to 5%. During the research period, the dry mass content averaged at 5.38%. Additionally, the amount of sludge supplied once to the installation often exceeded the designed value (90 m^3/d).

Examination of the dry mass content and its organic fraction before and after the process allows to determine the loss of organic matter content in the sludge, which can be a measure of the degree of its stabilization. As can be seen, the installation still ensures that this condition is met.

Of course, one should not forget about the temperature control of the sludge in the reactors, as the temperature of the thermophilic process guarantees the sanitation of the sludge. Achieving this temperature is possible, among other things, owing to the corresponding content of organic substances in the dry mass of the sludge.

Acknowledgement

The analyzes were carried out as part of the work No S/WBiIŚ /02/2014 and financed from the funds of the Ministry of Science and Higher Education.

REFERENCES

1. Bartkowska I. 2017. Autothermal thermophilic aerobic digestion of sewage sludge. Wydawnictwo Seidel-Przywecki Sp. z o. o., Warszawa (in Polish).

2. Bartkowska I., Dzienis L. 2007. Technical and economic aspects of autothermal thermophilic aerobic digestion exemplified by sewage treatment plant in Giżycko. Environment Protection Engineering, 33(2), 17–25.

3. Bartkowska I., Wawrentowicz D. 2011. Efficiency analysis of sewage sludge treatment by means of autoheated thermophilic aerobic digestion (ATAD) on the example of wastewater treatment plant in Giżycko. Inżynieria Ekologiczna, 25, 165–175 (in Polish).

4. Borowski S., Szopa J. 2007. Experiences with the dual digestion of municipal sewage sludge, Bioresource Technology, 98, 1199–1207.

5. Dąbrowski W., Żyłka R., Malinowski P. 2017. Evaluation of energy consumption during aerobic sewage sludge treatment in dairy wastewater treatment plant. Environmental Research, 153, 135–139.

6. Layden N.M. 2007. An evaluation of autothermal thermophilic aerobic digestion (ATAD) of municipal sludge in Ireland. Journal of Environmental Engineering and Science, 6, 19–29.

7. Movahedian A.H., Bina B., Moeinian K. 2005. Effects of aeration rate and detention time on thermophilic aerobic digestion of mixed sludge and its dewaterability. International Journal of Environmental Science and Technology, 2(2), 105–111.

8. Podedworna J., Heidrich Z. 2010. Directions of final disposal of sewage sludge. Gaz, Woda i Technika Sanitarna, 3, 25–28 (in Polish).

9. Rojas J., Zhelev T., Graells M. 2010. Energy Efficiency Optimization of Wastewater Treatment – Study of ATAD. Computer Aided Chemical Engineering, 28, 967–972

10. Shugen L., Nanwen Z., Ping N., Xudong G. 2012. The one-stage autothermal thermophilic aerobic digestion for sewage sludge treatment: Effects of temperature on stabilization process and sludge properties. The Chemical Engineering Journal, 197, 223–230.

11. Shugen L., Nanwen Z., Ping N., Xudong G. 2013. Semicontinuous Operation of One-Stage Autothermal Thermophilic Aerobic Digestion of Sewage Sludge: Effects of Retention Time. Journal of Environmental Engineering, 139(3), 422–427.

12. Song Yu-dong, Hu Hong-ying 2005. Autothermal Thermophilic Aerobic Digestion Technology for Sewage Sludge Treatment. China Water & Wastewater, 27(6).

13. Zhelev T., Jamniczky-Kaszas D., Brzyszcz B., Trévarain M., Kovacs R Vaklieva-Bancheva N. 2008. Energy efficiency improvement of waste-water treatment processes using process integration techniques. Report prepared for the Environmental Protection Agency by University of Limerick.

14. Zupančič G.D., Roš M. 2008. Aerobic and two-stage anaerobic–aerobic sludge digestion with pure oxygen and air aeration. Bioresource Technology, 99, 100–109.

23

Identification of Threats to the Soil and Water Environment on the Example of an Inactive Landfill Site

Agnieszka Pusz[1*], Magdalena Wiśniewska[1], Dominik Rogalski[1], Grzegorz Grzyb[1]

[1] Warsaw University of Technology, Faculty of Building Services, Hydro and Environmental Engineering, Nowowiejska 20, 00-653 Warsaw, Poland

[*] Corresponding author's e-mail: agnieszka.pusz@pw.edu.pl

ABSTRACT

Pollution migration is the main problem of the landfills that have been located, installed and operated without meeting basic environmental protection requirements. Undertaking effective reclamation treatments is conditioned by a good recognition of the object and its close and distant surroundings. The purpose of the work was to determine the potential hazard to the soil and water environment, based on selected factors, in the vicinity of a closed waste landfill at Głębocka Street in Warsaw. In the soil samples taken at the site from various depths, high concentrations of chlorides, EC, ammonium nitrogen and phosphates were found. Furthermore, the high ammonium nitrogen content correlated with the scent of ammonia, and the content of phosphates with the scent of organic compounds. These pollutants testify to the progressive decomposition of substances occurring in the mass of littering waste, which may pose a serious threat to groundwater. The landfill is also a potential geotechnical hazard because it is a dynamically changing object that can change stability, create caverns and landslides.

Keywords: landfill, soil and water environment, distribution of pollutants, available phosphorus, ammonium nitrogen, chlorides, EC, migration of pollutants

INTRODUCTION

Municipal waste and the waste from the economic sector deposited in landfills are a potential centre of environmental pollution. In the absence of proper sealing of the storage site, leached contaminants may cause degradation of the soil and water environment. Leachates constitute one of the sources causing the spread of pollutants in the soil and water environment. They arise due to the biochemical decomposition of organic compounds and through the leaching by rainwater and surface runoffs of soluble mineral and organic fractions that are contained in the stored waste [Tabor 2008]. The composition and amount of leachate generated depend on many factors, including the type of waste, the degree of their fragmentation, thickness and degree of waste compaction, age of the deposited waste, storage technology, the nature of biochemical and physicochemical changes, humidity and re-

tention capacity of the stored waste, the amount and intensity of atmospheric precipitation, evaporation rate, protection method (sealing of the surface of the landfill), the hydrogeological conditions of the substrate, as well as the type of vegetation growing on the land after reclamation [Tabor 2008; Marcinkowski 2009]. The range of pollutants spread in groundwater is affected, among others by the ground level of the landfill, groundwater flow rate, hydraulic loss of these waters, filtration rate of aquifers, concentration of pollutants, properties and amount of leachate and the sorption capacity of the ground [Tabor 2008]. Soil is the element that is relatively the most resistant to the impact of landfills. Most often, the soil contamination occurs in a close proximity to the object and mainly in the subsurface layers [Siuta 1995; Koda, Koper 2009]; however, as far as permeable works are concerned, the migration of pollutants may occur, which may pose a potential threat to the groundwater

[Grygolczuk-Patersons, Wiater 2012]. The problem particularly concerns the landfills that have not been designed in the right way. Undertaking effective recultivation measures [The Waste Act 2012] is conditioned by a proper diagnosis of the object and its immediate and further environment [Mizera 2007]. Conducting a cursory reclamation does not lead to full protection against the negative effects of impacts on the ground and water environment. Often, there are conditions conducive to the migration of toxic mineral substances and organic compounds that persist for many years [Koda, Koper 2009]. The purpose of the work was to determine the potential hazard to the soil and water environment, based on selected properties, in the vicinity of a closed waste landfill at Głębocka Street in Warsaw.

RESEARCH OBJECT

The research site was a landfill site located at Głębocka Street in Warsaw, in the municipality of Targówek on the western edge of the Forest Park "Bródno". It covers an area of about 4 ha, and the difference in altitude at the highest point in relation to the surrounding area is 12.5 m. The scarp from the north-east is steep, whereas in the southern part it is mild (Figure 1). In the eastern part of the landfill, there are wetlands that may indicate leachate and abnormal ground-water relations (Figure 2). The lack of documentation regarding the landfill, the amount and type of waste deposited and its history, causes problems related to its potential impact on the ground and water environment. The landfill mainly deposited the organic waste from Zakłady Farmaceutyczne "Polfa" and Zakłady Mięsne "Żerań".

The vegetation around the object under examination is mainly composed of *Betula L.* and *Pinus sylvestris L.* species. In the eastern part, wetlands are present, the following species can be found: *Alnus glutinosa, Betula pubescens, Sambucus nigra* and *Prunus padus*. The crown of the landfill is mainly covered with grasses, single trees and shrubs.

MATERIALS AND METHODS

Twenty wells were drilled to assess the land degradation. From each well, two samples were taken from two depths 0–0.3 m and 0.8–1.0 m (taking into account the differences in the height of the terrain) using a soil drill. Individual depths were marked with the letters "a" for 0–0.3 m and "b" for 0.8–1.0 m. The first area of research included the samples taken along the forest road (1), while the second area of research involved the samples taken on the scarp of storage sites (2). In each area, 10 wells from I to X were located (Figure 3, 4). The boreholes for both areas are placed at equal distance, every 15 m.

The collected samples were subjected to further preparations for the selected analyzes, according to PN-ISO 11464 standard. The chemical analysis was carried out in air-dry samples of known hygroscopic humidity. The tests were performed according to the following methodologies: type of smell, particle size distribution. sieve and Casagrande in modif. Prószyński, color, pH in 1M KCl according to PN-ISO 10390:

Figure 1. Landfill – view of the scarp from the east

Figure 2. Leachates at the eastern base of the landfill site

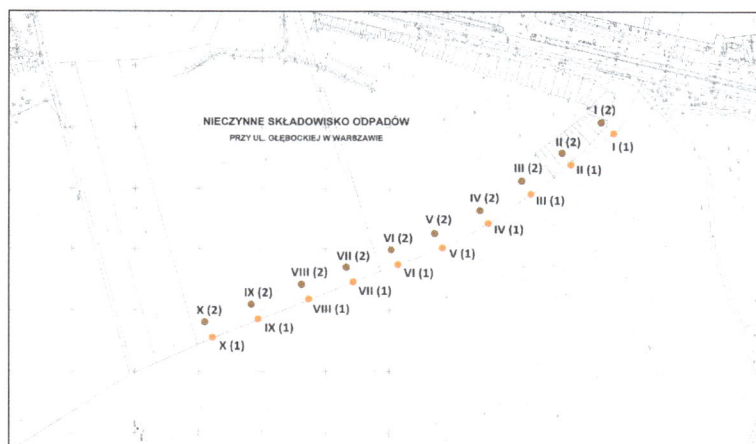

Figure 3. Location of wells along the forest road (1) and on the scarp of the landfill (2)

Figure 4. Soil sample from point VI (2) from 0.8–1.0 m layer

1997 standard; electrolytic conductivity (EC) according to PN-ISO 11265 standard + ACI: 1997, hydrolytic acidity (Hh) according to PN-R-04027:1997 standard, chloride content according to PN-ISO 9297: 1994 standard, organic carbon content according to 14235: 2003 standard, content of available nitrogen forms, according to PN-R-04028:1997 standard and, content of available phosphorus, according to [Ostrowska et al. 1991].

RESULTS AND DISCUSSION

The determination of the physical properties (Table 1) constituted a very important stage of the research work. On the scarp, at the depth of 0–0.3 m, loose sand was found, and at the depth of 0.8–1.0 m, the loamy sand was found. At the both depths studied, the dark-brown and the brown-yellow color predominates. In the top level, the soil was characterized by a sallow smell, in the level with a deeper, organic-specific odor, characteristic of ammonia. By the road in the upper level "a" loose sands, loamy loams, loamy light and strong ones were found, while in the middle of the deeper "b" level, medium clay was found in a single . In the "a" level, the dark and brown

colors prevailed, while in the deeper "b", the colors were lighter, yellow. At both depths, there was an earthy smell and in two samples at a depth of 0.8–1.0 m, a slightly specific odor was observed.

Table 2 presents the results of the performed analyses of pH value, hydrolytic acidity, electrolytic conductivity (EC) and chloride content. On the scarp (2), a larger difference in pH between the levels "a" and "b" was observed. By the road (1) the pH value increased with depth. This increase was smaller than on the scarp (2), ranging from about 0.5 to 1.0 pH value, which was caused by the presence of soil alkalizing substances at the depth above 1.0 m (Figure 5). In the samples taken on the scarp (2) a negative correlation was found between the pH value and the hydrolytic acidity (the higher the pH value, the lower the acidity). Therefore, the "a" level values were slightly higher than the "b" level. Along the road (1), the hydrolytic acidity in the upper level was characterized by an elevated value, typical for forest soils. It is clearly visible that the electrolytic conductivity (EC) on the scarp (2) is about 10 times higher in the "b" level compared to the "a" level, which indicates a significant amount of soluble salts in the examined soils [Karczewska 2012], affecting the degree of their degradation.

Table 1. Physical properties of soil taken from two layers along the road (1) and on a landfill site (2)

No of sample	Soil formation*		Munsell soil color, HUE 10YR		Scent (intensity)**	
	„a"	„b"	„a"	„b"	„a"	„b"
I (1)	s	s	dark grayish brown 4/2	yellow 8/6	earthly (3)	chemical (2)
II (1)	s	s	dark brown 3/3	yellow 7/6	chemical (1)	earthly (1)
III (1)	s	s	light brownish gray 6/2	brownish yellow 6/6	earthly (1)	earthly (1)
IV (1)	s	s	brown 4/3	yellow 7/6	earthly (2)	earthly (2)
V (1)	s	s	brown 4/3	brownish yellow 6/6	earthly (2)	earthly (1)
VI (1)	s	scl	yellowish brown 5/4	yellow 7/6	earthly (3)	chemical (2)
VII (1)	sl	ls	dark grayish brown 4/2	grayish brown 5/2	earthly (2)	earthly (1)
VIII (1)	ls	s	dark grayish brown 4/2	grayish brown 5/2	earthly (2)	earthly (1)
IX (1)	s	s	dark grayish brown 4/2	grayish brown 5/2	earthly (1)	earthly (1)
X (1)	ls	s	grayish brown 5/2	yellowish brown 5/6	earthly (1)	earthly (1)
I (2)	s	s	dark yellowish brown 4/6	dark brown 3/3	ammonia (1)	ammonia (3)
II (2)	s	s	dark yellowish brown 4/4	very dark gray 3/1	chemical (2)	chemical (3)
III (2)	s	s	dark yellowish brown 3/6	very dark gray 3/1	chemical (2)	chemical (3)
IV (2)	s	s	yellowish brown 5/8	dark grayish brown 4/2	ammonia (2)	ammonia (3)
V (2)	s	s	yellowish brown 5/8	very dark gray 3/1	organic (1)	organic (3)
VI (2)	s	s	yellowish brown 5/8	black 2/1	ammonia (2)	ammonia (3)
VII (2)	s	s	brownish yellow 6/8	very dark brown 2/2	rotten (1)	rotten (3)
VIII (2)	s	s	brownish yellow 6/8	black 2/1	ammonia (2)	ammonia (3)
IX (2)	s	s	yellowish brown 5/8	yellowish brown 5/4	organic (1)	organic (3)
X (2)	s	s	brownish yellow 6/6	very dark gray 3/1	ammonia (2)	ammonia (3)

***Soil formation**: s – sand, sl – sandy loam, ls – loam sand, scl – sandy clay loam; ****Intensity of scent**: weak (1), medium (2), strong (3)

Table 2. Comparison of pH, hydrolytic acid, EC and chloride acid results for samples taken along a forest road (1) and on a landfill site (2)

Nr odwiertu	pH 1M KCl		Hh [cmol(+)/kg d.m.]		EC [µS/cm]		Chloride [mg/kg d.m.]	
	„a"	„b"	„a"	„b"	„a"	„b"	„a"	„b"
I (1)	5.91	6.89	2.60	1.00	96.5	188.9	404.0	804.6
II (1)	7.31	6.90	0.60	0.80	247.6	338.6	203.0	361.6
III (1)	5.05	4.86	3.80	1.40	356.5	472.2	302.6	401.8
IV (1)	5.44	6.96	4.40	0.60	601.5	396.4	616.8	453.8
V (1)	6.8	6.97	2.00	3.80	413.5	409.1	50.9	402.7
VI (1)	5.3	5.63	3.80	1.40	56.8	43.8	909.1	602.6
VII (1)	5.65	5.97	4.00	1.40	152.0	226.4	508.0	501.2
VIII (1)	7.05	6.99	2.00	0.80	764.5	330.3	506.8	402.2
IX (1)	4.91	6.47	7.60	1.00	439.2	173.5	508.6	301.1
X (1)	4.9	5.62	6.40	1.60	159.6	426.5	805.4	1006.6
I (2)	5.39	7.92	3.80	0.40	125.3	1707.9	604.9	519.9
II (2)	5.48	7.57	1.80	1.40	83.6	2091.0	251.0	656.4
III (2)	6.75	7.61	2.00	2.20	141.1	2991.0	300.5	1856.0
IV (2)	5.3	7.79	2.00	1.00	278.4	3330.0	252.1	1874.7
V (2)	5.33	8.15	2.00	2.40	697.2	3487.0	620.0	2902.2
VI (2)	6.02	7.37	2.20	1.40	81.0	3328.0	705.1	2345.5
VII (2)	5.23	7.39	3.00	1.60	60.8	3025.0	704.7	2646.0
VIII (2)	5.37	7.87	2.40	1.60	91.2	1739.2	327.4	651.3
IX (2)	5.93	8.01	2.00	0.40	61.5	1543.5	403.0	829.0
X (2)	5.56	7.43	1.60	1.40	95.6	2533.0	802.0	2542.3

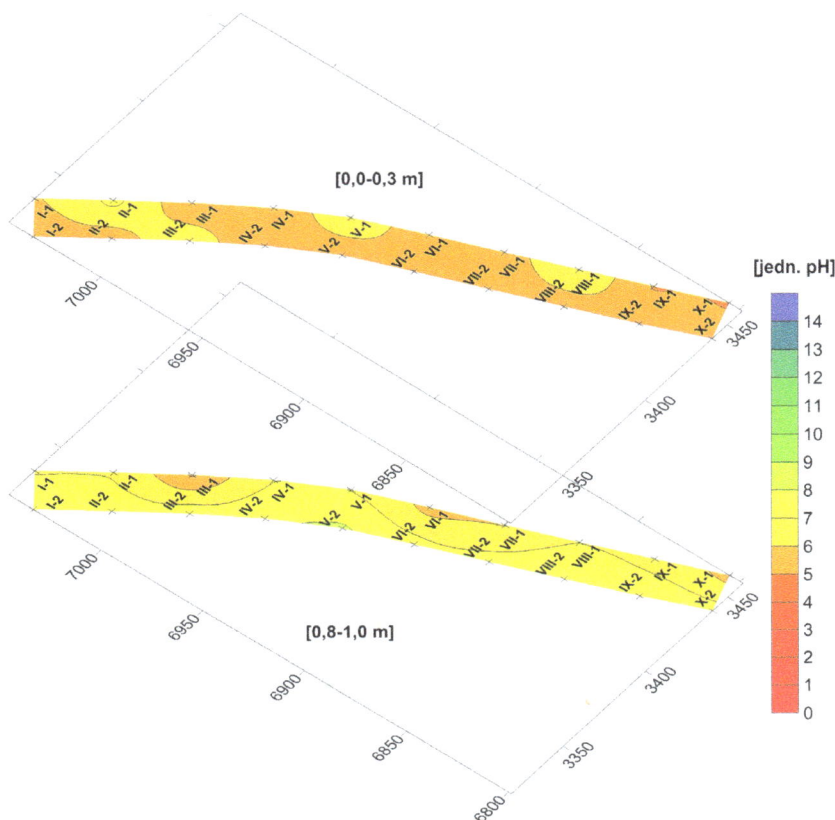

Figure 5. Simulation of the pH distribution between the individual wells along the forest road (1) and on the scarp of the landfill (2)

The soil collected by the road (1) in which low values of EC were found in both levels (Figure 6) may constitute the reference sample. The chloride content is equally high and correlates with the value of EC. The highest content of chlorides was found in the samples taken from the scarp (2) in the deeper level "b", which may pose a potential threat to groundwater (Figure 7).

Table 3 summarizes the results of analysis involving the organic carbon content, absorbed phosphorus and nitrogen forms. The obtained results of the organic carbon content along the forest road (1) show the correct tendency, as they are higher in the "a" level (0–0.3 m). At this level, the accumulation and decomposition of organic matter, characteristic of natural processes, occurs [Zawadzki 1999; Bednarek et al. 2005]. In the case of the boreholes located on the scarp's landfill (2), the results of the organic carbon content increase with depth (Figure 8). Such a high content of organic matter is caused by leaching it from the waste. The content of available phosphates on the scarp (2) at the depth "b" (0.8–1.0 m) reached very high values compared to the upper levels "a" on the scarp (2) and along the road (1) (Figure 9). This probably indicates a progres-

sive decomposition of the substance, occurring in the mass of littering waste, which may pose a serious threat to the groundwater. The high values of phosphates analyzed correlated with the perceptible specific smell of organic compounds.

The nitrate nitrogen content on the scarp (2) was significantly higher than on the road (1), especially at the "b" level (0.8–1.0 m) (Figure 10). The highest content of ammonium nitrogen was found in the border wells. The content of ammonium nitrogen on the scarp (2) was very high and in the sample III (2) in the level "b" it amounted to 1917.91 mg/kg d.m. The difference between the levels "a" and "b" was very clear, in each borehole a higher content of the examined feature was noted at a depth of 0.8–1.0 m (Figure 11). The high analyzed ammonium nitrogen content correlated with the perceived specific smell of ammonia.

Summing up, the vertical and horizontal migration of pollutants from the landfill may become a big threat to the soil and water environment. The landfill is also a potential geotechnical hazard because it is a dynamically changing object that can change stability, create caverns and landslides.

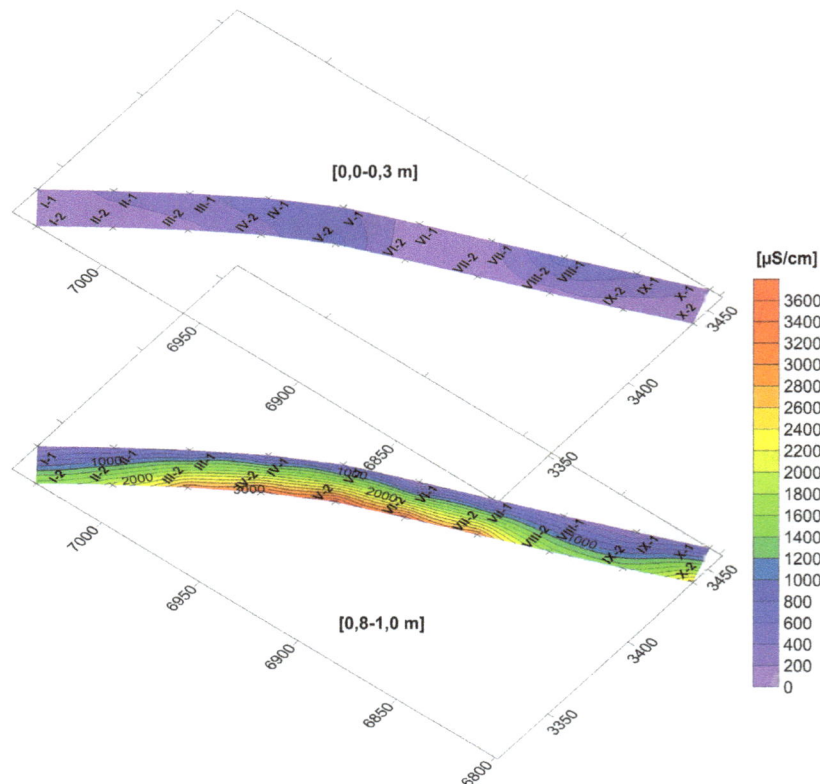

Figure 6. Simulation of the electrolytic conductivity (EC) distribution between individual wells along the forest road (1) and on the scarp of the landfill (2)

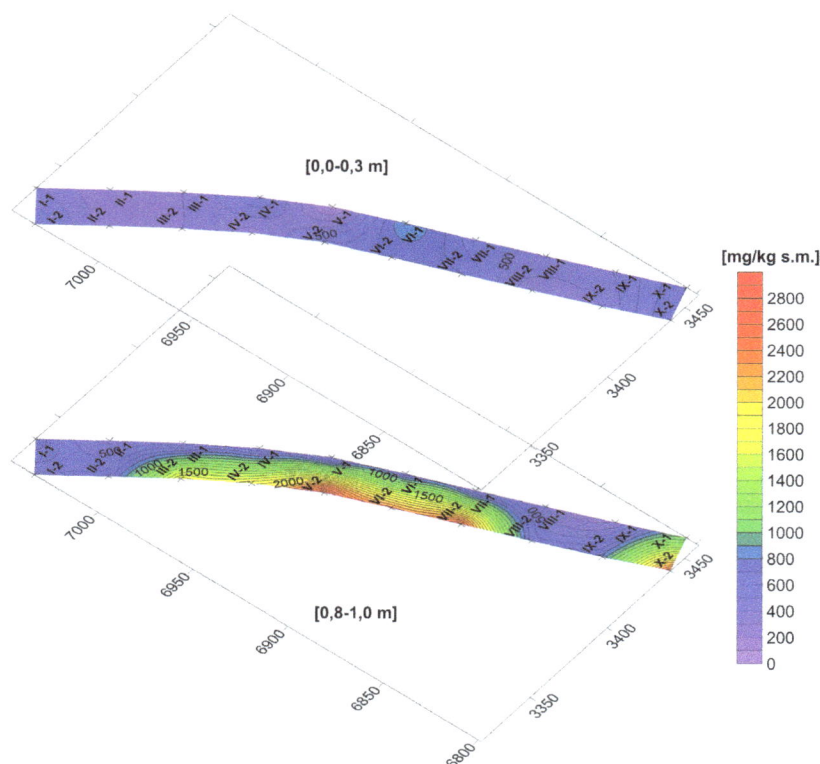

Figure 7. Simulation of the distribution of chloride content between individual wells along the forest road (1) and on the scarp of the landfill (2)

Table 3. Content of carbon, available phosphorus and nitrogen forms in the soil samples taken along the forest road (1) and on the scarp of the landfill (2)

No of sample	Organic carbon [%]		Available phosphorus [mgP$_2$O$_5$/kg d.m.]		Nitrogen [mg/kg d.m.]			
					ammonia		nitrate	
	„a"	„b"	„a"	„b"	„a"	„b"	„a"	„b"
I (1)	1.724	0.890	24.2	120.7	404.03	402.32	85.86	70.41
II (1)	0.836	0.119	162.4	41.3	274.04	69.74	5.07	2.07
III (1)	1.514	0.033	242.1	32.1	58.00	57.76	20.17	1.00
IV (1)	3.940	0.477	205.6	40.3	277.55	68.07	10.28	2.02
V (1)	1.542	0.349	407.2	40.3	274.84	171.14	2.04	5.03
VI (1)	1.602	0.596	202.0	24.1	22.73	20.09	5.05	1.51
VII (1)	1.546	0.356	203.2	24.1	33.02	92.72	3.05	18.04
VIII (1)	1.849	0.356	186.5	40.2	50.68	25.14	65.88	30.17
IX (1)	2.955	0.238	81.4	20.1	50.86	27.61	27.47	15.06
X (1)	1.723	1.945	80.5	32.2	90.61	40.26	5.54	110.72
I (2)	2.310	1.419	282.3	2495.7	90.73	779.91	30.24	129.99
II (2)	0.179	1.155	241.0	3675.6	77.81	1772.16	50.20	78.76
III (2)	0.497	7.671	240.4	2969.7	57.60	1917.91	5.01	61.87
IV (2)	0.716	3.280	201.7	2707.8	118.48	1405.99	5.04	26.04
V (2)	2.547	3.597	248.0	3009.7	320.35	994.27	7.23	2.15
VI (2)	1.135	5.547	137.0	2855.4	553.97	1682.64	5.04	20.40
VII (2)	1.479	2.187	88.6	2116.8	193.79	1068.58	4.03	20.35
VIII (2)	0.534	4.178	131.0	2822.2	62.76	705.55	2.18	3.80
IX (2)	0.834	1.769	120.9	2486.9	37.78	217.60	6.04	1.04
X (2)	1.123	3.396	120.3	1694.8	37.59	635.57	7.02	158.89

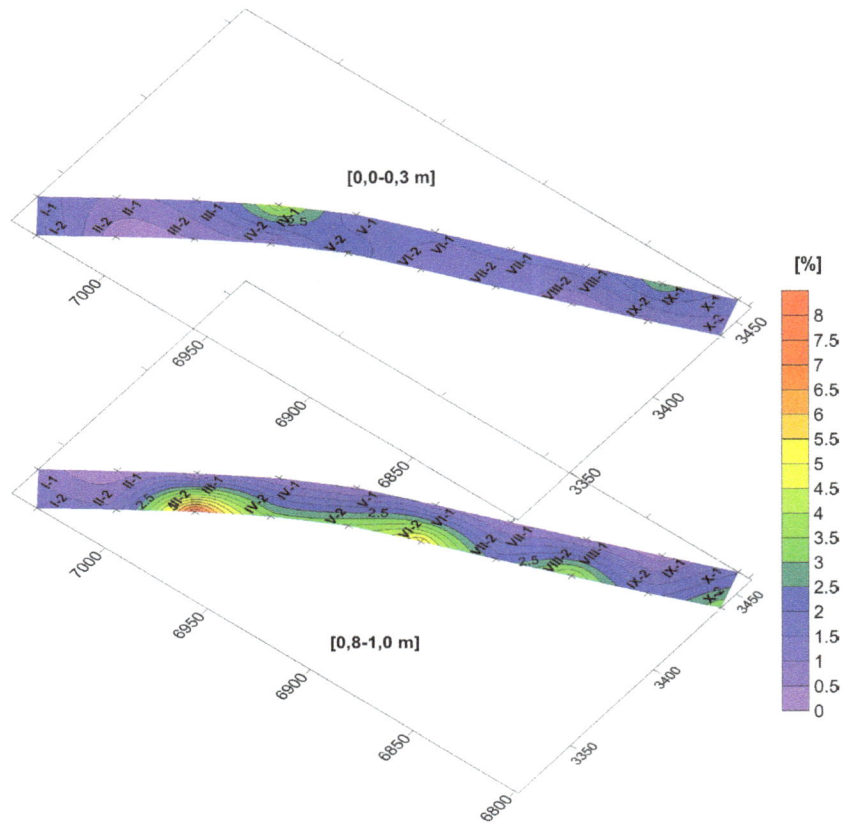

Figure 8. Simulation of the distribution of organic carbon content between individual wells along the forest road (1) and on the scarp of the landfill (2)

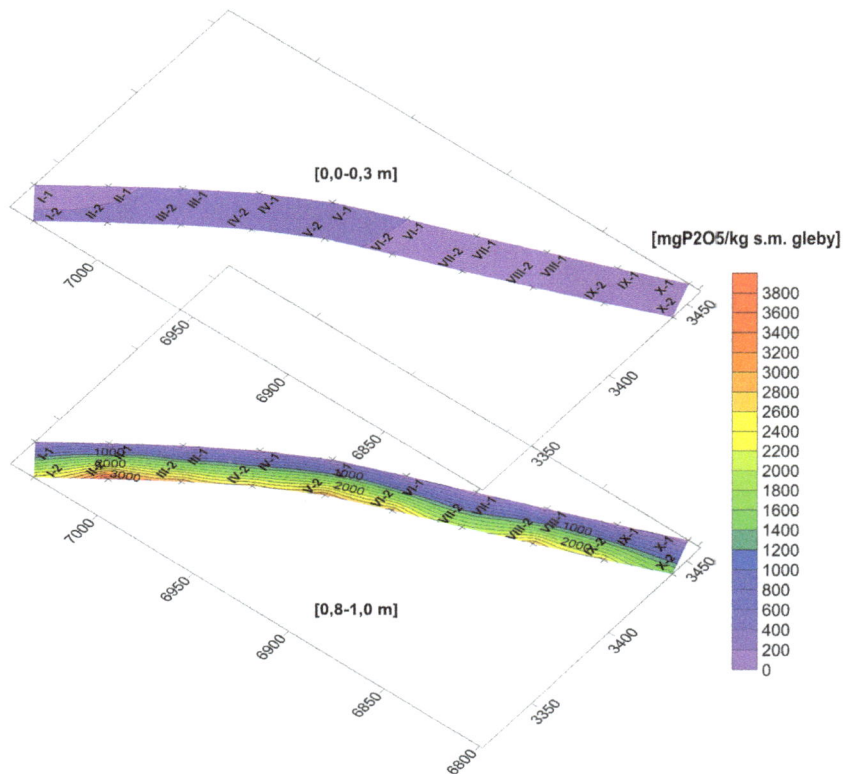

Figure 9. Simulation of the distribution of organic carbon content between individual wells along the forest road (1) and on the scarp of the landfill (2)

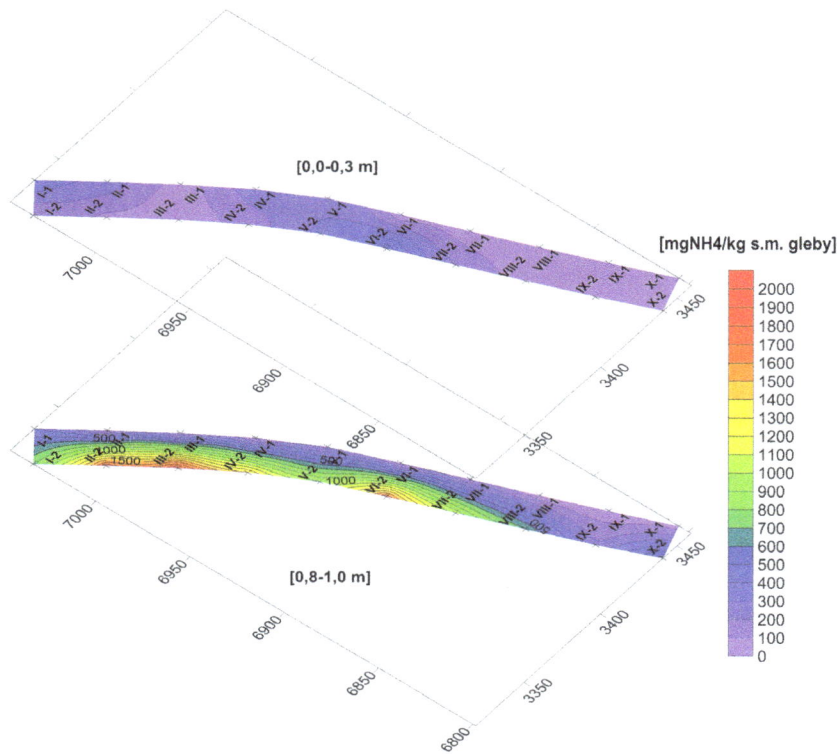

Figure 10. Simulation of the distribution of nitrate nitrogen between individual wells along the forest road (1) and on the scarp of the landfill (2)

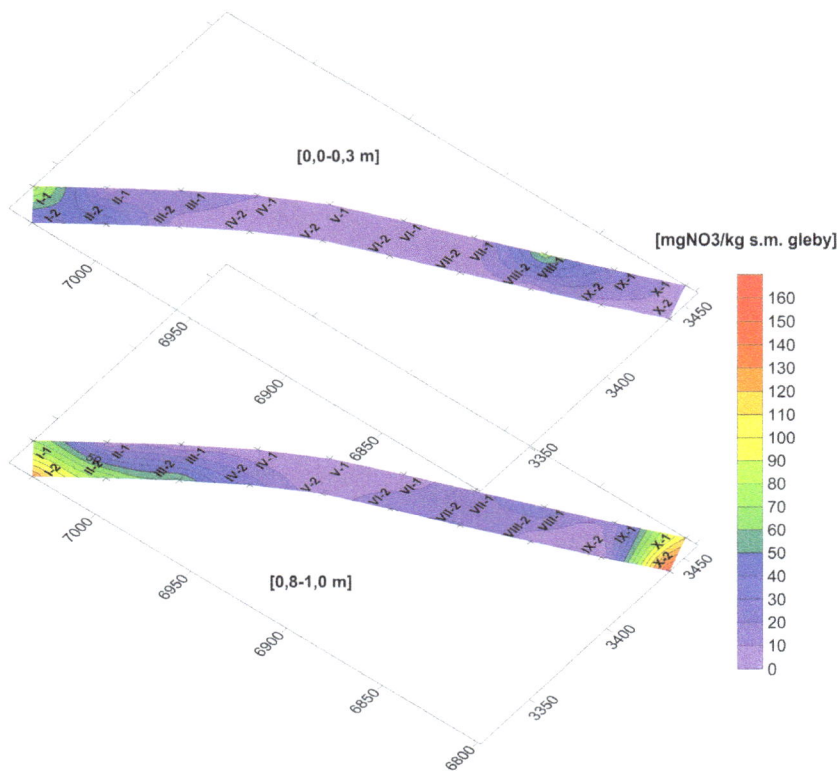

Figure 11. Simulation of the distribution of ammonium nitrogen content between individual wells along the forest road (1) and on the scarp of the landfill (2)

CONCLUSIONS

1. In the western part of the former landfill site, the area under the supervision of the Warsaw Forests, an increased content of EC, chlorides, ammonium and nitrate nitrogen and assimilable phosphorus in the layer of 0.8–1.0 m was observed. There was also a close relationship between the perceptible smell of ammonia, and high content of ammonium nitrogen.

2. An increased content of organic carbon and higher pH values were found in relation to the soils surrounding the landfill (acidic soils).

3. High content of the above-mentioned properties testifies to the anthropogenic pollution. The most likely cause of contamination of the investigated soils involves the leachates from the landfill, which pose a risk of groundwater contamination.

4. The samples taken on the scarp of the repository, especially from the depth of 0.8–1.0 m, were characterized by much higher values of the tested properties than the samples taken at a distance of about 5–10 m from the scarp.

5. The samples taken from the deeper levels on the scarp of the landfill were characterized by a sharp specific organic smell, which indicates the migration of leachate through the permeable layers deep into the ground.

6. The landfill site is to a large extent chemically degraded and constitutes a potential hazard to the groundwater, which should be tested and monitored.

REFERENCES

1. Bednarek R., Dziadowiec H., Pokojska U., Prusinkiewicz Z. 2005. Ecological and soil science research (in Polish). PWN, Warszawa.

2. Grygolczuk-Patersons E., Wiater J. 2012. Influence of municipal landfill site on the ground water quality (in Polish). Inżynieria Ekologiczna, No 31.

3. Karczewska A. 2012. Soil protection and reclamation of degraded areas (in Polish). Wydawnictwo Uniwersytetu Przyrodniczego we Wrocławiu, Wrocław.

4. Koda E., Koper M. 2009. Assesment of reclamation efficiency of old municipal landfills (in Polish). VIII Międzynarodowe Forum Gospodarki Odpadami „Integrated waste management". Poznań.

5. Marcinkowski (Ed.) 2009. Comprehensive waste management management (in Polish). Polskie Zrzeszenia Inżynierów i Techników Sanitarnych Oddział Wielkopolski, Poznań.

6. Mizera A. 2007. Soil – mechanisms of its degradation and methods of reclamation (in Polish). Green World – Ochrona Środowiska i Ekologia.

7. Ostrowska A., Gawliński S., Szczubiałka Z. 1991. Methods of analysis and assessment of soil and plant properties (in Polish), Katalog, Instytut Ochrony Środowiska, Warszawa.

8. PN-ISO 11464:1999 – Soil quality – Pretreatment of samples for physico-chemical analyses (in Polish).

9. PN-ISO 10390:1997 – Soil quality – Determination of pH (in Polish).

10. PN-ISO 11265 + ACI:1997 – Soil quality – Determination of the specific electrical conductivity (in Polish).

11. PN-ISO 9297:1994 – Water quality – Determination of chloride – Silver nitrate titration with chromate indicator (Mohr's method) (in Polish).

12. PN-ISO 14235:2003 – Soil quality – Determination of organic carbon by sulfochromic oxidation (in Polish).

13. PN-R-04027:1997 – Agrochemical soil analysis – Determination of hydrolytic acidity in mineral soils (in Polish).

14. PN-R-04028:1997 – Agrochemical soil analysis – Determination of nitrate and ammonium ions in the mineral soils (in Polish).

15. Siuta J. 1995. Soil – diagnosis of condition and threat (in Polish). Institute of Environmental Protection, Warszawa.

16. Tabor A., Generowicz A., Korzeniowska – Rejmer E., Sacharczuk J., Truś S. 2008. Waste management and land protection, Vol. III (in Polish). Centrum szkolenia i organizacji systemów jakości Politechniki Krakowskiej im. Tadeusza Kościuszki.

17. The Waste Act of 14 December 2012 (Dz.U. 2013 poz. 21, z późn. zm.) (in Polish).

18. Zawadzki S. 1999. Soil Science (in Polish). Państwowe Wydawnictwo Rolnicze i Leśne, Warszawa.

24

The Condition of Sanitary Infrastructure in the Parczew District and the Need for its Development

Agnieszka Micek[1], Michał Marzec[1*], Karolina Jóźwiakowska[2], Patrycja Pochwatka[1]

[1] Department of Environmental Engineering and Geodesy, University of Life Sciences in Lublin, Leszczyńskiego 7, 20-069 Lublin, Poland

[2] Warsaw University of Life Sciences, Nowoursynowska 166, 02-787 Warszawa, Poland

* Corresponding author's e-mail: michal.marzec@up.lublin.pl

ABSTRACT

The aim of this paper is to present the current state and the need for development of sanitary infrastructure in the communes of Parczew District, in Lublin Voivodeship. Parczew District encompasses seven communes: Parczew, Dębowa Kłoda, Jabłoń, Milanów, Podedwórze, Siemień, and Sosnowica. The present paper uses the data from the surveys conducted in these communes in 2016. On average, 88% of the population used the water supply system in the communes surveyed, while 48% of the inhabitants were connected to a sewerage system. Parczew District had 12 collective mechanical and biological wastewater treatment plants with a capacity exceeding 5 m³/d. The households which were not connected to the sewerage network discharged the wastewater mainly to non-return tanks. In the communes surveyed, 1,115 households had domestic wastewater treatment plants. All of them were systems with infiltration drainage, which do not ensure high efficiency of removing pollutions and may even contribute to the degradation of the groundwater quality. In order to solve the existing problems of sewage and water management in the communes of Parczew District, it is necessary to further develop the collective sewerage systems and equip the areas which have a dispersed development layout with highly efficient domestic treatment plants, such as constructed wetlands.

Keywords: sanitary infrastructure; commune; water supply system; sewerage system; wastewater treatment plant; septic tank

INTRODUCTION

After joining the European Union, Poland has undertaken to comply with the legal provisions regarding the sound management and protection of water. The changes that had to be introduced into the national law and the need to adjust this law to the requirements of the Water Framework Directive 2000/60/EC of 23 October 2000, as well as the Council Directive 91/271/EEC of 21 May 1991 concerning urban wastewater treatment, prompted local government bodies to take actions aimed at developing the water and sewerage infrastructure [Jóźwiakowski and Pytka 2010]. For this reason, in recent years there has been a significant increase in the investments into building water supply and sewage removal systems in urban and rural areas. This increase has been connected not only with the requirements of environmental protection, but above all with the better possibilities of financing the investments with the EU funds. In addition to the ecological and economic effects, the development of the technical and sanitary infrastructure brings many benefits of a social and economic nature, as it contributes to improving the living conditions of the inhabitants and the development of entrepreneurship.

The statistical data show that the length of the water supply and sewerage systems as well as the number of water supply and sewage disposal facilities in Poland have significantly grown in the recent years. The length of the water supply system increased from 245,600 km in 2005 to 301,000 km in 2016. In rural areas, the increase was from 190,700 km to 233,200 km, and the number of households connected to the system grew by 25.3%. In the period of 2005–2016, the length of the sewerage network increased by 73,900 km (92.2%), reaching 154,000 km in 2016. In rural areas, the length of the sewerage network increased by 53,600 km (146%). The increase was larger than in cities, where 20,300 km of new pipelines were laid (an increase by 46.8%) [GUS 2017a].

In 2016, almost 91.9% of the total population used the water supply network (a 5.8% increase compared to 2005). In urban areas, over 95.5% of inhabitants had access to a water supply system (an increase of 1.2% compared to 2005). In rural areas, the proportion of people using the water supply network in 2016 was 85%. As far as the sewerage systems are concerned, in the period of 2005–2016, the percentage of people having access to a sanitation network increased from 59.2% to 70.2% (an 11.0% increase). In cities, 90.0% of the population used the sewage system (a 5.2% increase), whereas in rural areas, the proportion was 40.3% (a 21.5% increase) [GUS 2017a].

In 2016, there were 3,319 collective sewage treatment plants in Poland, including 763 facilities located in cities and towns, and 2,556 in the countryside [GUS 2017b]. In villages with dispersed housing, wastewater was discharged into non-return tanks or domestic wastewater treatment plants. In 2016, there were 2,333,000 such facilities, out of which about 91% were non-return tanks, and about 9% were domestic wastewater treatment plants. For several years now, a systematic decrease in the number of non-return tanks has been observed, while the number of domestic sewage treatment plants has been on the increase. In 2016, the number of non-return tanks was 2,117,000, whereas the number of domestic wastewater treatment plants amounted to 217,000 [GUS 2017a].

These data show that after the Poland's accession to the European Union, the increase in water and sewerage infrastructure investments in towns and villages was much larger than in the early 1990s [Jóźwiakowski and Pytka 2010]. Nevertheless, there are still great disproportions among Polish cities and villages regarding the access to water supply and sewage removal systems. These disparities are most prominent in rural areas.

The aim of the present study was to evaluate the state of the development pertaining to the sanitary infrastructure and the need for expanding this infrastructure in Parczew District (county), which constitutes one of the 24 districts of Lublin Voivodeship. The analyses were carried out using the official statistical data and the results of surveys conducted in 2016 in seven communes of Parczew District by the employees of the Department of Environmental Engineering and Geodesy of the University of Life Sciences in Lublin. The surveys identified the percentage of the population having access to public water supply and sanitation systems and the number of non-return tanks as well as collective and domestic wastewater treatment plants.

CHARACTERISTICS OF THE PARCZEW DISTRICT

Parczew District is located in the northeastern part of Lublin Voivodeship and occupies an area of 952 km^2, i.e. 3.8% of the area of the province (Figure 1). The district is inhabited by 36,147 people, who constitute 1.7% of the province's population. It borders with the districts of Lubartów, Radzyń, Biała Podlaska, Włodawa, and Łęczna. The following communes belong to Parczew District: Parczew, Milanów, Jabłoń, Podedwórze, Siemień, Dębowa Kłoda, andi Sosnowica (Figure 2) [SYNERGIA. Advisory Office, 2013]. The largest area is occupied by Dębowa Kłoda Commune, whereas the smallest by Podedwórze (Table 1). The population density in the county varies over a very wide range from 16 people/km^2 in the communes of Podedwórze and Sosnowica to 101 people/km^2 in Parczew Commune (Table 1).

In terms of the physical-and-geographical units, Parczew District extends over the area of three regions of the Western Polesie macroregion: Równina Parczewska (the Parczew Plain, northeastern part of the district), Pojezierze Łęczyńsko-Włodawskie (the Łęczna-Włodawa Lake District, southern part of the district) and Garb Włodawski (the Włodawa Elevation, between the Parczew Plain and the Łęczna-Włodawa Lake District) [SYNERGY. Advisory Office, 2013].

Parczew District is a typically agricultural area, which, at the same time, uses its landscape and natural resources to create a welcoming environment for tourism and leisure activities. The agricultural land constitutes about 67% of the total area of the district; about 26% of the district

Fig. 1. Geographical location of Parczew District

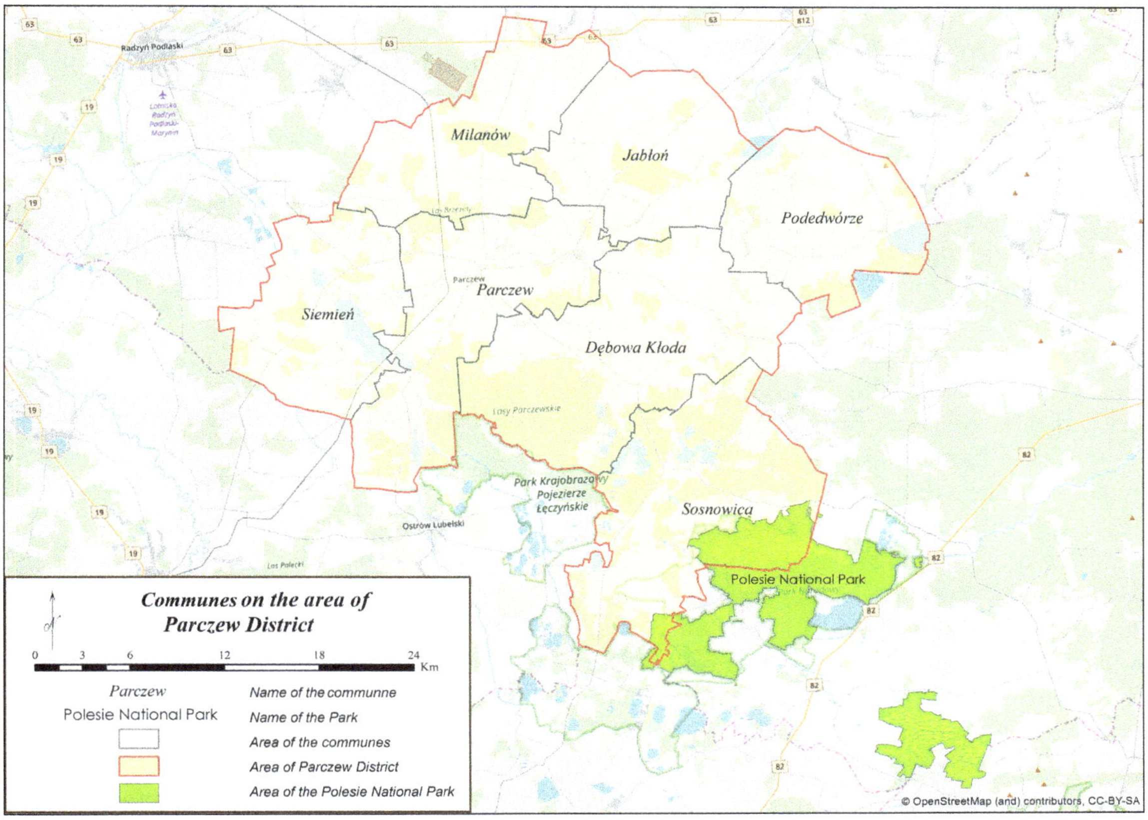

Fig. 2. The communes of Parczew District

Table 1. Characteristics of the communes of Parczew District

Name of commune	Area [km²]	Population	Population density [people/km²]	Proportion of district area occupied by the commune [%]
Parczew	147	14 860	101	15.42
Dębowa Kłoda	188	4009	21	19.72
Jabłoń	110	3947	36	11.54
Milanów	117	4052	35	12.27
Podedwórze	107	1749	16	11.22
Siemień	111	4802	43	11.64
Sosnowica	171	2728	16	17.94

[SYNERGY. Advisory Office, 2013]

is occupied by woodlands, which are considered to be a touristically valuable element of the landscape and ecosystem. The district's varied soil types (quality-wise) provide a basis for growing not only cereals, but, increasingly often, also herbs (chamomile, mint, valerian). The prevailing soil valuation classes are arable land (63%) and permanent grasslands (30%) [US Lublin 2017].

Parczew District boasts many highly valuable natural features. The most important of them is part of the Polesie National Park, along with the surrounding areas, covering vast stretches of meadows, forest tundra and peat bogs, a group of ponds in Siemień, retention reservoirs in the Podedwórze commune and the Parczew Forests (Lasy Parczewskie) complex. Unfortunately, many valuable boggy ecosystems and peat bogs have been destroyed as a result of large-scale water drainage interventions carried out in the recent decades [SYNERGY. Advisory Office, 2013].

Because of its landscape and natural heritage, a significant part of Parczew District has been placed under legal protection. No new investments which might pose threat to the natural environment can be made in such areas. All human interference is legally prohibited to protect the valuable natural features. The most important forms of nature conservation are the Polesie National Park, the Polesie Landscape Park, and the Polesie Area of Protected Landscape. In addition, there are six nature reserves in the district (Czarny Las, Lasy Parczewskie, Warzewo, Królowa Droga, Jezioro Obradowskie, and Torfowisko przy Jeziorze Czarnym), as well as 44 natural monuments. After the Poland's accession to the European Union, numerous parts of Parczew District have been incorporated into the Natura 2000 European network of protected areas [SYNERGIA. Advisory Office, 2013].

The clean environment and the little-changed landscape of Parczew District provide favorable conditions for the development of tourism and recreation as well as the production of healthy food. Apart from enjoying the natural assets of the district, tourists can also visit some historical buildings, including churches, manor houses and palaces as well as former Uniate Orthodox churches [http1].

In order to keep the district's natural environment, with its numerous conservation areas and tourist attractions, in good shape, it is necessary to invest in the water supply and sewage removal infrastructure [Kaczor et al. 2015; Jóźwiakowski et al. 2017a].

The development strategy update for Parczew District for the years 2014–2020 defines the following objectives and priority environmental actions:

- reduction of emissions of harmful substances and energy,
- protection of natural environment and landscape resources,
- sound management of the environment,
- increasing the civic activity and the environmental awareness of the society [SYNERGY. Advisory Office, 2013].

RESULTS AND DISCUSION

The condition of sanitary infrastructure in Parczew District was analyzed on the basis of results obtained from the surveys carried out in each commune of the district in 2016. The surveys covered the following issues: length of the water supply and sewerage network in a commune, number and capacity of collective wastewater treatment plants of over 5 m³/d, number of domestic wastewater treatment plants by type of technological design used. The survey data were supplemented with official statistics.

WATER SUPPLY AND SEWAGE REMOVAL SYSTEMS

The status of the sanitary infrastructure depends on the length of the water and sewerage network, and, to a greater extent, on the percentage of residents who are connected to this network. In this respect, the water supply system in Parczew District is quite well developed. The total length of the water supply network in 2016 was 630 km and constituted about 2.9% of the length of the water supply network in Lublin Voivodeship. The proportion of people using the network exceeded 87% (Figure 3). The longest water supply network was that of the Siemień commune (126.7 km); it supplied water to 4,550 people, who represented approximately 95.5% of the commune's population. A similarly high percentage of the population had access to the collective water supply system in the communities of Milanów (98.3%) and Podedwórze (93.3%); however, the water supply networks in those administrative areas were substantially shorter, at 82.2 km and 60 km, respectively (Fig. 4). This was mainly due to the smaller population of these communes and a greater concentration of settlements. The second longest water supply network was that in the urban-rural Parczew Commune (100.1 km). Because of the urban character of Parczew Commune and its compact development layout, this network provided water to nearly 13,000 people, or approximately 86.4% of the commune's population. In the communes of Dębowa Kłoda and Jabłoń, the length of the water supply network in 2016 was approx. 90 km, which allowed water to be supplied to 85.3% and 86.1% of the inhab-

itants of these communes, respectively. In Sosnowica Commune, the length of the water supply network was 77.8 km, but it supplied water to only 60% of the inhabitants, a percentage that was very low and significantly deviated from the average for the entire district (Figure 3).

The condition (extent) of sewerage infrastructure in Parczew District was much worse than that of the water supply network. The survey showed that there was a very large disproportion between the development of the sanitation and water supply systems (Figures 3 and 4).

The total length of the sewage network in the district was 164.4 km in 2016, which constituted only 26.1% of the length of the water supply network. This meant that an average of 48.2% of the commune's population was connected to a sewage removal system. Most of the people who used the collective sewage removal systems came from the urban-rural Parczew Commune, which is inhabited by over 40% of the district's population and provides good access to sanitation services (Figure 3). In this commune, a 76.3 km long sanitation system collected the sewage from over 12,300 people (88.3% of the population of the commune). In other communes of the district, the situation was less favorable. In the communes of Siemień, Sosnowica, Dębowa Kłoda, and Jabłoń, the percentage of inhabitants discharging wastewater to the sewerage systems fluctuated within a quite narrow range from 26.8% to 32.8%. The length of the sewerage network in each of these communes was approx. 12–15 km, with the exception of Siemień Commune, in which it exceeded 32 km (Figure 4). This may indicate that Siemień has a more dispersed development pat-

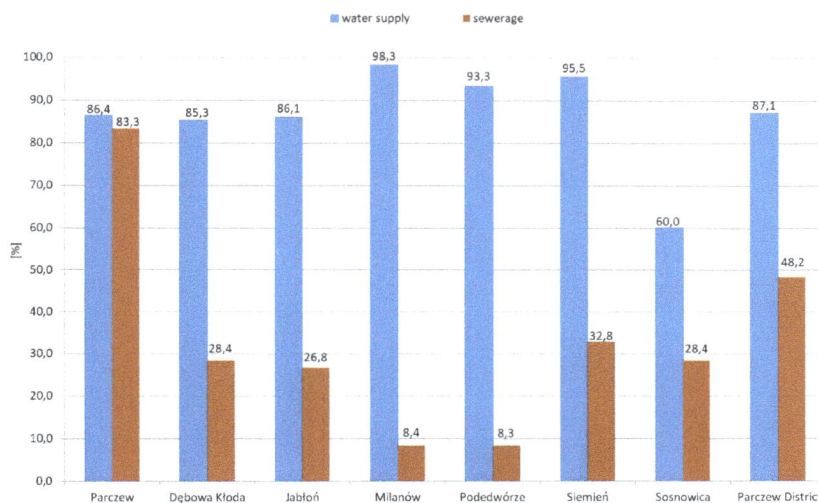

Fig. 3. Percentage of residents of Parczew District with access to the sanitary infrastructure in 2016

tern, requiring a larger scale of investment to meet the needs of a specific group of residents. A similar situation was found in Milanów Commune, in which the sewerage network was not particularly shorter than in the communes mentioned above (12.6 km), yet it served only 8.4% of residents. The lowest percentage of users of collective sewage disposal systems was recorded in Podedwórze Commune (8.3%), which still seems to be a good result, taking into account that the length of the sewerage network in that commune in 2016 was only 0.5 km (Figure 4) .

The data presented above point to some unfavorable phenomena regarding the development of the sewerage network in Parczew District, including the very large disproportion between the length of this network in relation to the length of the water supply network and lower than average (for Lublin voivodeship – about 57% in 2016) percentage of population using the collective sewage disposal and treatment systems. These problems have many causes, including the environmental, social, technical and, above all, economic ones. The sewerage network is strictly dependent for its development on the existence of a water supply system. It is also characterised by a greater technical complexity and capital intensity of investment compared to the latter type of system [US in Lublin 2017]. In addition, in rural communes, especially in their peripheral parts, where the population or economic activity are not sufficiently concentrated, it is difficult to find justification for constructing sewerage systems; hence, the dominant role of scattered systems consisting of domestic sewage treatment plants or non-return tanks (cesspools). An equally important factor affecting the rate and scale of sewerage investments are the financial capacities of the individual communes and local public utility companies.

COLLECTIVE AND DOMESTIC WASTEWATER TREATMENT PLANTS

The survey conducted in 2016 shows that 12 collective wastewater treatment plants with a capacity of more than 5 $m^3 \cdot d^{-1}$ operated in Parczew District at that time. They were all biological treatment plants. The largest plant was located in the urban-rural Parczew Commune and served mainly the area of Parczew Town. The facility had been designed to treat about 2,400 m^3 of sewage per day (as of 2016). At present, the plant is undergoing a thorough modernization, the aim of which is to improve the efficiency of the wastewater treatment and reduce the consumption of the energy used in the process. Larger facilities, with a capacity of over 100 m^3/d, are located in the communes of Milanów, Jabłoń, Siemień and Sosnowica (Table 2).

The capacity of the remaining facilities did not exceed 70 m^3/d. The structure and operation of collective wastewater treatment plants are closely related to the existence of collective sewerage networks, which is why they are most often located in the settlements with the highest concentration of population, where the construction of the network is justified. The only exception includes the facilities intended for neutralization of the sewage coming from non-return tanks, which is brought to the sewage treatment plant by gully emptiers. An example of a wastewater

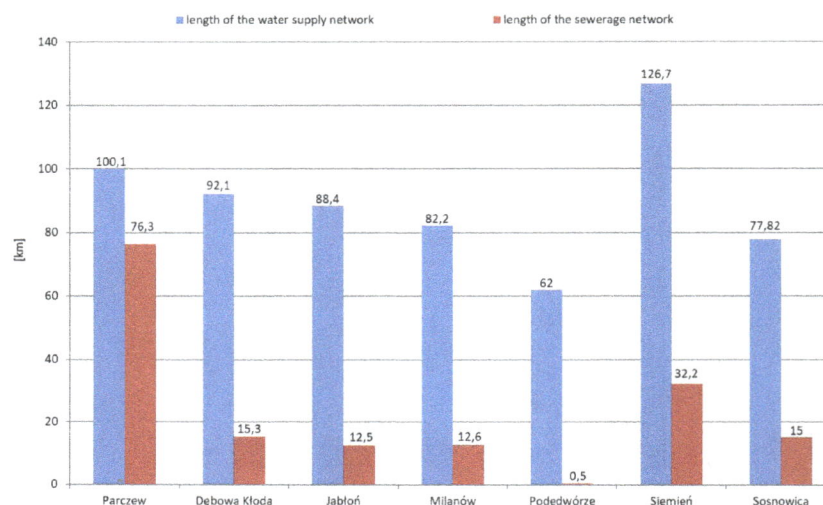

Fig. 4. The length of the water supply and sewerage network in Parczew District in 2016

Table 2. Collective wastewater treatment plants with a capacity of more than 5 m³/d in the communes of Parczew District

Name of treatment plant (commune)	Capacity [m³/d]
Parczew (Parczew)	2400
Leitnie (Dębowa Kłoda)	70
Wyhalew (Dębowa Kłoda)	25
Uhnin (Dębowa Kłoda)	70
Jabłoń (Jabłoń)	193
Kalinka (Jabłoń)	60
Milanów (Milanów)	220
Nowe Mosty (Podedwórze)	30
Siemień (Siemień)	120
Sosnowica (Sosnowica)	120
Zienki (Sosnowica)	60
Turno (Sosnowica)	30

neutralization plant in Parczew District, is the treatment plant in the village of Leitnie (Dębowa Kłoda Commune).

According to the National Municipal Wastewater Treatment Program (NPMWW), the construction of a centralized (collective) sewerage system is justified when there are no fewer than 120 inhabitants per kilometer of sewerage network (excluding the sewer laterals). This means that the unit length of the sewerage network should not exceed about 8 m/inhab. [Heidrich and Stańko 2008; AKPOŚK 2010]. Many settlements in rural areas do not meet this criterion, and, therefore, the use of domestic sewage disposal systems is recommended for those places. The non-return tanks are the most commonly used domestic sewage disposal systems, mainly due to the low investment costs. Because many of them are inaccurately built and used beyond their useful life, they do not provide adequate tightness, with some of the contaminants migrating into the ground and groundwater. According to the survey data, in 2016, there were 3,331 operating non-return tanks in Parczew District. Unfortunately, the data cover only four of the district's seven communes. The largest numbers of tanks were recorded in the communes of Dębowa Kłoda (1,702) and Parczew (1,028).

An alternative to non-return tanks are domestic wastewater treatment plants. They are by far cheaper in use than cesspools, which is why more and more of them are built every year [Karolinczak et al. 2015]. The term 'domestic wastewater treatment plant' is not officially recognized by the existing Polish legal acts. The Polish PN-

EN 12566–3:2016–10 Standard [2016] defines domestic wastewater treatment plants as the facilities that can be used to serve up to 50 inhabitants. Pursuant to the Water Law Act [2001], the limit capacity of this type of facilities is 5m³/d, while the Construction Law [2003] sets the limit at 7.5 m³/d. They are built as two-stage facilities, consisting of a mechanical and a biological waste removal stage [Pawęska et al. 2011].

The following technological systems can be used in domestic wastewater treatment plants [Pawełek and Bugajski 2017]:
- a septic tank with infiltration drainage,
- a septic tank with a sand filter,
- a containerized mobile wastewater treatment plant with activated sludge,
- a containerized mobile wastewater treatment plant with biological bed
- constructed wetlands

While choosing a domestic wastewater treatment plant, one should pay special attention to its reliability as well as the ecological and technical properties, and be less concerned with the economic aspects of the investment [Mucha & Mikosz 2009, Jóźwiakowski et al. 2015]. Unfortunately, in practice, the investment costs are the basic criterion for selecting a wastewater treatment technology, which means the largest number of domestic wastewater treatment plants are the cheapest structures that do not provide sufficient cleaning efficiency.

In 2016, Parczew District had 1,115 domestic sewage treatment plants. The largest numbers of domestic facilities were found in the communes of Siemień (422 plants), Podedwórze (322 plants), and Jabłoń (229 plants). In each of the remaining communes, there were fewer than fifty domestic sewage treatment plants (Figure 5).

Considering the large number of non-return tanks, a significant part of which can potentially be replaced by domestic wastewater treatment plants, the possibilities of developing this form of sewage disposal are still very large. All domestic wastewater treatment plants operating in Parczew District use a technological system with a septic tank and infiltration drainage, which is highly undesirable. According to many authors, the drainage treatment plants should not be approved for common use, because they do not allow to control the quality of treated sewage, and discharge of mechanically treated wastewater to the soil and water environment can cause its degradation

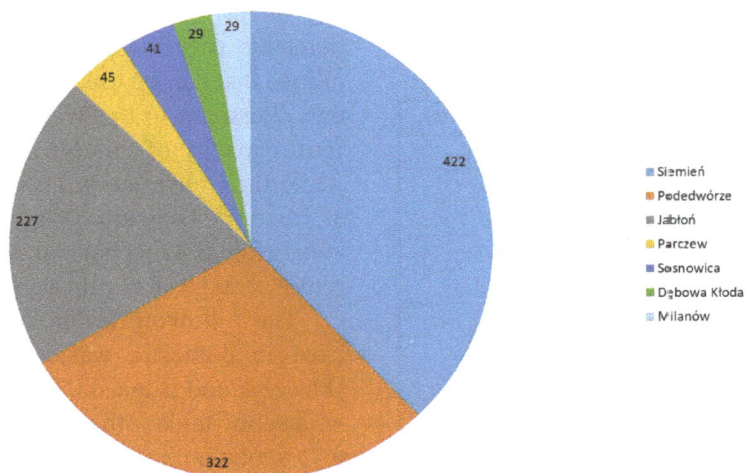

Fig. 5. Number of domestic sewage treatment plants in the communes
of Parczew District in 2016

[Jucherski and Walczowski 2001; Jóźwiakowski et al. 2014; Pawełek and Bugajski 2017]. Therefore, emphasis should be placed on promoting and implementing solutions that are most beneficial for the environment, thus minimizing the possibility of environmental pollution. In this context, the possibility of using constructed wetlands, which are highly reliable and efficient in removing contaminants from sewage, should be mentioned [Dębska et al. 2015; Gajewska et al. 2015; Jóźwiakowski et al. 2015; Gizińska et al. 2016; Jóźwiakowski et al. 2017b; Jóźwiakowski et al. 2018; Jucherski et al. 2017].

CONCLUSIONS

1. The water supply system in the district of Parczew is quite well developed, both in terms of the length of the water supply pipelines (630 km), and the proportion of inhabitants using the network (over 87%).
2. Compared to the water supply infrastructure, the sanitation infrastructure is being developed very slowly. There are very large disparities between the two systems. The length of the water supply network exceeds the length of the sewerage network almost four times.
3. Large differences in access to the sewerage network were found between the urban-rural commune of Parczew and the remaining communes of the district.
4. In 2016, twelve collective wastewater treatment plants with a total capacity of approximately 3400 m³/d operated in the district.

5. In the communes of Parczew District which were not connected to a collective sewerage system, wastewater was discharged into septic tanks or transported by gully emptiers to collective wastewater treatment plants.
6. The network of domestic wastewater treatment plants in Parczew District is made up entirely of systems with infiltration drainage, which may pose a serious threat to the soil and water environment.
7. The condition of the sewerage infrastructure in Parczew District, especially in the rural communes, is unsatisfactory and requires the investment activities aimed at expanding and modernizing the existing collective sewage disposal and treatment systems as well as developing a network of reliable, high-efficiency domestic wastewater treatment plants.

Acknowledgements

Publication supported by the Polish Ministry of Science and Higher Education as a part of the program of activities disseminating science from the project „Organization of the First International Science Conference – Ecological and Environmental Engineering", 26-29 June 2018, Kraków.

REFERENCES

1. AKPOŚK 2017. Update of the National Municipal Wastewater Treatment Program, Państwowe Gospodarstwo Wodne Wody Polskie, p. 42 (in Polish).
2. Dębska A., Jóźwiakowski K., Gizińska-Górna M.,

Pytka A., Marzec M., Sosnowska B., Pieńko A. 2015. The efficiency of pollution removal from domestic wastewater in constructed wetland systems with vertical flow with Common reed and Glyceria maxima. Journal of Ecological Engineering 16 (5) 110–118.

3. Gajewska M., Jóźwiakowski K., Ghrabi A., Masi F. 2015. Impact of influent wastewater quality on nitrogen removal rates in multistage treatment wetlands. Environ. Sci. Pollut. Res. 22, 12840–1284.

4. Gizińska-Górna M., Czekała W., Jóźwiakowski K., Lewicki A., Dach J., Marzec M., Pytka A., Janczak D., Kowalczyk-Juśko A., Listosz A. 2016. The possibility of using plants from hybrid constructed wetland wastewater treatment plants for energy purposes. Ecological Engineering 95, 534–541.

5. GUS 2017a. Municipal infrastructure in 2016. Główny Urząd Statystyczny, Warszawa, p. 35 (in Polish).

6. GUS 2017b. Environmental Protection. Główny Urząd Statystyczny, Warszawa, p. 551 (in Polish).

7. Heidrich Z,, Stańko M. 2008. Directions of solutions of wastewater treatment plants for rural settlement units. Infrastruktura i Ekologia Terenów Wiejskich 5/2008, 169–177 (in Polish).

8. http1. http://www.parczew.pl/p,81,walory-turystyczne

9. Jóźwiakowski K., Pytka A. 2010. Development of water and sewage management in rural areas in Poland in 1990–2008. Gospodarka Odpadami Komunalnymi. Monografia Komitetu Chemii Analitycznej PAN (red. K. Szymański), VI, 31–39 (in Polish).

10. Jóźwiakowski K., Steszuk A., Pieńko A., Marzec M., Pytka A., Gizińska M., Sosnowska B., Ozonek J. 2014. Evaluation of the impact of wastewater treatment plants with drainage system on the quality of groundwater in dug and deep wells. Inżynieria Ekologiczna 39, 74–84 (in Polish).

11. Jóźwiakowski K., Mucha Z., Generowicz A., Baran S., Bielińska J., Wójcik W. 2015. The use of multi-criteria analysis for selection of technology for a household WWTP compatible with sustainable development. Archives of Environmental Protection, 41 (3), 76–82.

12. Jóźwiakowski K., Podbrożna D., Kopczacka K., Marzec M., Listosz A., Pochwatka P., Kowalczyk-Juśko A., Malik A. 2017a. The state of water and wastewater management in the municipalities of the Polesie National Park. Journal of Ecological Engineering 18 (6), 192–199.

13. Jóźwiakowski K., Bugajski P., Mucha Z., Wójcik W., Jucherski A., Natawny M. Siwiec T., Mazur A., Obroślak R., Gajewska M. 2017b. Reliability of pollutions removal processes during long-term operation of one-stage constructed wetland with horizontal flow. Separation and Purification Technology 187, 60–66.

14. Jóźwiakowski K., Bugajski P., Kurek K., Nunes de Carvalho M. F, Araújo Almeida M. A., Siwiec T., Borowski G., Czekała W, Dach J, Gajewska M. 2018. The efficiency and technological reliability of biogenic compounds removal during long-term operation of a one-stage subsurface horizontal flow constructed wetland. Separation and Purification Technology 202, 216–226.

15. Jucherski A, Walczowski A. 2001. Drainage systems. Cleaning or discharging untreated sewage into the soil. Wiadomości Melioracyjne i Łąkarskie 3 (390), 131–132 (in Polish).

16. Jucherski A, Nastawny M., Walczowski A., Jóźwiakowski K., Gajewska M. 2017. Assessment of the technological reliability of a hybrid constructed wetland for wastewater treatment in a mountain eco-tourist farm in Poland. Water Sci. Technol. 75 (11), 2649–2658.

17. Kaczor G., Bergel T., Bugajski P., Pijanowski J. 2015. Aspects of sewage disposal from tourist facilities in national parks and other protected areas. Pol. J. Environ. Stud. 24 (1), 107–114.

18. Karolinczak B., Miłaszewski R., Sztuk A. 2015. Cost-effectiveness analysis of different technological variants of single-house sewage treatment plants. Rocznik Ochrona Środowiska, 17, 726–746 (in Polish).

19. Mucha Z., Mikosz J. 2009. Rational use of small wastewater treatment plants taking into account the sustainability criteria. Czas. Techn. Środowisko 106 (2), 91–100 (in Polish).

20. Pawełek J., Bugajski P. 2017. The development of household wastewater treatment plants in Poland – advantages and disadvantages. Acta Scientiarum Polonorum, Formatio Circumiectus, 16 (2), 3–14 (in Polish).

21. Pawęska K., Pulikowski K., Strzelczyk M., Rajmund A. 2011. Septic tank – basic element of household treatment plant. Infrastruktura i Ekologia Terenów Wiejskich, 10, 43–53 (in Polish).

22. Polish Norm PN-EN 12566–3:2016–10. 2016. Small wastewater treatment plants for a population calculation (OLM) up to 50 – Part 3: Container and / or home sewage treatment plants on site (in Polish).

23. SYNERGIA. Kancelaria doradcza 2013. Update of the Development Strategy of Parczew District on the years 2014–2020, 96 (in Polish).

24. US w Lublinie 2017. Statistical Vademecum of the Local Government. Parczew District – A statistical portrait of the territory. Lublin, 4.

25. Construction Law Act of March 27, 2003 [Dz.U. nr 80/03, poz. 718, art.29 ustęp 1, pkt. 3] (in Polish).

26. Water Law Act of 18 July 2001, [Dz.U. nr 115/01, poz. 1229, art. nr 36, 39, 42] (in Polish).

Adsorption of Phenol from Water on Natural Minerals

Alicja Puszkarewicz[1*], Jadwiga Kaleta[1], Dorota Papciak[1]

[1] Rzeszow University of Technology, The Faculty of Civil and Environmental Engineering and Architecture, Department of Water Purification and Protection, Poznańska 2, 35-084 Rzeszów, Poland

* Corresponding author's e-mail: apuszkar@prz.edu.pl

ABSTRACT

Phenol and its derivatives (chlorophenol, nitrophenol, methylphenol, cresol etc.) belong to highly toxic contaminants, and their occurrence in industrial and municipal sewage as well as in groundwater carries a high threat to the environment and human health. Elimination of such contaminants is one of the major challenges in solving the global environmental problems. Implementation of pro-ecological methods of water treatment is associated with the use of natural, cheap and unprocessed materials, with the possibility of their repeated use. The article presents the results of the studies on the use of powdery adsorbents for the removal of phenol from aqueous solutions. The following natural minerals were used: attapulgite – Abso'net Superior Special (ASS) and alganite – Abso'net Multisorb (AM). Tests were performed under non-flowing conditions, in series, depending on the type and dose of adsorbents. Tests were conducted on a model solution of phenol with the initial concentration of $C_0 = 20$ mg /dm³, at the temp. of 20° C. Alganite mineral (AM) proved to be effective in adsorption of phenol. Maximum adsorption capacity P = 0.21 g/g, was obtained for a dose 10 mg/dm³. Almost complete removal of phenol (99.9%) was obtained for a dose of 500 mg/dm³. For natural attapulgite – Abso'net Superior Special (ASS) the maximum adsorption capacity (at a dose 5 mg/dm³) amounted to P = 0.15 g/g. The efficiency of phenol removal at the level 99% was obtained at a dose of 1000 mg/dm³).

Keywords: phenol, adsorption, attapulgite, alganite

INTRODUCTION

The presence of phenol in the aquatic environment may cause a decrease in the quality of water, as well as lead to the death of aquatic organisms. Almost all phenols are toxic, and some of them are carcinogenic to humans. If its concentration exceeds the value of 5.6 g/m³, it disrupts the water self-purification processes. In contrast, at the concentrations above 30 g/m³, it completely inhibited photosynthesis [Al-Khalid et al. 2012, Bazrafshan et al. 2013].

The presence of phenols in ecosystems is associated with the production and degradation of many pesticides, as well as the production of industrial and municipal sewage. Large quantities (up to several g/m³) are found in wastewater from various industries, mainly coking plants, gas plants, plastic mills, synthetic fibres, gasification plants for solid fuels, processing of crude oil, production of dyes, plant protection chemicals, and

pharmaceutical plants. Phenols are also found in the wastewater from other industrial plants in which phenol is used as a raw material [Ahmaruzzaman 2007, Michałowicz et al. 2007].

Phenol is biodegradable in an aqueous environment; however, only when it is absent in the concentrations toxic to microorganisms [Indu Nair et. al. 2008]. Therefore, it is necessary to develop effective and time-saving methods for removing phenols from wastewater. Photocatalytic oxidation can be used to remove phenols from aqueous solutions and adsorption [Alam et al. 2014, Bielecka-Daszkiewicz 2008]. Various materials such as activated carbon, natural minerals, synthetic polymers and waste materials are used as absorbents [Dąbrowski et al. 2005]. Research was also carried out to assess the effectiveness of cheaper adsorbents, such as peat, fly ash, bentonite and other materials for which high phenol removal efficiency was obtained. An additional advantage of adsorption is the ability to burn and reuse the adsorbent

[Bizerea Spiridon et al. 2013, Kaleta 2005, Puszkarewicz 2010, Viraraghavan et al. 1998].

This article focuses on determining the effectiveness of phenol removal using natural adsorbents: attapulgite – Abso'net Superior Special (ASS) and alganite – Abso'net Multisorb (AM).

MATERIALS AND METHODS

Materials

The subject of the research was a model aqueous solution of phenol (C_6H_5OH). It was prepared on the basis of distilled water, to which phenol compounds were added. The initial concentration of phenol in water was $C_0 = 20$ mg/dm³.

In laboratory tests, dusty fractions of the following raw materials were used as adsorbents:

- mineral Abso'Net Superior Special (ASS),
- mineral Abso'net Multisorb (AM),

Characteristics of Abso'Net Superior Special (ASS) adsorption material – catholytic card – Horpol:

- bulk density – 500 g/dm³
- produced from finely ground mineral called attapulgite, during the activation process at high temperature,
- it is a natural product (it does not contain any chemical additives, it is not modified),
- it is not harmful to people and animals,
- it is resistant to abrasion when subjected to mechanical action,
- after absorbing the fluid it does not decompose,
- has NIH approval.

Attapulgite is a filbrillar silicate with the chemical formula of $Mg_5[Al]Si_8O_{20}(OH_2)_4$, showing a relatively large sorption capacity oscillating in the range of 0.15 – 0.30 mval/g. Like zeolites, it is characterized by features of molecular sieve [Rybiński et. al. 2013].

Characteristics of the Abso'net Multisorb adsorp-tion material (AM) – catholytic card – Horpol:

- made of finely ground mineral called Alganite,
- bulk density – 500 g/dm³
- it is resistant to abrasion when subjected to mechanical action,
- after absorbing the fluid it does not decompose,
- designed to remove all types of leaks, from industrial fluids, aqueous suspensions, through

hydrocarbons (oil, gasoline, oils), organic solvents, to alkaline compounds;

- has NIH approval.

Alganite is made of oiled mineral slates after their deoiling in the process (extraction), resulting in a porous, absorbent, reticular fabric material. It is one of the few materials the production of which is ecological and does not increase the emission of carbon dioxide. During production, more energy is gained than it consumes.

Alganite is a hydrophobic product, resistant to bacterial colonization according to the catalogue card [MKM Holdings LTD group 2014].

Method of conducting research

The adsorption process was carried out using an adsorbent dried at 105°C. The phenol content was determined indirectly by indicating the organic carbon content with the GE Analytical Instruments TEST Sievers InnovOx laboratory analyser. The tests were carried out under static (non-flow) conditions. They were made in series depending on the dose of the dusty adsorbent.

Laboratory tests were carried out in two stages:

Stage I –tests to determine the initial conditions for the adsorption process – the effect of pH and process kinetics

Stage II – determination of the effect of the adsorbent dose on the effectiveness of adsorption and determination of Freundlich adsorption isotherms.

Effect of pH on the adsorption process

The tests were performed for pH: 2, 3, 4, 5, 6, 7, 8, 9, 10, 11, 12, 13And involved adding 12.5 mg of adsorbent to 250 ml adsorptive , shaking for 0.5 h and after sedimentation (24h), the phenol in solutionwas determined.

Effect of contact time on the adsorption process under static conditions (adsorption kinetics)

The effect of the contact time on the adsorption process was determined at a sorbent dose of 50 mg/dm³. The initial concentration of phenol in the water tested was $C_0 = 20$ mg/dm³, pH = 7. Five adsorbent weights were poured into five conical flasks, containing 250 cm³ each. Then, each flask was shaken successively for 0.5; 1.0; 1.5; 2.0; 2.5; 3.0; 3,5; 4.0, 4.5, 5.0, hours and left for 24 hours of sedimentation. After this time, the samples were filtered and phenol determinations were made.

Adsorption of phenol (adsorption capacity) was determined from the equation:

$$q_r = \frac{V \times (C_0 - C)}{m} \text{ (g/g)} \tag{1}$$

where: V – volume of adsorptive (dm^3)

C_0 and C – initial and equilibrium (final) concentration of phenol (g/m^3),

m – mass of adsorbent (g)

Effect of adsorbent dose on adsorption efficiency

Effect of adsorbent dose on adsorption efficiency of phenol was conducted at pH = 7, contact time 3.5 h, for the doses of AM and ASS from $1 - 1000$ mg/dm^3.

Adsorption isotherms

The separation of the adsorbate between the solution and the adsorbent at steady state was described by Freundlich adsorption isotherms, which represent the relationship between the amount of adsorbed substance and the equilibrium concentration.

In exponential form, the equation has the following form:

$$y/m = K \cdot C^{1/n} \tag{2}$$

where: y/m – adsorption capacity,

K – constant isotherm,

n – constant isotherm,

C – equilibrium concentration

The tests were carried out in vessels with the content of 0.5 dm^3 of adsorption, for the concentration of phenol $C_0 = 20$ mg/dm^3, at pH 7 and for variable doses of tested adsorbents from

$1 - 1000$ mg/dm^3. The content of the adsorb at steady state q_r, after 3.5 hours of shaking and 24 hours of sedimentation for the tested carbons, was calculated with formula (1).

RESULTS AND DISCUSSION

Effect of pH

The effect of the pH of the solution on phenol adsorption is shown in Figure 1. The phenol particles showed much greater affinity for the adsorbents at a pH in the range of 1–4, with the phenol adsorption efficiency for AM being higher than for ASS and fluctuating in the range of 14–16%. This can be explained by the fact that the pH of the solution may affect the adsorption process also due to the change in the surface properties of the adsorbent. For many solids, the potential-forming ions are H+ and OH- ions, from which it follows that the surface potential can be influenced by changing the pH of the solution in which the given solid is located, and the adsorption proceeds with greater intensity when the surface charge of the adsorbent and potential the electric adsorbate are opposite. In the case of very strong adsorption of potential-forming ions on the surface of a solid (which may have occurred in the tests carried out), it may happen that the charge of the adsorption layer changes its sign. Therefore, depending on the pH of the solution in which the adsorbent is located, it will take up a positive or negative electric charge of the surface [Janusz 2000]. Very low removal of phenol for the AM and ASS (1–3%) at high pH could be the result of competition between the OH- ions and dissociated phenol anions [Liu et. al. 2010]. The effectiveness of phenol adsorption for AM was higher than for ASS and fluctuated in the range of 14–16%.

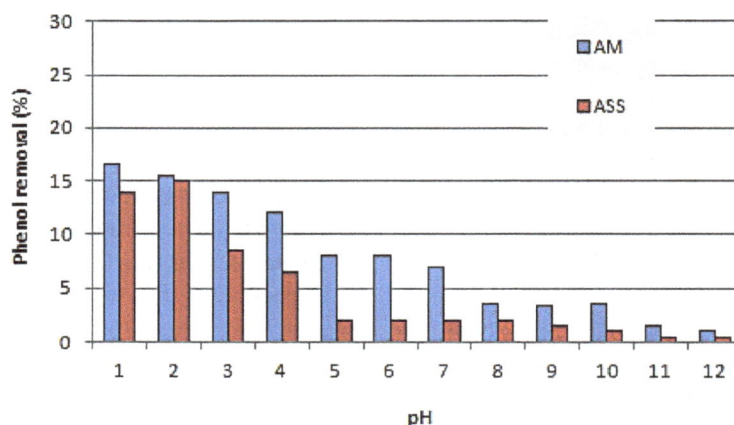

Fig. 1. Effect of pH on the adsorption of phenol

For the AM tested, the increase in the adsorption efficiency at the acid reaction was not significant enough to introduce a correction of water pH and despite visible differences for ASS, further tests were carried out at pH = 7.

Adsorption kinetics

The process of phenol absorption consists of several stages. The first stage is the transport of the adsorbed particle from the solution's bulk phase to the adsorbent boundary layer, followed by external diffusion (on the adsorbent surface), internal diffusion in the adsorbent structure and the last stage involves the reactions from the adsorbate to the active site. The reaction in the active sites is the fastest, but the adsorption kinetics limit the slowest processes. Under static (non-flow) conditions, these phenomena are diffusion at the interface and inside the pores of the adsorbent. Solutions diluted at the adsorption rate are affected by the mixing and shaking time. The concentration of phenol close to the adsorption equilibrium (adopted as equilibrium) for ASS and AM was obtained after 3.5 hours. The effectiveness of adsorption expressed as C/C_0 is illustrated in Figure 2.

In order to describe the adsorption kinetics, two kinetic models are used most often: the pseudo-first-order (PFO) model defined by the Lagergren equation and the secondary model – pseudo-second-order (PSO) popularized by Ho et al. [1999]. The modelling process comes down to matching the empirical results with these equations and choosing the one that better correlates the data. Choosing the right model will not explain the mechanisms controlling the adsorption rate in the system, but it can be helpful in determining the factors limiting the process speed [Płaziński et al. 2011].

The PFO equation, in a differential form, has the following form:

$$\frac{dq(t)}{dt} = k_1(q_e - q(t)) \qquad (3)$$

where: t is the time,
  q the amount of adsorbate bound by the adsorbent (this amount may depend on time),
  q_e corresponds to the value of q in equilibrium, i.e. $q\,(\,t \rightarrow \infty) = q_e$;
  k_1 is a constant, called the PFO constant.

During data analysis, the line form is more commonly used:

$$\ln(q_e - q(t)) = \ln q_e - k_1 t \qquad (4)$$

The kinetic curves for the adsorbents tested, made in function f(t) = log(qe – q(t)) are shown in Figure 3. On the basis of simple equations determined with the least squares method, the values of k_1 (1/h) speed constants and adsorption capacities in equilibrium were calculated and are given in Table 1.

The pseudo-second order equation (PSO) in a differential form is as follows:

$$\frac{dq(t)}{dt} = k_2(q_e - q(t))^2 \qquad (5)$$

where: k_2 is a constant.

In order to correlate experimental data, the linear representation using which the constant k_2 and q_e are most often relevant is:

$$\frac{t}{q(t)} = \frac{1}{k_2 q_e^2} + \frac{t}{q} \qquad (6)$$

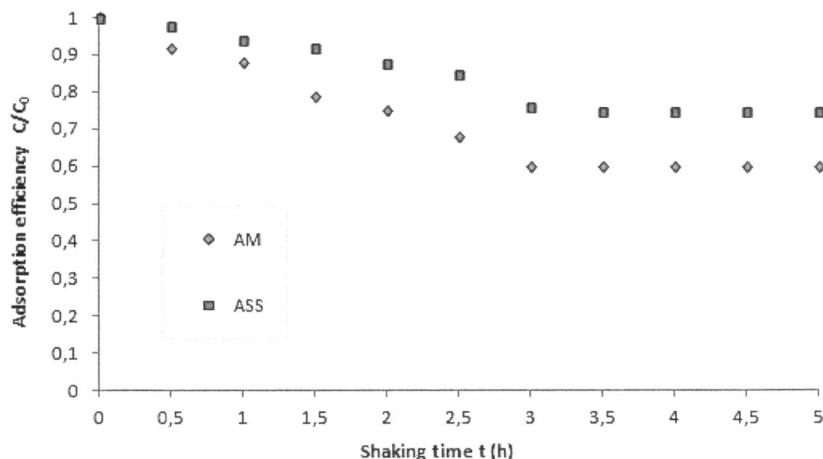

Fig. 2. Influence of adsorption time on the effectiveness of the removal of phenol

The graphs for PSO were prepared in the function $f(t) = t/q(t)$ and are shown in Figure 4. Rate constants k_2 (g/g·h) were calculated from the inclination and displacement coefficients of straight lines. All calculated k_2 rate constants and adsorption capacities are shown in Table 1.

With regard to AM and ASS adsorbents, the adsorption kinetics were approximately in line with the PFO model, because the determination coefficients were 0.885 and 0.881, respectively. We can say that the model is weaker to match the experimental values, which is also indicated by large discrepancies (between calculated and obtained in tests) in the adsorption capacity (several times). Quicker adsorption was performed for AM and the slowest for ASS, as evidenced by k_1 rate constants of 1.540 and 0.720 1/h, respectively.

The presented dependences show that the adsorption efficiency increased along with the contact time of the water with the adsorbent, but its speed (the highest in the first 30 minutes) dropped sharply after 3 hours. The adsorption process was much slower and its continuation did not significantly affect the phenol removal efficiency. In further studies (adsorption isotherms and dose effect), due to slight changes in the adsorption kinetics, 3.5 hours was assumed as the mixing time for all adsorbents.

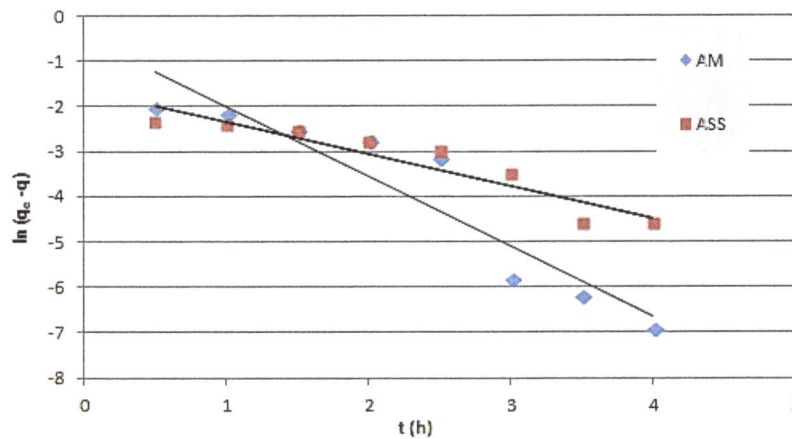

Fig. 3. The course of adsorption kinetics of phenol described pseudo first-order equation

Table 1. Rate constants for PFO and PSO of phenol adsorption on AM and ASS

Adsorbent	q_r (g/g)	Pseudo-first-order (PFO)			Pseudo-second-order (PSO)		
		k_1 (h^{-1})	q_e (g/g)	R^2	k_2 (g/g·h)	q_e (g/g)	R^2
AM	0.163	1.540	0.199	0.885	0.192	0.37	0.800
ASS	0.105	0.720	0.619	0.881	772	4.21	0.005

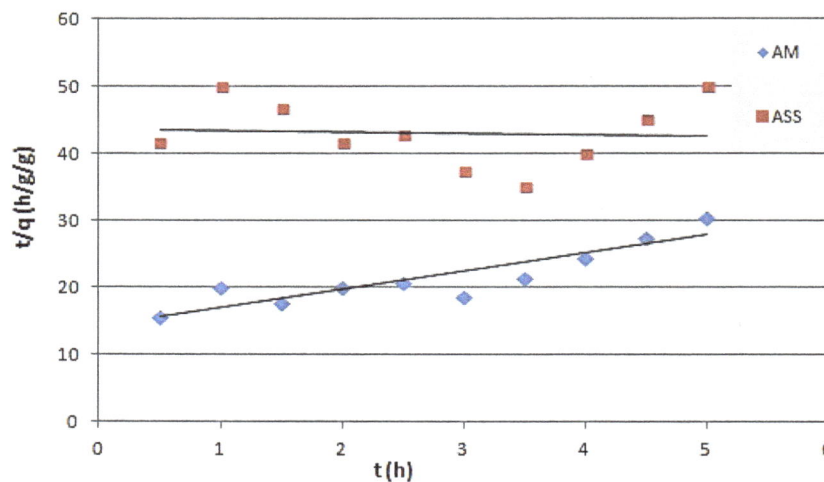

Fig. 4. The course of adsorption kinetics of phenol described pseudo second-order equation

Adsorption isotherms

The phenol adsorption isotherms for the adsorbents tested are shown in Figure 5. The determined values of K in constants, together with the values of the determination coefficients R^2 are presented in Table 2. The course of isotherms (determined under static conditions) indicates that the maximum adsorptive capacity for AM (at a dose of 10 mg/dm³) was $q_r = 0.21$ g/g. For ASS, the maximum adsorption capacity was much lower and amounted to $q_r = 0.15$ g/g (at a dose of 5 mg/dm³). The K values expressing the maximum adsorption on the adsorbent surface are 0.0862, and 0.0321 for AM and ASS respectively.

The highest value of the correlation coefficient observed for AM (0.9427) indicates a good adjustment of the theoretical model to the experimental isotherm. Slightly smaller fitting (but still good) of experimental data to the model was demonstrated by ASS (0.9214). The n-factor for the adsorbents tested is greater than 1, which may mean that the

Table 2. Values of constants n and K of Freudlich isotherms

Adsorbent	Constants of isotherms		Factor of determination R^2
	n	K	
AM	3.77	0.0862	0.9427
ASS	2.16	0.0321	0.9214

surface concentration of the adsorbate increases more slowly than its concentration in the solution. In principle, no full saturation is achieved, because there are always places with high adsorption energy on the surface. In the analyzed adsorbents, the lowest factor n is for ASS, which may indicate the possibility of reaching a higher saturation state with the adsorbate than in the case of AM.

Effect of dose on adsorption

The phenol adsorption efficiency also depended on the amount of adsorbent used. Figure 6

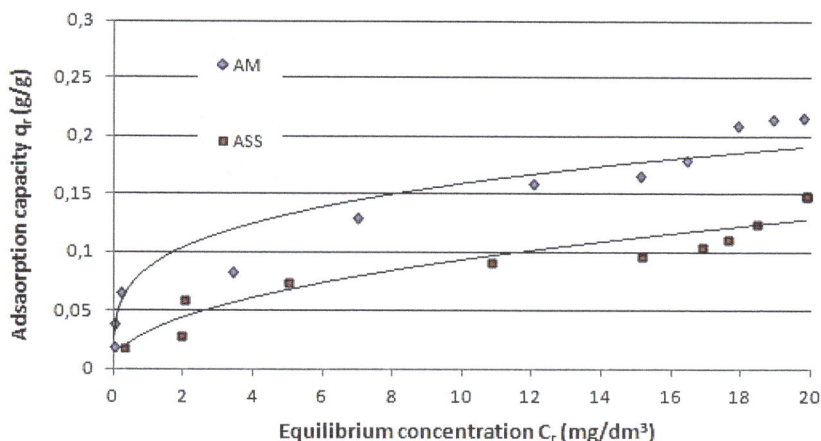

Fig. 5. Adsorption isotherms of phenol onto adsorbent (AM, ASS), temp. 20° C

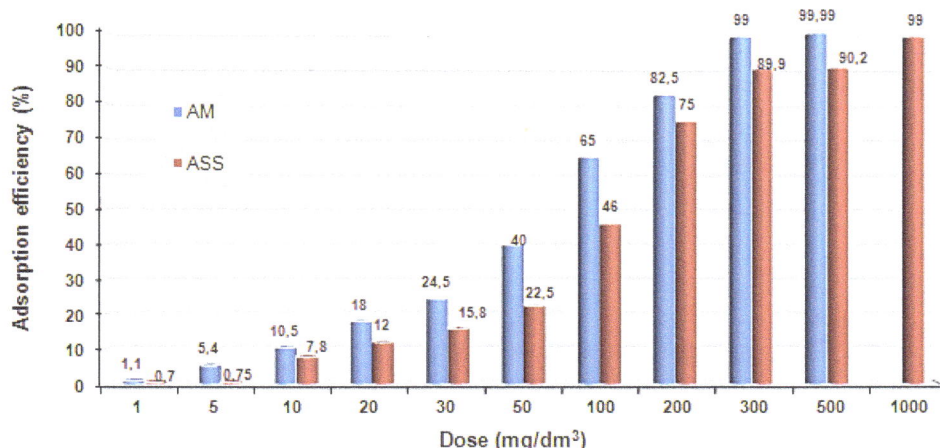

Fig. 6. Effect of dose on the effectiveness of the removal of phenol

depicts the effect of the adsorbent dose on the percent removal of phenol from the solution. It is clearly seen that the higher dose of adsorbent gave better final efficiency of the process. The efficiency of phenol removal for ASS at the dose of 1000 mg/dm^3 was 99%, yielding the equilibrium concentration $C_r = 0.28$ mg/dm^3. For AM at a dose of 500 mg/dm^3, the equilibrium concentration of phenol was $C_r = 0.02$ mg/dm^3, which gave 99.9% adsorption efficiency.

CONCLUSIONS

1. The pH of the solution affected the adsorptive capacity of the adsorbents tested. The phenol particles showed much greater affinity for the adsorbents at a pH in the range of 1–4, with the phenol adsorption efficiency for AM being higher than for ASS.
2. The adsorption rate changed as a function of the contact time (mixing) of the adsorbents with the adsorbate and dropped sharply after 3 hours. With respect to the adsorbents AM and ASS, the adsorption kinetics proceeded according to the PFO (Pseudo-first-order) model, and the determination coefficients were 0.885 and 0.881, respectively.
3. Adsorption of phenol for all tested adsorbents was well-described with the Freudlich model. The highest value of the correlation coefficient was obtained for AM (0.9427).
4. Analysis pertaining to the effect of the adsorbents dose showed that the adsorption efficiency in phenol removal clearly increased along with the dose of the adsorbent. For AM at a dose of 500 mg/dm^3, 99.9% efficiency was obtained.
5. In the light of the conducted research, the mineral adsorbent AM effectively removed phenol from water, which enables to use it in water treatment systems. Studies have shown that a natural adsorbent – Abso'net Multisorb (AM), with granulation <0.1 mm, could be an alternative to activated carbons for effective pre-treatment of phenol-containing solutions.

REFERENCES

1. Ahmaruzzaman M. 2008. Adsorption of phenolic compounds on low-cost adsorbents: A review, Advances in Colloid and Interface Science (143), 48–67.
2. Alalm M. G. Tawfik A. 2014. Solar photocatalytic degradation of phenol in aqueous solutions using titanium dioxide, International Journal of Chemical, Materials Science and Engineering 8(2), 43–46.
3. Al-Khalid T., El-Naas M. H. 2012. Aerobic Biodegradation of Phenols: A Comprehensive Review, Critical Reviews in Environmental Science and Technology 42, 1631–1690.
4. Bazrafshan E., Mostafapour F. K., Mansourian H. J. 2013. Phenolic Compounds: Health Effects and Its Removal From Aqueous Environments by Low Cost Adsorbents, Health Scope. 2(2), 65–66.
5. Bielicka-Daszkiewicz K. 2008. Removing phenol from wastewaters by oxidation, Przemysł Chemiczny 87(1), 24–32.
6. Bizerea Spiridon O., Preda E., Botez A., Pitulice L. 2013. Phenol removal from wastewater by adsorption on zeolitic composite, Environ Sci. Pollut. Res. 20, 6367–6381.
7. Dąbrowski A., Podkościelny P., Hubicki Z., Barczak Z., Robens E. 2005. Adsorption of phenols on activated carbon, Wiadomości Chemiczne, 59(7–8), 631–694.
8. Ho Y.S., McKay G. 1999. Pseudo-second-order model for sorption processes, Process Biochemistry 34, 451–465.
9. Indu Nair C., Jayachandran K., Shashidhar S. 2008. Biodegradation of phenol, African Journal of Biotechnology 7(25), 4951–4958.
10. Janusz W. 2000. The electrical double layer parameters for the group 4 metal oxide/electrolyte system, Adsorpt. Sci. Technol.18 (2), 117–134.
11. Kaleta J. 2005. Phenols in water medium, Ecology and Technique 73(1), 3–11.
12. Liu Q-S., Zheng T., Wang P., Jiang J.P., Li N. 2010. Adsorption isotherm, kinetic and mechanism studies of some substituted phenols on activated carbon fibers, Chemical Engineering Journal 157, (2–3), 348–356.
13. Michałowicz J., Duda W. 2007. Phenols – Sources and Toxicity, Polish J. of Environ. Stud. 16(3), 347–362.
14. MKM Holdings Ltd Group, 2014. Catalog card.
15. Płaziński W., Rudziński W. 2011. Adsorption kinetics at solid /solution interfaces . The meaning of the pseudo -first – and pseudo -second -order equations, Journal of Polish Chemical Society 65, 11–12, 1055-1067.
16. Puszkarewicz A., 2010. Analysis of phenol adsorption on raw and modified carpathian diatomite, Chemistry-Didactics-Ecology-Metrology 15(2), 189–192.
17. Rybiński, P., Janowska G. 2013. Flammability and other properties of elastomeric materials and nanomaterials. Part II. Nanocomposites of elastomers with attapulgite, nanosilica, nanofibres and carbon nanotubes, Polimery 7–8, 533–542.
18. Viraraghavan T., Alfaro F. M. 1998. Adsorption of phenol from wastewater by peat, fly ash and bentonite, Journal of Hazardous Materials 57, 59–70.

Permissions

List of Contributors

Kairat T. Ospanov and Zhanar Kudaiberdi
Kazakh National Research Technical University Named After K.I. Satpayev, Satpayev's Street 22, Almaty, Kazakhstan

Aleksandr V. Demchenko
State Municipal Enterprise "Tospa Su", Shubazeva's Str. 128, Alamty Region, Yhapek Batyr Village, Kazakhstan

Marta Kisielewska, Marcin Dębowski, Marcin Zieliński and Mirosław Krzemieniewski
University of Warmia and Mazury in Olsztyn, Department of Environmental Engineering, ul. Warszawska 117, 10-950 Olsztyn, Poland

Joanna Szulżyk-Cieplak, Aneta Tarnogórska and Zygmunt Lenik
Faculty of Fundamentals of Technology, Lublin University of Technology, ul. Nadbystrzycka 38, 20-618 Lublin, Poland

Agnieszka Micek, Michał Marzec and Patrycja Pochwatka
Department of Environmental Engineering and Geodesy, University of Life Sciences in Lublin, Leszczyńskiego 7, 20-069 Lublin, Poland

Karolina Jóźwiakowska
Warsaw University of Life Sciences, Nowoursynowska 166, 02-787 Warszawa, Poland

Bieby Voijant Tangahu, Adhi Triatmojo, Ipung Fitri Purwanti and Setyo Budi Kurniawan
Department of Environmental Engineering, Faculty of Civil, Environmental and Geo Engineering, Institut Teknologi Sepuluh Nopember, Jalan Raya ITS, Kampus ITS Sukolilo, Surabaya 60111, Indonesia

Michelle Peñaherrera
Universidad de Las Américas (UDLA), Faculty of Engineering and Agrarian Sciences, Environmental Engineering, Av. de los Granados and José Queri, 59302, Quito, Ecuador

Elvito Villegas
Universidad Nacional Mayor de San Marcos (UNMSM), 07001, Lima, Perú

Paola Posligua
Universidad de Las Américas (UDLA), Faculty of Engineering and Agrarian Sciences, Environmental Engineering, Av. de los Granados and José Queri, 59302, Quito, Ecuador
Universidad Nacional Mayor de San Marcos (UNMSM), 07001, Lima, Perú
Instituto Antártico Ecuatoriano (INAE), 59316, 9 de Octubre y Chile, Guayaquil, Ecuador

Carlos Banchón
Universidad Agraria del Ecuador (UAE), Environmental Engineering School, Faculty of Agrarian Sciences, Av. 25 de Julio and P. Jaramillo, 59304, Guayaquil, Ecuador

Agnieszka Pusz, Magdalena Wiśniewska, Dominik Rogalski and Grzegorz Grzyb
Warsaw University of Technology, Faculty of Building Services, Hydro and Environmental Engineering, Nowowiejska 20, 00-653 Warsaw, Poland

Agnieszka Cydzik-Kwiatkowska, Michał Podlasek, Dawid Nosek and Beata Jaskulska
University of Warmia and Mazury in Olsztyn, Department of Environmental Biotechnology, Słoneczna 45G, 10-709 Olsztyn, Poland

Radosław Żyłka and Wojciech Dąbrowski
Bialystok University of Technology, Faculty of Building and Environmental Engineering, ul. Wiejska 45E, 15-351 Białystok, Poland

Elena Gogina and Olga Yancen
Moscow State University of Civil Engineering, Yaroslawskoyoe Shosse, Moscow, Russia

Zaidun Naji Abudi
Environmental Engineering Department, Faculty of Engineering, Al-Mustansiryiah University, Baghdad, Iraq

Piotr Maciołek
Municipal Water Supply and Sewerage Systems Co. Ltd., ul. Wojska Polskiego 14, 75-711 Koszalin, Poland

Kazimierz Szymański
Faculty of Civil Engineering, Environmental and Geodetic Sciences, Koszalin University of Technology, ul. Śniadeckich 2, 75-453 Koszalin, Poland

Rafał Schmidt
Municipal Public Utilities Company Co. Ltd., 1 Maja 1, 76-150 Darłowo, Poland

Robert Malmur and Maciej Mrowiec
Czestochowa University of Technology, Faculty of Infrastructure and Environment, Institute of Environmental Engineering, ul. Brzeźnicka 60a, 42-200 Częstochowa, Poland

Anatoli Hurynovich
Faculty of Civil and Environmental Engineering, Bialystok University of Technology, ul. Wiejska 45A, 15-351 Bialystok, Poland

Uladzimir Maroz
Faculty of Environmental Engineering Brest State Technical University, Moskowskaja 265, 224017 Brest, Belarus

Hanna Bauman-Kaszubska and Mikołaj Sikorski
Faculty of Civil Engineering, Mechanics and Petrochemistry, Warsaw University of Technology, Łukasiewicza 17, 09-400 Płock, Poland

Izabella Kłodowska, Joanna Rodziewicz and Wojciech Janczukowicz
University of Warmia and Mazury in Olsztyn, Faculty of Environmental Sciences, Department of Environment Engineering, Warszawska 117a, 10-719 Olsztyn, Poland

Piotr Stachowski, Karolina Kraczkowska, Daniel Liberacki and Anna Oliskiewicz-Krzywicka
Poznań University of Life Sciences, Institute of Land Improvement, Environment Development and Geodesy, Piątkowska 94, 60-649 Poznań, Poland

Ali Hadi Ghawi
Department of Civil Engineering, Collage of Engineering, University of Al-Qadisiyah, Iraq

Mohamed F. Eida, Osama M. Darwesh and Ibrahim A. Matter
Agricultural Microbiology Department, National Research Centre, Giza, Egypt

Dariusz Boruszko
Faculty of Civil and Environmental Engineering, Bialystok University of Technology, 45A Wiejska St., 15-351 Bialystok, Poland

Justyna Zamorska
Department of Water Purification and Protection, Rzeszow University of Technology, Al. Powstańców War-szawy 6, 35-959 Rzeszów, Poland

Tomasz Ciesielczuk and Katarzyna Łuczak
Department of Land Protection, Opole University, ul. Oleska 22, 45-052 Opole, Poland

Czesława Rosik Dulewska
Institute of Environmental Engineering of the Polish Academy of Sciences, ul. Skłodowskiej-Curie 4, 41–819 Zabrze, Poland

Joanna Poluszyńska and Ewelina Ślęzak
Institute of Ceramics and Building Materials, ul. Oswiecimska 21, 45-641 Opole, Poland

Paweł Malinowski
Bialystok University of Technology, Faculty of Building and Environmental Engineering, Wiejska 45E, 15-351 Białystok, Poland

Beata Karolinczak
Warsaw University of Technology, Faculty of Building Service, Hydro and Environmental Engineering, Nowowiejska 20, 00-653 Warsaw, Poland

Marcin Zieliński, Marcin Dębowski, Mirosław Krzemieniewski, Paulina Rusanowska and Agata Głowacka-Gil
Department of Environment Engineering, Faculty of Environmental Sciences, University of Warmia and Mazury in Olsztyn, ul. Warszawska 117, 10-720 Olsztyn, Poland

Magdalena Zielińska
Department of Environmental Biotechnology, Faculty of Environmental Sciences, University of Warmia and Mazury in Olsztyn, ul. Słoneczna 45G, 10-709 Olsztyn, Poland

Joanna Struk-Sokolowska and Justyna Tkaczuk
Bialystok University of Technology, Faculty of Civil and Environmental Engineering, Department of Technology and Systems in Environmental Engineering, ul. Wiejska 45E, 15-351 Bialystok, Poland

Izabela Bartkowska and Dariusz Wawrentowicz
Bialystok University of Technology, Department of Technology and Environmental Engineering Systems, ul. Wiejska 45 A, 15-351 Bialystok, Poland

Agnieszka Pusz, Magdalena Wiśniewska, Dominik Rogalski and Grzegorz Grzyb
Warsaw University of Technology, Faculty of Building Services, Hydro and Environmental Engineering, Nowowiejska 20, 00-653 Warsaw, Poland

Agnieszka Micek, Michał Marzec and Patrycja Pochwatka
Department of Environmental Engineering and Geodesy, University of Life Sciences in Lublin, Leszczyńskiego 7, 20-069 Lublin, Poland

Karolina Jóźwiakowska
Warsaw University of Life Sciences, Nowoursynowska 166, 02-787 Warszawa, Poland

Alicja Puszkarewicz, Jadwiga Kaleta and Dorota Papciak
Rzeszow University of Technology, The Faculty of Civil and Environmental Engineering and Architecture, Department of Water Purification and Protection, Poznańska 2, 35-084 Rzeszów, Poland

Index